AF832517

Global Perspectives on Antiviral Drug Development

Muhammad Shahzad Aslam
Xiamen University, Malaysia

Bancha Yingngam
Ubon Ratchathani University, Thailand

Tariq Javed
GC University, Pakistan

Sana Riaz
Pakistan Council of Science and Industrial Research, Pakistan

IGI Global
Scientific Publishing
Publishing Tomorrow's Research Today

Vice President of Editorial	Melissa Wagner
Director of Acquisitions	Mikaela Felty
Director of Book Development	Jocelynn Hessler
Production Manager	Mike Brehm
Cover Design	Jose Rosado

Published in the United States of America by
IGI Global Scientific Publishing
701 East Chocolate Avenue
Hershey, PA, 17033, USA
Tel: 717-533-8845
Fax: 717-533-7115
Website: https://www.igi-global.com E-mail: cust@igi-global.com

Copyright © 2025 by IGI Global Scientific Publishing. All rights reserved. No part of this publication may be reproduced, stored or distributed in any form or by any means, electronic or mechanical, including photocopying, without written permission from the publisher.

Product or company names used in this set are for identification purposes only. Inclusion of the names of the products or companies does not indicate a claim of ownership by IGI Global Scientific Publishing of the trademark or registered trademark.

Library of Congress Cataloging-in-Publication Data

Names: Aslam, Muhammad Shahzad, 1985- editor | Yingngam, Bancha editor | Javed, Tariq, 1981- editor | Riaz, Sana, 1984- editor
Title: Global perspectives on antiviral drug development / edited by Muhammad Shahzad Aslam, Bancha Yingngam, Tariq Javed, Sana Riaz.
Description: Hershey, PA : IGI Global Scientific Publishing, [2025] | Includes bibliographical references and index. | Summary: "This book offers an evidence-based understanding of global antiviral strategies by integrating scientific innovation with policy analysis. By providing in-depth case studies, exploring cutting-edge technologies such as CRISPR and AI-driven drug discovery, and analyzing regulatory landscapes and public health implications, the book aims to inform and guide professionals in the field"-- Provided by publisher.
Identifiers: LCCN 2024055943 (print) | LCCN 2024055944 (ebook) | ISBN 9798369392768 h/c | ISBN 9798369392775 s/c | ISBN 9798369392782 eISBN
Subjects: LCSH: Antiviral agents--Development
Classification: LCC RS431.A66 G56 2025 (print) | LCC RS431.A66 (ebook) | DDC 615.7/924--dc23/eng/20250312
LC record available at https://lccn.loc.gov/2024055943
LC ebook record available at https://lccn.loc.gov/2024055944
British Cataloguing in Publication Data
A Cataloguing in Publication record for this book is available from the British Library.

All work contributed to this book is new, previously-unpublished material.
The views expressed in this book are those of the authors, but not necessarily of the publisher.
This book contains information sourced from authentic and highly regarded references, with reasonable efforts made to ensure the reliability of the data and information presented. The authors, editors, and publisher believe the information in this book to be accurate and true as of the date of publication. Every effort has been made to trace and credit the copyright holders of all materials included. However, the authors, editors, and publisher cannot assume responsibility for the validity of all materials or the consequences of their use. Should any copyright material be found unacknowledged, please inform the publisher so that corrections may be made in future reprints.

Editorial Advisory Board

Editorial Advisory Board

Yun Jin Kim, *School of Traditional Chinese Medicine, Xiamen University Malaysia, Malaysia*

Qian Linchao, *School of Traditional Chinese Medicine, Xiamen University Malaysia, Malaysia*

Lisha Yi, *School of Chemistry and Molecular Bioscience, The University of Queensland Brisbane, Australia*

S. M. Farid Hasan, *Department of Pharmaceutics, Karachi University, Karachi, Pakistan*

Qurratul ain Leghari, *Department of Pharmaceutical Chemistry, Faculty of Pharmacy, Hamdard University, Karachi, Pakistan*

Abhiruj Navabhatra, *Department of Pharmacology, College of Pharmacy, Rangsit University, Pathum Thani, Thailand*

Muhammad Afzal, *Department: Bionanosciences, BOKU University, Vienna, Austria*

Muhammad Ali Riaz, *Department of ophthalmology, Johns Hopkins University, Baltimore, USA*

Mahtab Ahmad Khan, *Faculty of Pharmaceutical Sciences, University of Central Punjab, Lahore, Pakistan*

Muhammad Ali, *Department of Biotechnology, Quaid-i-Azam University, Islamabad, Pakistan*

Aamir Mushtaq, *Department of Pharmaceutical Sciences, GC University Lahore, Pakistan*

Muhammad Tariq, *Faculty of Pharmaceutical Sciences, Times Institute, Multan, Pakistan*

Maida Aslam, *School of Biochemistry, Minhaj University, Lahore, Pakistan*

Saba Riaz, *Department of Microbiology and Molecular Genetics (MMG), University of the Punjab, Lahore, Pakistan*

Sara Junaid, *Department of Microbiology and Molecular Genetics (MMG), Women University, Multan, Pakistan*

Bakhtawar Bukhari, *PCSIR, Lahore, Pakistan*

Ishrat Parveen, *PCSIR, Lahore, Pakistan*

Hafiz Abdul Khaliq, *Faculty of Pharmacy Bahauddin Zakariya University, Multan, Pakistan*

Kanza Tariq, *Department of Pharmaceutical Sciences, GC University, Lahore, Pakistan*

Muhammad Ali, *Department of Biotechnology, Quaid-I-Azam University, Islamabad, Pakistan*

Rana Rashad Mahmood, *Khan Institute of Chemical and Life Sciences, GC University, Lahore, Pakistan*

Humera Kausar, *Department of Biotechnology, Kinnaird College for Women, Lahore, Pakistan*

Mahwish Aftab, *Department of Biotechnology LCW University, Lahore, Pakistan*

Ghulam Mustafa, *Lahore University of Applied and Biological Sciences (UBAS), Pakistan*

Table of Contents

Preface xi

Chapter 1
The Role of Vaccines in Complementing Antiviral Treatments and Reducing Viral Spread 1
Faiza Akhtar, Department of Pharmaceutical Chemistry, Faculty of Pharmacy, Jinnah University for Women, Karachi, Pakistan & Ziauddin University, Karachi, Pakistan
Somia Gul, Department of Pharmaceutical Chemistry, Faculty of Pharmacy, Jinnah University for Women, Karachi, Pakistan

Chapter 2
Targeted Delivery and Nanotechnology: Harnessing Nanoparticles for Precision Medicine in Viral Infection 31
Krishnaveni Manubolu, Narayana Pharmacy College, Nellore, Jawaharlal Nehru Technological University, Anantapur, India
Raveesha Peeriga, V.V. Institute of Pharmaceutical Sciences, India

Chapter 3
Personalized Antiviral Therapy TailoringTreatments to Genetic Environmental and Lifestyle Factors 63
Reyed M. Reyed, Genetic Engineering and Biotechnology Research Institute (GEBRI), City of Scientific Research and Technological Applications (SRTA-City), Alexandria, Egypt
Pranav Kumar Prabhakar, Research and Development Cell, Parul University, India

Chapter 4
Antiviral Drug Discovery in Emerging Economics: Challenges and Opportunities 139
Abhiruj Navabhatra, College of Pharmacy, Rangsit University, Thailand
Chonnipha Yangchaiya, College of Pharmacy, Rangsit University, Thailand

Chapter 5
Pioneering Technologies to Revolutionize Antiviral Treatment 175
Raveesha Peeriga, V.V. Institute of Pharmaceutical Sciences, India
Krishnaveni Manubolu, Narayana Pharmacy College, India

Chapter 6
Antiviral Therapeutics for Emerging Viruses 217
Odangowei Inetiminebi Ogidi, Bayelsa Medical University, Nigeria
Thomas Aye, Bayelsa Medical University, Nigeria
Iniebiyo Felagha, University of Africa, Nigeria

Chapter 7
The Battle Against Emerging Viruses: Current and Future Antiviral Therapeutics 261
Shafeeq Ur Rehman, Cholistan University of Veterinary and Animal Sciences, Bahawalpur, Pakistan
Ayesha Jahangir, The Islamia University of Bahawalpur, Pakistan
Minal Hussain, Cholistan University of Veterinary and Animal Sciences, Bahawalpur, Pakistan

Compilation of References 297

About The Contributors 381

Index 389

Detailed Table of Contents

Preface xi

Chapter 1
The Role of Vaccines in Complementing Antiviral Treatments and Reducing Viral Spread 1
Faiza Akhtar, Department of Pharmaceutical Chemistry, Faculty of Pharmacy, Jinnah University for Women, Karachi, Pakistan & Ziauddin University, Karachi, Pakistan
Somia Gul, Department of Pharmaceutical Chemistry, Faculty of Pharmacy, Jinnah University for Women, Karachi, Pakistan

This chapter highlighted the details of different types of vaccines available for different viral diseases. Right from origin and background of viral diseases to detailed epidemiology as well as their causative agent, symptoms and treatment available for these viral diseases. It is further emphasized on currently available viral vaccines, their types, dosing strategies, and population-specific recommendations, common side effects associated with various vaccines and as well as their approval status. Lastly, this chapter also decked challenges, mitigation strategies and future directions in vaccine access.

Chapter 2
Targeted Delivery and Nanotechnology: Harnessing Nanoparticles for Precision Medicine in Viral Infection 31
Krishnaveni Manubolu, Narayana Pharmacy College, Nellore, Jawaharlal Nehru Technological University, Anantapur, India
Raveesha Peeriga, V.V. Institute of Pharmaceutical Sciences, India

The emergence of nanotechnology has profoundly impacted diverse fields of science and medicine, specifically in the area of targeted drug delivery systems. This proposal will investigate the revolutionary ways nanotechnology has impacted antiviral drug therapies by discussing the advantages offered by targeted delivery systems and the upcoming precision medicine. The development of a nanocarrier and a targeting mechanism is anticipated to address the current hurdles responsible for unsatisfactory antiviral therapy outcomes, including underwhelming efficacies and adverse side effects. In this paper, we will focus on reviewing present paradigms in technologies, their applications, and future considerations in developing antiviral therapy.

Chapter 3
Personalized Antiviral Therapy TailoringTreatments to Genetic Environmental and Lifestyle Factors 63
Reyed M. Reyed, Genetic Engineering and Biotechnology Research Institute (GEBRI), City of Scientific Research and Technological Applications (SRTA-City), Alexandria, Egypt
Pranav Kumar Prabhakar, Research and Development Cell, Parul University, India

This chapter aims to discuss the individualized approach to antiviral therapy targeting genetic, environmental, and lifestyle issues. It talks about how pharmacogenomics, AI, and computational modeling help identify better drug targets and reduce the odds of resistance. These comprise genetic screening in personalized medicine, biomarkers, and clinical trials in therapeutics. It also examines infection control, the axis between the gut microbiome and the host, and probiotics as novel biotherapeutics. Positioned under the umbrella of a one-health concept, it entails concierge wellness nutrition and precision medicine to optimize therapeutic interventions, combat antiviral resistance, and advance patient satisfaction. The prospects provided are intended to open new avenues in the composition of antiviral drugs.

Chapter 4
Antiviral Drug Discovery in Emerging Economics: Challenges and Opportunities 139
Abhiruj Navabhatra, College of Pharmacy, Rangsit University, Thailand
Chonnipha Yangchaiya, College of Pharmacy, Rangsit University, Thailand

Emerging infectious diseases underscore the urgent need for new antiviral drugs because of their significant impact on global health. Although important advancements such as HAART for HIV and DAAs for hepatitis C have been made, challenges such as drug resistance, limited vaccine effectiveness, and complex regulatory landscapes persist. Traditional drug discovery methods are resource intensive, prompting a shift toward innovative techniques such as PROTACs, RIBOTACs, CRISPR-based strategies, and nanotechnologies, which address issues of resistance and drug delivery. Financial and regulatory challenges remain critical, particularly in developing countries where neglected diseases are prevalent. Collaborative efforts, including public–private partnerships, have successfully addressed funding gaps, as demonstrated during the COVID-19 pandemic. This chapter explores innovative antiviral approaches, including RNA-based therapies and biomaterials, while emphasizing sustainable development, regulatory reforms, and capacity building to address challenges in resource-limited settings.

Chapter 5
Pioneering Technologies to Revolutionize Antiviral Treatment 175
Raveesha Peeriga, V.V. Institute of Pharmaceutical Sciences, India
Krishnaveni Manubolu, Narayana Pharmacy College, India

Antiviral care is changing by the minute due to technological advancements that improve diagnosis, treatment and prevention of viral diseases. AI, ML, bioinformatics are tools that have allowed for rapid pathogen identification, outbreak models, and drug optiims. CRISPR technologies and next generation sequencing (NGS) allow for diagnostics and therapies to move into the next generation, uncovering viruses and their evolution and resistance particularly in underserved areas. The efficacy of antiviral therapies is enhanced through innovative drug delivery systems including nanotechnology and an mRNA based vaccine develops new standards for a rapid response to viral threats. In this chapter we explore how these cutting edge tools can be integrated toward overcoming existing limitations, increasing care precision, and fostering equitable access to the required antiviral treatments. These technologies will allow us to reduce the global burden of viral diseases and bring us closer to more effective, personalized, and more readily available antiviral healthcare.

Chapter 6
Antiviral Therapeutics for Emerging Viruses 217
Odangowei Inetiminebi Ogidi, Bayelsa Medical University, Nigeria
Thomas Aye, Bayelsa Medical University, Nigeria
Iniebiyo Felagha, University of Africa, Nigeria

Efficient pharmaceuticals are required to address the rise of new viral infections. HIV and HCV infections serve as examples of the past ineffectiveness of single-medication treatments for chronic viral diseases. Significant advancements in clinical outcomes have been observed through the implementation of combination therapy for a range of diseases, resulting in notable decreases in viral load, morbidity, and mortality. Several newly discovered and resurfacing viruses can be more efficiently treated using combinations of drugs, which exhibit additive and, ideally, synergistic effects. Viruses such as influenza, SARS-CoV, MERS-CoV, Ebola, Zika, and SARS-CoV-2 are included in this list. Discovering new drugs through traditional pipelines should always be the primary focus. However, drug-repurposing screens offer a rapid way to identify potential medication candidates with synergistic effects. This approach allows for more time to conduct clinical trials and potentially obtain emergency use authorisations. This chapter discusses antiviral treatments for emerging viruses.

Chapter 7
The Battle Against Emerging Viruses: Current and Future Antiviral Therapeutics 261
Shafeeq Ur Rehman, Cholistan University of Veterinary and Animal Sciences, Bahawalpur, Pakistan
Ayesha Jahangir, The Islamia University of Bahawalpur, Pakistan
Minal Hussain, Cholistan University of Veterinary and Animal Sciences, Bahawalpur, Pakistan

Emerging viruses pose a significant threat to global health due to their widespread illness, mortality, and economic impact. These zoonotic pathogens, often transmitted from animals to humans, can cause economic disruption and overburden healthcare systems. Infections like HIV, Ebola, SARS, MERS, and COVID-19 demonstrate how viruses can be linked to human behavior and environmental change. Antiviral therapeutics have been developed to combat viral infections, targeting different points in the viral life cycle. However, treatments still face challenges such as drug resistance, side effects, and the inability to reach latent viral reservoirs. To address these challenges, innovation in drug design, targeting less mutable, conserved viral structures, and exploiting combination therapies is needed. Drug repurposing, immunotherapy, and RNA-based therapies are promising approaches to address current and future viral threats. Antiviral therapeutics are expected to evolve in a multitiered framework, combining targeted and broad-spectrum agents to overcome the complexity of viruses.

Compilation of References 297

About The Contributors 381

Index 389

Preface

Despite all the progress in modern medicine, the world remains deeply vulnerable to viral threats. The pandemic reminded us that we are always one mutation, outbreak, or logistical failure away from crisis. However, it also reminded us of what is possible when science, policy, and global cooperation align.

Global Perspectives on Antiviral Drug Development is a collection of ideas and reflections shaped by this urgency. It does not aim to offer all the answers. Instead, it brings together different voices—from vaccine research and nanomedicine to personalized care and policy reform—to examine where we stand and where we might be heading.

The first chapter looks at vaccines not only as protective shields but as tools that need better systems behind them—better distribution, better communication, and better public trust. Other chapters take us into newer territory: nanocarriers delivering antiviral drugs with precision, therapies tuned to a person's genetics and environment, and technologies that can analyze patterns faster than we can see them.

The book also acknowledges the gaps. Many regions still lack the infrastructure to produce or access cutting-edge treatments. Innovation is moving fast, but without thoughtful policy and fair implementation, progress risks being unevenly shared. There's also the question of sustainability—how to develop treatments that don't just work today but remain effective in the long run, especially in places facing financial and logistical challenges.

This is not just a book for scientists. It's for anyone thinking seriously about reducing the burden of viral diseases worldwide. Each chapter offers something different—a concept, a case study, a future scenario—but all are linked by a belief that the future of antiviral medicine must be collaborative, adaptive, and rooted in both science and equity.

The ideas here will spark further conversation, research, and action. The challenges are complex, but so are the tools at our disposal. With the right questions and the right partnerships, there is still a lot we can do.

Chapter 1
The Role of Vaccines in Complementing Antiviral Treatments and Reducing Viral Spread

Faiza Akhtar
https://orcid.org/0000-0003-4953-6685
Department of Pharmaceutical Chemistry, Faculty of Pharmacy, Jinnah University for Women, Karachi, Pakistan & Ziauddin University, Karachi, Pakistan

Somia Gul
https://orcid.org/0000-0002-9742-8753
Department of Pharmaceutical Chemistry, Faculty of Pharmacy, Jinnah University for Women, Karachi, Pakistan

ABSTRACT

This chapter highlighted the details of different types of vaccines available for different viral diseases. Right from origin and background of viral diseases to detailed epidemiology as well as their causative agent, symptoms and treatment available for these viral diseases. It is further emphasized on currently available viral vaccines, their types, dosing strategies, and population-specific recommendations, common side effects associated with various vaccines and as well as their approval status. Lastly, this chapter also decked challenges, mitigation strategies and future directions in vaccine access.

DOI: 10.4018/979-8-3693-9276-8.ch001

Copyright © 2025, IGI Global Scientific Publishing. Copying or distributing in print or electronic forms without written permission of IGI Global Scientific Publishing is prohibited.

INTRODUCTION

Antiviral therapies and vaccines continue to be at the very forefront of global initiatives aimed at controlling and eradicating viral diseases, and they indeed have succeeded in reducing morbidity and mortality burdens around the world. Disease eradication through vaccination strategies has led to the virtually eliminated diseases of smallpox, measles, and polio. However, though there have been great strides made in many different directions, viral diseases continue to be not only a major but an extremely persistent threat because of ever-rapidly mutating viruses, inequities in global vaccine access and distribution, and increasing public attitudes against immunization. The COVID-19 pandemic also exposed rather it was swept away showing both assertiveness's and weaknesses of global vaccination efforts, deep inequities about access; logistical matters exceedingly influencing manifestations; as well as how profoundly misinformation can cut through multiple public health policies.

This chapter will go into detail about the importance of antiviral drugs and vaccines in managing viral diseases as well as their limitations that prevent them from being more effective. Some of the major topics to be covered include: scientific principles underlying vaccine development; effectiveness of current antiviral therapies; and the social, economic, and political barriers that provide or withhold supports for vaccine accessibility and use. The general discussion above continues with good consideration outside mere scientific facts about wider social, economic, and political determinants of vaccine availability and acceptability. This includes issues related to vaccine reluctance which is historically influenced, clinically misguided, and a generally misplaced suspicion of health systems. Its footprint is deeply engraved both in under-immunization in high-income countries as well as low-income countries. To bring these issues to life, the chapter examines real-world case studies, such as India's successful eradication of polio, the resurgence of measles linked to declining vaccination rates, and the global fight against COVID-19. These examples help illustrate the intricate challenges of vaccine distribution, public health policies, and the ever-evolving landscape of disease prevention.

Studies about vaccine effectiveness alongside antiviral medications exist in large numbers but scientists still lack clear knowledge about prolonged immunity mechanisms and fair vaccine sharing procedures and optimal ways to combat vaccine refusal among the general population and public officials. The study reviews recent research findings to generate a complete analysis of these vaccine-related issues. The chapter contributes the concepts to deliver valuable knowledge about strengthening global immunization approaches while building better trust in vaccines and increasing readiness for future viral outbreaks.

VIRAL DISEASES

Despite being microscopic and yet simple structured, virus demonstrate remarkable power that affects human health and affects the natural systems and environmental surroundings. Their method of infecting host cells together with their resource usage to reproduce makes them dangerous pathogens that cause diseases. Communication systems with host cells enable viruses to cause two ends of the spectrum of disease severity that ranges from benign colds to lethal diseases like Ebola and HIV/AIDS. Healthcare researchers face great challenges when addressing these pathogens because viruses vary extensively between their genetic composition and their pathogenic properties and the ways they spread infection. Knowing viral characteristics along with their corresponding diseases helps scientists develop proper disease prevention methods and control strategies.

The respiratory infection known as influenza spreads through airborne droplet, manifest with fever, cough, and fatigue. The combined use of vaccination and hand hygiene along with antiviral oseltamivir enables flu control (Guglani, L., 2018). The worldwide changes to public health practices emerged from the COVID-19 virus which stems from Severe Acute Respiratory Syndrome Coronavirus 2 (SARS-CoV-2) has transformed public health practices worldwide, with vaccination, masks, and symptomatic treatment, including antivirals, central to managing this highly transmissible respiratory disease (Prather, K. A et al., 2020). A et al., 2020). Two viruses that affect children are the Respiratory Syncytial Virus (RSV) which affects respiratory tissue specifically in infants and mumps with its hallmark of swell glands has become rare due to Measles Mumps Rubella (MMR) vaccination. One more respiratory illness commonly known as cold is mainly caused by rhinoviruses, that is easily transferable from surface and human contacts. It's only treatment is to manage its symptoms. (Smith et al.,2021).

Shifting the focus on chronic or life threating disease, Human immunodeficiency virus (HIV), a chronic virus that progresses to AIDS if untreated, compromises the immune system and spreads through bodily fluids. Preventive practices and antiretroviral therapies significantly improve life quality and control spread (CDC, 2022). Turning to virus that primarily target the liver is Hepatitis viruses, with each type (A, B, C, etc.) transmitted through unique pathways, from contaminated food to blood. Vaccines are available for hepatitis A and B, and antiviral medications help manage symptoms.

Considering the vector borne disease, the Zika virus, transmitted by Aedes mosquitoes, is a public health concern, especially during pregnancy due to the risk of severe birth defects (Chen, H. L., & Tang, R. B., 2016), while the Dengue virus spread by Aedes mosquitoes causes high fever, intense headaches, and joint pain, with mosquito control as the primary defense (WHO, 2023). Shifting to a disease that is instigated by dog bite and known for its rapid fatal progression is Rabies, almost universally fatal once symptoms appear, requires immediate post-exposure prophylaxis to prevent progression. Considering another virus with high death rate is Ebola virus, infamous for its high mortality rate, causes severe hemorrhagic fever and requires rapid isolation and supportive care (WHO, 2023).

Addressing the virus that impacts on skin and mucous membrane is Herpes Simplex Virus (HSV), which can cause cold sores or genital lesions, is transmitted by close contact, A new emerging concern is Monkeypox, with flu-like symptoms and a characteristic rash, gained attention in 2022 when the WHO declared it a public health emergency (WHO, 2022).

Reflecting on a historic milestone in global health is Smallpox, a once-devastating disease eradicated through vaccination, presented with high fever and a characteristic rash. Its elimination stands as a milestone in global health (Fineberg & Wilson, 2023). Another virus that is nearly eradicated through global efforts is Polio virus, transmitted through contaminated food and water, is preventable through vaccination and, thanks to global immunization efforts, is now rare in many regions (CDC, 2022).

The global burden of viral diseases varies significantly, from endemic conditions like dengue to pandemics like COVID-19. Table 01 provides an overview of major viral diseases, their causative agents, symptoms, and global impact.

Table 1. Overview of Viral Diseases, Global Burden and Their Management

Disease	Causative Agent	Symptoms	Global Burden (Annual Cases/Deaths)	Treatment
Influenza (Flu)	Influenza virus	Fever, cough, body aches, fatigue	1 billion cases, 290,000–650,000 deaths (WHO, 2023)	Antiviral drugs (oseltamivir, zanamivir), supportive care
COVID-19	SARS-CoV-2	Fever, cough, loss of taste/smell	689 million cases, 6.8 million deaths (as of 2023) (Johns Hopkins, 2023)	Antivirals (remdesivir), monoclonal antibodies, supportive care
HIV/AIDS	Human Immunodeficiency Virus (HIV)	Weak immune system, weight loss	39 million living with HIV, 630,000 deaths (UNAIDS, 2023)	Antiretroviral therapy (ART)
Dengue	Dengue virus	High fever, severe headache, joint pain	400 million infections, 40,000 deaths (WHO, 2023)	Supportive care, hydration
Rabies	Rabies virus	Fever, paralysis, hydrophobia	59,000 deaths (WHO, 2023)	Post-exposure prophylaxis (PEP), rabies immunoglobulin

continued on following page

Table 1. Continued

Disease	Causative Agent	Symptoms	Global Burden (Annual Cases/Deaths)	Treatment
Polio	Poliovirus	Muscle weakness, paralysis	5 cases (wild polio, 2023); 800 vaccine-derived cases (GPEI, 2023)	Supportive care, physical therapy
Ebola	Ebola virus	Fever, severe bleeding, organ failure	30 outbreaks since 1976, 50–90% mortality rate	Supportive care, investigational treatments (antivirals)

EPIDEMIOLOGY

Understanding the epidemiology of viral diseases is crucial for developing targeted prevention and treatment strategies. The following section examines the global burden of viral diseases, their transmission patterns, and their impact on public health.

Despite the perception that virus infects only small proportion of population, they contribute significantly to global health burdens causing millions of deaths annually. Viral diseases encompass a wide spectrum from easily transmittable infections like influenza and COVID-19 to geographically restricted or vector borne diseases such as dengue, zika and rabies (WHO, 2023). Factors such as seasonality, population density, mobility, geography, and human contact plays a crucial role in spread and severity of these diseases (Guglani, 2018). For instance, tropical climates facilitate the transmission of vector borne diseases while densely populated urban centers often become hotspots for viral outbreaks due to close human interaction (WHO, 2024).

Influenza and COVID-19 are examples of respiratory viruses, and they impact people globally, with influenza leading to epidemic like seasonal diseases, which results to 3 to 5 million severe illness and up to 290,000 to 650,000 respiratory deaths (WHO, 2024). The SARS-CoV-2 virus that led to COVID-19 changed the world healthcare system dramatically: over 680 million people received a diagnosis, and over 6.8 million died by 2023 (Johns Hopkins University, 2023). Such respiratory viruses are very relevant in the cases of pandemic and the need to put in place strategies to control spread.

Vector-borne diseases that are transmitted through mosquitos, flees or ticks are another major global concern in tropical and subtropical regions. The incidence of vector-borne viral diseases including dengue, Zika, and chikungunya has risen significantly by effects of climate change, increasing population density in urban areas, and intercontinental travel. Some diseases such as dengue affect about 390 million people every year out of which only 96 million experience the disease symptoms

(CDC, 2022, European Centre for Disease Prevention and Control, 2023). Vector control remains the cornerstone of dengue prevention, focusing on eliminating mosquito breeding sites, using insecticide-treated nets, and deploying Wolbachia-infected mosquitoes to reduce viral transmission (Ritchie et al., 2022). Zika virus which was quickly spread in 2015- 2016 is a mosquito- borne virus that has become established in several new areas and caused public health concerns (WHO, 2022).

Rabies is a fatal zoonotic disease with an estimated 59,000 human deaths annually, primarily in Asia and Africa, where access to post-exposure prophylaxis (PEP) remains limited (Hampson et al., 2022). The virus is transmitted through the bite or scratch of infected animals, predominantly domestic dogs, which account for 99% of human rabies cases (WHO, 2023). In contrast, wildlife reservoirs (e.g., bats, foxes) contribute to transmission in high-income countries.

Despite being 100% preventable through timely PEP and canine vaccination, rabies persists due to low awareness, inadequate vaccine availability, and insufficient surveillance. The "Zero by 30" initiative, led by WHO and partners, aims to eliminate human rabies deaths by 2030 through mass dog vaccination, improved access to PEP, and public education (WHO, 2018). Countries like India and the Philippines have successfully reduced rabies incidence through nationwide vaccination campaigns. However, challenges such as cost, cold-chain logistics, and rural healthcare access continue to hinder elimination efforts.

However, high-risk communicable diseases that are now preventable by vaccination include children's measles, mumps, rubella, and hepatitis B diseases still pose a problem in areas of low overall vaccination rates. They can reinfect, especially where immunization rates have dropped. For example, measles led to the death of more than 140000 people around the world in 2018 particularly in children who had not been immunized (WHO, 2019).

Polio a dreadful disease costing millions of people their mobility – has been almost wiped out by vaccinations. Up till now only Afghanistan and Pakistan have not been polio free however few cases of outbreaks of vaccine-derived poliovirus have been reported in other countries in areas where no or very little vaccination has taken place (Global Polio Eradication Initiative, 2022). The global eradication of smallpox in late 1980s through vaccination is a noble example of what vaccination could achieve in the world, as this alone has saved millions of lives (WHO, 2020). Efforts to eradicate polio continue through routine immunization campaigns, environmental surveillance, and targeted outbreak response. However, challenges such as conflict, vaccine hesitancy, and gaps in immunization coverage pose barriers to complete eradication. Recent outbreaks of VDPV in countries like Nigeria and the Democratic Republic of Congo highlight the need for sustained immunization efforts and continued surveillance (GPEI, 2022).

Some more viral diseases, for example, Ebola and SARS remain workable threats to man, especially in the Third World where facilities are limited. The west Africa Ebola viral disease outbreak, which begun in 2014 and ended in 2016 affected more than 28000 people and lead to 11225 deaths; the most affected countries were Guinea, Liberia, and Sierra Leone (WHO, 2021). The Democratic Republic of the Congo (DRC) continues to experience recurrent Ebola outbreaks, largely due to persistent zoonotic transmission from bats and great apes, as well as human-to-human spread in healthcare settings. Recent advancements, such as the rVSV-ZEBOV vaccine (Ervebo), have significantly improved outbreak containment by providing rapid protection for high-risk populations (Henao-Restrepo et al., 2017).

In the same manner, pointed out that viral diseases are one of the leading concerns to public health globally. Vaccination, vector control, and global surveillance have played a role to reducing and or preventing most of these diseases though new viral threats call for increased investment in researching and strengthening healthcare systems, and public health preparedness.

Understanding the epidemiology of viral diseases is crucial for developing effective treatment strategies. As viral outbreaks evolve, antiviral therapies play a pivotal role in managing infections and reducing disease burden.

BACKGROUND AND THERAPEUTIC APPROACHES OF VIRAL DISEASES

Antiviral treatments targets different stages of viral replication cycle to inhibit or reduce viral activity within the host cell. In the below sections, we are going to discuss the main antiviral agents for different viral diseases, exploring the mechanisms and challenges in viral treatments.

Antivirals are designed to inhibit the viral replication, have achieved the varying degree of success. For example: oseltamivir (Tamiflu) is a neuraminidase inhibitor, halts the influenza virus release from the infected cell, thereby reducing the viral spread in the body. Similarly, nucleoside analogs such as acyclovir, is effective against herpes virus by integrating into viral DNA and inhibit the replication process (Brown et al., 2021). For HIV. Antiretroviral therapy (ART) combines drug targeting multiple viral processes that includes, protease activity and reverse transcription to achieve the viral suppression (Saag et al., 2020). Despite these advancement issues of viral resistance, particularly for HIV and hepatitis, it is necessary to continue the development of novel treatments (Temereanca & Ruta, 2023).

Combination therapies have proven particularly effective in treating chronic infections like HIV and hepatitis B. For instance, highly active antiretroviral therapy (HAART) has transformed HIV from fetal to disease to a manageable chronic condition, significantly reducing the viral load enhancing patients' quality of life. However, these therapies require high adherence rate to avoid resistance and sustain effectiveness (Adewuyi et al., 2024). In hepatitis C, combination therapies with direct0acting antivirals (DAAs) have achieved cure rate exceeding 95%, setting the example for potential targeted antiviral treatment (Nguyen et al., 2023).

While antiviral drugs help manage and treat viral infections, they do not always provide complete immunity. This limitation has driven the development of vaccines, which offer long-term protection and play a critical role in global disease eradication efforts. To understand the progress in vaccine development, it is essential to examine their origins and how they have evolved over time.

Origin of Viral Vaccines

The origin of viral vaccines traces back to the early attempts at smallpox prevention through variolation, later redefined into vaccine development by Edward Jenner in the 18th century. The discoveries made by Jenner established modern immunology as a field that pushed future developments in vaccine science. The 20th century saw historical progress when scientists developed inactivated and live attenuated vaccines which allowed successful prevention programs of polio and measles among other diseases (CDC, 2023).

The recent Scientific progressions in 21st century have announced mRNA vaccines and recombinant vector vaccines. These technologies emerged with the Ebola and COVID-19 pandemics. DNA and protein subunit vaccines as seen with hepatitis B and Human papillomavirus (HPV), further expand the arsenal, tailoring immunity based on viral structure and infection pathways (Wang, J. et al., 2022). The success of mRNA vaccines against COVID-19, developed by Pfizer-BioNTech and Moderna, demonstrated the adaptability and efficacy of this approach, offering a promising platform for future rapid response vaccines (Pardi et. al., 2018). These innovations underscore the field's evolution towards precision targeted vaccines capable of responding to emerging viral threats.

Currently available viral vaccines

The early breakthroughs in vaccine science laid the foundation for the development of numerous life-saving vaccines. Today, a variety of vaccines exist to combat the most prevalent and dangerous viral infections. The following section provides an overview of the key vaccines currently in use and their role in global health.

Several vaccines against viral infections are available in the global market, providing immunity against some of the most prevalent and dangerous viruses. These include vaccines for diseases like measles, influenza, hepatitis, and COVID-19, which have significantly reduced morbidity and mortality rates. Continued advancements in vaccine technology, such as mRNA and vector-based platforms, are expanding the scope for preventing emerging viral threats.

Vaccines are tailored to specific population based on the epidemiology of the disease and the immune system's response in different stages of life. For example, the influenza vaccine is available in both inactivated and live-attenuated forms is recommended annually for all ages particularly high-risk groups such elder and young children. This is mainly due to influenza evolves rapidly requiring yearly updates to vaccine composition to match circulating strains (CDC, 2023).

The recombinant hepatitis B virus vaccine, including brands like Engerix-B, Recombivax HB is administered in 3 doses scheduled over six months. It is essential for infants, healthcare workers and other high patients to prevent chronic hepatitis which may lead to liver cirrhosis or cancer (WHO 2023, CDC 2023). Similarly, the recombinant Human Papilloma Virus vaccines, Gardasil, Cervarix targets pre-teens and young adults. This population is prioritized because the vaccine is most effective before the exposure to the virus, typically through sexual activity (Markowitz et al., 2018)

COVID-19 vaccines, including mRNA viral vaccines like Pfizer, Moderna and adenoviral vector vaccines like Johnson & Johnson are administered in all age individuals. These vaccines play a critical role in reducing severe disease and transmission rates with additional booster doses recommended for maintaining the immunity in high-risk groups (Weinreich et al., 2021; Haidar et al., 2021).

The live attenuated measles, mumps and rubella vaccines are routinely given to children in two dose regimens to achieve long lasting immunity. Susceptible adults without prior vaccination or immunity are also advised to receive the MMR vaccines to prevent outbreaks of these highly contagious diseases (Plotkin, 2014). Additionally, the varicella-zoster vaccines, Varivax and Shingrix, are administered to children for chicken pox prevention (CDC, 2023).

These vaccine recommendations are shaped by the diseases burden, transmission dynamics and the ability to achieve optimal immune protection within the targeted population. Table 02 gives a detailed idea about the prominent viral vaccines, their types and target populations:

Table 2. Viral Vaccines Classification, Administration and Target Demographics

Virus	Vaccine Type	Vaccine Name(s)	Dosage Regimen	Target Population
Influenza	Inactivated/Live-Attenuated	Fluzone, FluMist	Annual	All ages, high risk (elderly, young children)
Hepatitis B	Recombinant	Engerix-B, Recombivax HB	3 doses over 6 months	Infants, healthcare workers, at-risk adults
Human Papillomavirus (HPV)	Recombinant	Gardasil, Cervarix	2-3 doses (age-based)	Pre-teens, young adults
COVID-19	mRNA, Adenoviral Vector	Comirnaty (Pfizer), Moderna, J&J	2 doses (+ boosters)	All eligible individuals
Measles, Mumps, Rubella (MMR)	Live-Attenuated	MMR	2 doses	Children, susceptible adults
Varicella-Zoster	Live-Attenuated	Varivax, Shingrix	2 doses	Children, adults (Shingles in older adults)

Ref: CDC (2023); WHO (2023)

The global success of these vaccines is demonstrated by the near eradication of disease like polio, with only a few remaining cases in isolated regions (Global Polio Eradication Initiatives, 2020). HPV vaccines have also contributed significantly to reduce cervical cancer rates, while influenza vaccines are revised annually to address circulating strains effectively (Gul, Somia, et al, 2024, Schiller & Lowy, 2012). The breadth of available vaccines illustrates the impact of immunization on public health, particularly in controlling preventable viral infections (WHO, 2024).

The impact extends beyond individual protection, contributing to population-level immunity, which helps to protect vulnerable groups such as the elderly, immunocompromised and those unable to relieve vaccines. In countries with high vaccination coverage, the virus's reproduction number (RO) declined significantly, reducing the frequency and scale of outbreaks. This has been further supported by targeted booster campaigns which have sustained immunity over time and against variants of concerns.

However, the success of vaccination programs has varied globally due to disparities in vaccine access, logistical challenges in distribution, vaccine hesitancy and public health policies (Ghamdi, E., et al., 2023). Higher income countries were able to procure and distribute vaccines more rapidly achieving earlier widespread coverage, while low- and middle-income countries faced delays due to supply chain issues. Vaccine hesitancy fueled by misinformation and cultural factors has posed additional challenges, emphasizing the need for tailored strategies and community engagement to address fears and misinformation (Tiwana, M. H., & Smith, J., 2024).

Side effects of viral vaccines

Vaccines, while highly beneficial may carry side effects ranging from mild to moderate, and in rare cases severe. Typical side effects include pain at site of administration, mild fever, and muscle aches. These mild symptoms usually resolve without intervention (CDC, 2023). Serious adverse reactions are rare but can include allergic responses which may be managed through pre vaccination screening protocols (Wen et.al., 2021). For example, rare cases of pericarditis and myocarditis have been reported in young males after receiving the second dose of Pfizer -BioNTech and Moderna-mRNA vaccines (CDC, 2023; Oster, M. E. et. al., 2022). Similarly, in women predominately under the age of 50, few but serious cases of thrombosis with thrombocytopenia syndrome (TTS) were also reported who received Johnson and Johnson (Shimabukuro, T. T., Cole, M., & Su, J. R., 2021).

Vaccines are generally safe and effective; however, they can cause side effects, which vary in frequency and severity. Table 03 categorizes common vaccine side effects and their public health relevance to provide a clearer understanding of vaccine safety

Table 3. Common Vaccine Side Effects and Their Public Health Relevance

Vaccine	Side Effect	Public Health Relevance
COVID-19 (Pfizer, Moderna - mRNA)	Pain at Injection Site	Normal immune response, self-limiting
COVID-19 (Pfizer, Moderna - mRNA)	Fatigue	Can affect daily activities, resolves in 1–2 days
COVID-19 (J&J - Adenoviral Vector)	Blood Clotting Issues	Higher risk in younger women, monitoring advised
MMR (Measles, Mumps, Rubella - Live-Attenuated)	Rash, Fever	Short-lived, no long-term effects
HPV (Gardasil, Cervarix)	Fainting (Vasovagal Response)	Requires observation post-vaccination
Varicella (Chickenpox - Live-Attenuated)	Mild Rash	No major health impact
Hepatitis B (Recombinant)	Anaphylaxis	Requires emergency intervention
COVID-19 (Pfizer, Moderna - mRNA)	Pain at Injection Site	Normal immune response, self-limiting
COVID-19 (Pfizer, Moderna - mRNA)	Fatigue	Can affect daily activities, resolves in 1–2 days
COVID-19 (J&J - Adenoviral Vector)	Blood Clotting Issues	Higher risk in younger women, monitoring advised

continued on following page

Table 3. Continued

Vaccine	Side Effect	Public Health Relevance
MMR (Measles, Mumps, Rubella - Live-Attenuated)	Rash, Fever	Short-lived, no long-term effects
HPV (Gardasil, Cervarix)	Fainting (Vasovagal Response)	Requires observation post-vaccination
Varicella (Chickenpox - Live-Attenuated)	Mild Rash	No major health impact

While most vaccine side effects are mild and self-limiting, rare severe reactions reinforce the importance of post-vaccination monitoring and patient education.

Among the various types of vaccines globally available today, the COVID-19 vaccines have garnered particular attention due to the global impact of pandemic The Figure 01 shows how major global regulatory authorities including FDA and WHO handle vaccine approval timelines for COVID-19 vaccines.

Figure 1. Major Global Regulatory Authorities Vaccine Approval Timelines For COVID-19 Vaccines

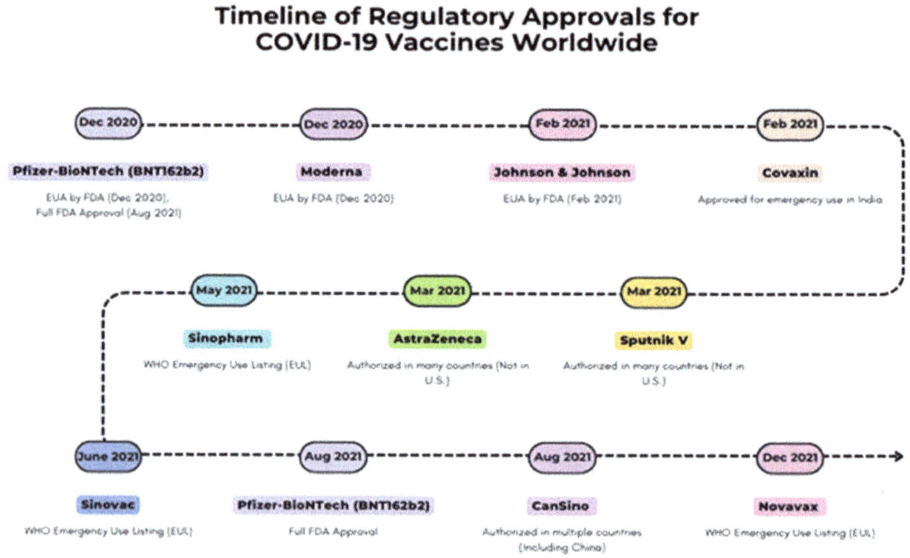

VACCINE HESITANCY; MISINFORMATION AND PUBLIC DISTRUST

Vaccine hesitancy remains a significant public health challenge, fueled by misinformation, historical distrust, and personal beliefs. The World Health Organization (WHO) has identified vaccine hesitancy as one of the top ten global health threats, as it continues to undermine immunization efforts and contribute to the resurgence of preventable diseases (WHO, 2023). False claims about vaccine safety, potential infertility risks, and conspiracy theories such as microchips being implanted through vaccines have significantly weakened public trust, particularly during the COVID-19 pandemic.

The extent and impact of vaccine hesitancy vary across regions, often influenced by healthcare infrastructure, government policies, and public attitudes. Some countries struggle with public distrust despite strong healthcare systems, while others face logistical barriers that exacerbate hesitancy. Regional approaches to vaccine implementation reflect these challenges, with differences in strategies, success factors, and obstacles shaping immunization efforts. The following table outlines key vaccine implementation strategies across different regions, highlighting the factors that contribute to success and the challenges that hinder widespread acceptance. An overview of regional vaccine strategies with their associated obstacles and enablers are listed in Table 04.

Table 4. Different Vaccine Strategies in Various Regions

Region	Vaccine Implementation Strategy	Success Factors	Challenges
North America (USA, Canada)	High reliance on mRNA vaccines (Pfizer, Moderna)	Strong public health funding, rapid rollout	Vaccine hesitancy, misinformation
Europe	Hybrid approach (mRNA, vector-based, inactivated vaccines)	Universal healthcare ensures accessibility	MMR hesitancy in some regions
Africa	Focus on routine immunization programs & COVAX	International aid, WHO support	Limited cold-chain infrastructure
Asia (India, China)	Mass campaigns & domestic vaccine production	High population coverage, local manufacturing	Rural accessibility issues
Latin America	Mixed strategy (national programs + COVAX aid)	High public trust in vaccination	Economic constraints, supply chain issues

Moon, S., et al. (2023). Overcoming vaccine inequity: Lessons from COVID-19. Nature Medicine, 29(1), 15-19.; WHO. (2023). Global vaccine distribution and regional strategies.

Challenges in Vaccine Accessibility and Distribution

Inadequate vaccine access continues to be the worst form of disparity in global health, with stark contrasts between high and low-income countries. While the COVID-19 pandemic was unfolding, the relatively wealthier nations pre-ordered and secured vaccines from various suppliers quite early, leaving the low-income countries grappling with a delayed burden of insufficient quantities (Moon et al., 2023). Besides supply deficiencies, logistical factors such as cold chain requirements for vaccines particularly those ultra-cold mRNA-based vaccines have rendered intra-country distributions extremely problematic for rural and less developed areas. Moreover, due to intellectual property rights constraining production in low-income countries, they were unable to produce their own supplies and thereby further extended lag times in equitably vaccinating the world.

To overcome these hurdles, COVAX (a global initiative led by Gavi, the Vaccine Alliance, the Coalition for Epidemic Preparedness Innovations (CEPI), the World Health Organization (WHO) and UNICEF) and regional manufacturing hubs have been initiated to enhance fair vaccine access. However, such initiatives struggle to fulfill their massive demand for vaccine, especially in resource-constrained settings. Increasing vax production capacity in regions like Africa, Asia, and Latin America can be a more sustainable remedy whereby they will not depend on external suppliers and ensure faster distribution among themselves in case of any public health emergency.

Historical Vaccine Controversies and Their Impact on Public Perception

Vaccine hesitancy is not a new phenomenon; it has pretty much been molded by past debates and public health blunders that go on to affect how people trust vaccines now. Misinformation about vaccines, alongside safety concerns and violations of ethics, have all added to the skepticism; getting top-notch immunization coverage in some communities becomes a challenge. These debates left behind long-lasting effects; even when scientific proof of the safety and efficacy of vaccines comes up, some populations still hold their hesitancy because of past experiences.

Events that have taken place and been put in a table above demonstrate how errors in vaccine production, fraudulent scientific claims, and misinformation campaigns have bent public perception over so many years. While some controversies did result in incidents like the Cutter Incident and perhaps deservedly so led to improvements in the quality control of vaccines (Offit, 2005), others like the MMR-autism myth (Li, N., Stroud, N. J., & Jamieson, K. H., 2017) have far longer-term tragically resulting-in-lower-vaccination-rates-and-outbreaks-of-preventable-diseases consequences.

In many instances, misinformation has been magnified by social media and digital platforms through which false narratives gain rapid traction to alter confidence in vaccination on a global scale (Broniatowski et al., 2018; Wilson & Wiysonge, 2020). Addressing these historical concerns requires transparent communication, stronger public health education, and proactive measures to counter misinformation, ensuring that vaccine hesitancy does not continue to hinder immunization efforts. Table 05 gives brief details about significant vaccine related controversies and their impact on public perception.

Table 5. Major Historical Vaccine Controversies and Their Impact on Public Perception

Event	Incident Details	Impact on Public Perception
The Cutter Incident (1955)	A faulty batch of the polio vaccine led to 40,000 cases of polio due to improper inactivation.	Increased fears about vaccine safety; led to stricter vaccine quality control.
MMR Vaccine and Autism Myth (1998)	A fraudulent study falsely linked the MMR vaccine to autism.	Despite being debunked, this claim led to a decline in MMR vaccination rates.
Sterilization Rumors (Africa, 1990s–2000s)	False claims that polio vaccines contained sterilizing agents.	Contributed to vaccine refusal in Nigeria, delaying polio eradication efforts.
COVID-19 Vaccine Misinformation (2020s)	Social media-driven myths about infertility, DNA alteration, and microchips in vaccines.	Created widespread distrust, slowing vaccine uptake globally.

Offit, P. A. (2005); Centers for Disease Control and Prevention, 2024);
Jegede, A. S. (2007), World Health Organization (2019)

Such historical incidents underscore the importance of proactive public engagement, ethical transparency, and rapid response to misinformation to maintain public confidence in vaccination programs (Dubé et al., 2022).

Case studies on vaccines rollouts

To better understand the impact of different vaccine implementation strategies, it is essential to examine real-world case studies. These examples highlight both the achievements and ongoing challenges in global vaccination efforts.

In India, the Pulse Polio Immunization Program transformed the public health landscape by reducing polio cases from over 200,000 per year in the 1990s to achieving a polio-free status by 2014. This remarkable success was driven by comprehensive mass vaccination campaigns that reached even the most remote areas, coupled with rigorous surveillance and innovative strategies such as house-to-house vaccination drives. Community engagement played a critical role in overcoming cultural and logistical barriers, ensuring high vaccine uptake, and sustaining immunity across diverse populations (WHO,2021; Gavi, 2021). This case illustrates how persistent efforts and coordinated public health initiatives can eradicate a once-devastating disease.

In contrast, the resurgence of measles in parts of the United States and Europe underscores the consequences of declining vaccination rates. Despite the availability of the highly effective MMR vaccine with a two-dose regimen offering approximately 97% protection vaccine hesitancy fueled by misinformation and historical controversies has led to significant outbreaks. In 2019, the United States experienced an outbreak with around 1,250 confirmed cases, while several European nations reported over 20,000 cases in 2022. These outbreaks not only expose the vulnerability of populations when herd immunity is compromised but also highlight the critical need for proactive public health communication and community-based interventions to rebuild trust in vaccination programs (CDC, 2020; WHO, 2025).

The global rollout of COVID-19 vaccines revealed deep disparities in vaccine access between wealthier and lower-income nations. High-income countries, leveraging substantial resources and early pre-purchase agreements, secured large vaccine stocks, which enabled rapid mass immunization. Conversely, many low-income countries faced severe challenges due to limited supply, inadequate cold chain infrastructure, and logistical hurdles that delayed widespread vaccination. This inequitable distribution exacerbated existing health disparities and hampered efforts to control the pandemic on a global scale. The experience underscored the urgent need for coordinated international strategies—such as technology transfer, flexible intellectual property policies, and strengthening local manufacturing capacity to ensure fair and timely access to vaccines for all populations (Moon et al., 2023; Yamey et al., 2023).

GAPS IN RESEARCH, AVAILABILITY, AND EFFICACY FOR SPECIFIC VIRAL TREATMENTS

The lessons learned from past vaccine rollouts have significantly influenced ongoing research and development efforts. Innovations in vaccine technology continue to address the challenges highlighted in these case studies, ensuring more effective and accessible immunization strategies. Despite advancements, viral treatment research remains uneven across viruses, with some diseases lacking effective antiviral therapies altogether (CDC 2023). For example, dengue virus (DENV) poses a unique challenge as it has four distinct serotypes that can infect the person making the vaccine and antiviral development. Currently, there is no specific antiviral for dengue and treatment primarily involves supportive care to manage symptoms and prevent complications like dengue induced hemorrhagic fever. While the vaccine Dengvaxia has been developed, it is only recommended for the individuals with a prior dengue infection and is not widely used due to safety concerns especially in individuals who have not been previously infected (Wilder-Smith, A., 2020). The lack of universal and broadly applicable antiviral treatment leaves millions vulnerable particularly in tropical regions where dengue is endemic.

However, several other issues have also been found to affect treatment in consideration to antiviral research. There are certainly research gaps unique to antiviral treatments but there are also broader issues of accessibility and resistance to antiviral therapies. Societal factors that interrupt with the administration of the antiviral medications include the cost of medications, absence of healthcare facilities, geographical disparities, lack of equitable access to healthcare amenities. Pre-access barriers to effective treatments persist in low- and middle-class income regions leading to greater health disparities among populations in these nations. For example, treatment with direct acting antivirals (DAAs) are available for hepatitis C, but they are expensive and not accessible in most parts of the world and as a result hepatitis C continues being spread and affecting people's health (WHO, 2022).

Another problem that lies ahead is antiviral resistance. It can develop through the Genetic mutations of viruses that permit the virus to be resistant from the effects of drugs. This can clearly be seen in HIV and influenza whereby the treatment failures associated with the kind of resistance have been reported. In HIV therapy, drug resistance of the virus strains poses subsequent problem on treatment, thus there is the constant invention of the new antiretroviral drugs is needed. Likewise, the influenza virus has a seasonal pattern as well as develops subtypes making existing vaccines and antiviral drugs less effective; this underlines the importance of constant research on resistance (Paules et al., 2022).

vaccine resistance is another major problem, which also contributes to poor coverage and subsequent poor consequence while conducting viral disease prevention. Hesitancy ends up due to factors such as; misperception, low confidence in the healthcare system and doubts over the efficacy of vaccines (Dube et al., 2021; Katz et al., 2019). The consequence of vaccine hesitancy can be high and includes decreased vaccination rates, which in turn means that the easily preventable diseases break out. Measles is one of the brightest examples of how hesitancy can reverse the progress achieved over the years in the field of public health (Gastanaduy et al., 2019).

In background of COVID-19, some gaps and challenges evident in COVID-19 context where evident progress has been achieved in terms of vaccine and antiviral treatment development. The one which exists to some extent is the gap in COVID-19 vaccine distribution to different countries of the world and its access to low middle income countries have been challenged by low vaccine uptake leading to slow vaccination than high income countries due to insufficient supply of vaccines (Chaudhari et al., 2021). The main hindrances which have been attributed to delayed dispersal of vaccine include manufacturing restrictions, logistic tasks, and intellectual property issues. Even the initiatives like in case of COVAX vaccines which meant to ensure equitable access, logistical and financial constraints are still a challenge in unleashing vaccines in many untouched areas hence continue to expose the more vulnerable populations to the deadly diseases (Gavi, 2021). These gaps must be filled not only through global commitment and assistance but also through the enhancement of efforts in developing countries to increase immunization coverage and access in hard-to-reach populations (Usher, et al. 2021).

Furthermore, the appearance of SARs-COV-2 variants including Delta and Omicron has therefore created a new major concern on effectiveness of existing vaccines and treatments. Variants with mutations in the spike protein have demonstrated some ability to evade immunity which may lead to reduced vaccines effectiveness, and increase the need of booster doses to maintain protective immunity (Wang et al., 2021). Research indicates that while m-RNA vaccines Pfizer-Biotech and Moderna show highly efficacy against greater transmissibility (Tregoning et al. 2021). This dynamic highlight the need for ongoing surveillance and rapid adaptation in vaccine formation to respond the emerging variants, along with continued investment in research to develop more broadly neutralizing antiviral agents and next generation vaccines that offer durable immunity across variants (Sadoff et al., 2021). An analysis of challenges and their proposed solutions for improving the access of vaccines across different regions of the world are summarized in Table 06.

Table 6. Challenges and Mitigation Strategies in Vaccine Access

Challenges	Details	Potential Solutions
Global Inequity	Vaccine shortages and delays in low- and middle-income countries, widening health disparities.	Expand COVAX, boost local production capacity
Manufacturing & Distribution	Limited production, shipping issues, and IP restrictions hamper timely access in many regions.	Increase production, relax IP for faster distribution
Variant Evolution	New variants (Delta, Omicron) reduce vaccine efficacy, increasing demand for booster doses.	Rapid vaccine adaptation, ongoing research
Antiviral Options	Limited availability of broad-spectrum antivirals to cover emerging variants and severe cases.	Investment in next-gen antivirals, variant-resistant vaccines

continued on following page

Table 6. Continued

Challenges	Details	Potential Solutions
Vaccine Hesitancy	Misinformation and mistrust hinder vaccine uptake, especially in certain communities.	Strengthen public education, community engagement
Ongoing Surveillance Need	Continuous tracking needed to monitor variant spread and vaccine effectiveness over time.	Enhance global tracking systems, flexible vaccine platforms

WHO, 2022

CURRENT TRENDS AND FUTURE DIRECTIONS ON VIRAL VACCINES

Despite these global initiatives, significant challenges persist in achieving universal vaccine coverage. Addressing emerging variants, overcoming distribution barriers, and tackling misinformation remain critical priorities for the future.

The field of viral vaccines has received considerable advancements particularly with the emergence of newest technologies that diversify coverage and enhance the effectiveness of vaccines. These include viral vector and mRNA technologies which are classified as revolutionary inventions and the COVID-19 pandemic accelerated the use and advancement of the technology. Moreover, trends that are emerging the future of viral immunization include, the development of nanoparticle vaccines and discovery of Universal vaccines. This heading will discuss these current trends and prospects of viral vaccines, including the exciting novelties in delivery systems that are being developed with the probabilities of enhancing vaccine delivery.

Moderna and Pfizer-BioNTech COVID-19 vaccines are the most pioneering types of mRNA vaccines that are predicted to bring a revolutionary shift in present and future vaccine development. These vaccines carry messenger RNA within the body to prompt cells to manufacture a harmless fragment of the spike protein on the surface of the SARS CoV-2 virus, thereby causing the desired immune reaction (Polack et al., 2020). Another significant strength of the mRNA vaccines, the fact that they can easily be created in a more regimen of weeks thus they can be of great help in case of a new emerging infectious diseases. This technology is now being used in virus diseases such as influenza, the Zika virus and HIV with current clinical trials on evaluating the safety and efficacy (Gunter et al., 2021).

Another example is viral vector vaccines which involved a harmless virus to introduce genetic material carrying an antigen that belongs to the target virus. Two vaccines made using the viral vector technology are the Oxford AstraZeneca and the Johnson & Johnson. Measuring between 10 and 25 nanometers, these vaccines use an adenovirus vector to trigger immunity without causing illness (Voysey et al., 2021). These viral vectors render easy adaptation in response to host-pathogen challenges, making them the future vaccine development area for Nipah and Hendra emergence virus (Gazal, S., et al., 2022).

In addition to mRNA and viral vector technologies, recent advancements have come up with nanoparticle vaccines in the vaccine catalogue. These vaccines use chemically synthesized nanoparticles to display viral antigens that correctly mimic the structure of viruses to stimulate the immune response (Baker & Wang et al., 2021). Nanoparticles can enhance the stability and the efficiency of vaccines and boost immunogenicity. For instance, the development of nanoparticle-based vaccines against flu and other respiratory viruses, including COVID 19 is in progress, which began to demonstrate the immune effect (Baker et al., 2021). This technology may open the door to the development of better vaccines against numerous viral diseases.

A breakthrough in vaccine studies is the development of universal vaccines that will target different disease strains. An example of these is the search for broad influenza vaccines which offers long-term cross-reacting immunity against many strains of the influenza virus hence eliminating the need to take the flu vaccine seasonally (Beans, C., 2022). There are existing strategies of designing a broad-spectrum vaccine that would target conserved regions of the various RNA virus, which are the least likely to mutate (Wright et al., 2021). Likewise, researchers are working on next generation HIV vaccines which should trigger broadly neutralizing the antibodies that can target various strains of the viruses. Although, HIV vaccine research continues to present difficulties because of viral genetic variability and immunopathogenesis, the advancement in other areas of vaccine development offers a potential avenue toward the search for the effective HIV vaccine (Sanders et al. 2018).

Advancements in vaccines are also revolutionizing the way the vaccines are delivered. Traditional intramuscular injections may be unappealing to some people, thus hampering the overall coverage of immunizations. To this end, new delivery systems are under consideration and planning to improve patient compliance and utility. For example, nasal vaccines offer non-parenteral delivery with the ability to trigger mucosal immunity desirable for respiratory infections. The work to improve the nasal formulations for the vaccines against influenza and COVID-19 is on the way (Sarkar et al. 2022).

Another frontier of the vaccine area is the oral vaccines. To ease vaccination, especially in the developing nations, scientists work in developing vaccines that can be taken orally. Currently, there is steady growth in formulation technologies as a prerequisite to the development of oral vaccines that can survive stomach acid besides stimulating an immune response (Katz et al., 2020).

Microneedle patches are another innovative approach that seems to be causing considerable interest. These patches are in the form of tiny needles that do not cause any pain and pierce the skin to administer the vaccine. Besides, it improves patient satisfaction and can invite for self-vaccination, which might help to increase the people's immunization rates (Matsumoto et al., 2021). The fact that these patches can be transported with no need for refrigeration expands their utility in conditions with limited resources.

CONCLUSION

Vaccines have played an instrumental role in controlling and, in some cases, eliminating viral diseases, significantly reducing global morbidity and mortality. Success from vaccination programs continuously demonstrates how immunization transforms public health outcomes for smallpox, polio, and measles infections. Several ongoing challenges continue to impact vaccine programs including vaccine hesitancy, inequitable distribution, and logistical barriers particularly in low outcome regions of the world.

The continuing spread of misinformation together with distrust prevents people from receiving vaccines despite living in high-income countries. Efforts to reduce vaccine reluctance require more than distributing information because purposeful outreach needs community-based programs that resolve cultural social concerns. The obstacles to vaccine access increase since economic inequalities and geopolitical factors have made access challenging. While initiatives like COVAX have attempted to bridge the gap, the COVID-19 pandemic exposed the fragility of global vaccine distribution systems, emphasizing the need for more sustainable and decentralized manufacturing capabilities.

For making vaccines more accessible in rural areas, future research should focus on improving delivery methods and storage systems. Advances in mRNA technology and the development of universal vaccines that could revolutionize immunization via synthesizing the vaccines more adaptable and effective against emerging variants. Furthermore, additional studies are needed to discover how the data analytics and artificial intelligence (AI) can help to predict outbreaks and improve vaccination efforts, guaranteeing a more proactive and effective public health response.

From a policy perspective, governments and international organizations should encourage regional manufacturing centers and establish flexible intellectual property regulations during global health crises to prioritize fair vaccine availability. To overcome current obstacles and increase immunization coverage worldwide, it will be essential to fortify public health infrastructure, guarantee open communication, and promote international cooperation.

Ultimately, while vaccines remain one of the most powerful tools in combating viral diseases, their full potential can only be realized through a combination of scientific innovation, strategic policymaking, and proactive public engagement. The future of global health depends on a more integrated and inclusive approach to immunization, ensuring that no population is left behind in the fight against preventable diseases.

REFERENCES

Adewuyi, O. S., Orevba, J., Yaria, J., Bamgboye, E., & Dairo, D. (2024). Factors that determine the level of non-adherence to HAART among HIV-positive pregnant women in Kaduna State. *World Journal of Biology Pharmacy and Health Sciences*, *19*(03), 458–471. DOI: 10.30574/wjbphs.2024.19.3.0653

Ahmed, R., & Hu, Z. (2021). The emerging role of mRNA vaccines: Lessons learned from the COVID-19 pandemic. *Nature Reviews. Immunology*, *21*(9), 569–570.

Baden, L. R., El Sahly, H. M., Essink, B., Kotloff, K., Frey, S., Novak, R., Diemert, D., Spector, S. A., Rouphael, N., Creech, C. B., McGettigan, J., Khetan, S., Segall, N., Solis, J., Brosz, A., Fierro, C., Schwartz, H., Neuzil, K., Corey, L., & Zaks, T. (2021). Efficacy and safety of the mRNA-1273 SARS-CoV-2 vaccine. *The New England Journal of Medicine*, *384*(5), 403–416. DOI: 10.1056/NEJMoa2035389 PMID: 33378609

Baker, J. R., Dubacheva, G. V., & Gentry, S. (2021). Nanoparticle vaccines: A review of the latest advancements. *Advanced Drug Delivery Reviews*, *170*, 107–123. DOI: 10.1016/j.addr.2020.10.009

Baker, J. R., & Wang, F. (2021). Nanoparticle vaccines for the prevention of infectious diseases. *Frontiers in Immunology*, *12*, 648–658.

Beans, C. (2022). Researchers getting closer to a "universal" flu vaccine. *Proceedings of the National Academy of Sciences of the United States of America*, *119*(5), e2123477119. DOI: 10.1073/pnas.2123477119 PMID: 35082157

Broniatowski, D. A., Jamison, A. M., Qi, S. H., AlKulaib, L., Chen, T., Benton, A., Quinn, S. C., & Dredze, M. (2018). Weaponized health communication: Twitter bots and Russian trolls amplify the vaccine debate. *American Journal of Public Health*, *108*(10), 1378–1384. DOI: 10.2105/AJPH.2018.304567 PMID: 30138075

Browne, M., Thomson, P., & Mijan, J.. (2021). Vaccine hesitancy: A systematic review of quantitative studies. *International Journal of Public Health*, *66*, 584–592. DOI: 10.3389/ijph.2021.00059

CDC. (2023). Vaccine side effects. Centers for Disease Control and Prevention. Retrieved from CDC website.

Centers for Disease Control and Prevention. (2020). Measles Cases and Outbreaks. Retrieved March 28, 2025, from https://www.cdc.gov/measles/data-research/index .html

Centers for Disease Control and Prevention. (2022). *HIV/AIDS: Basics*. https://www.cdc.gov/hiv/basics

Centers for Disease Control and Prevention. (2023). History of measles. Retrieved March 28, 2025, from https://www.cdc.gov/measles/about/history.html

Centers for Disease Control and Prevention. (2023). *Influenza (Flu)*. https://www.cdc.gov/flu

Centers for Disease Control and Prevention. (2023). Underutilization of Influenza Antiviral Treatment Among Children and Adolescents—United States, 2023–24. *Morbidity and Mortality Weekly Report*, *73*(45), 1012–1016. Retrieved March 28, 2025, from https://www.cdc.gov/mmwr/volumes/73/wr/mm7345a2.htm

Centers for Disease Control and Prevention. (2023). *Possible side effects after getting a COVID-19 vaccine*. https://www.cdc.gov/coronavirus/2019-ncov/vaccines/expect/after.html

Centers for Disease Control and Prevention (CDC). (2023). Possible Side Effects After Vaccines. Retrieved from https://www.cdc.gov

Centers for Disease Control and Prevention (CDC). (2023). Varicella and Shingles Vaccines. Retrieved from https://www.cdc.gov

Dube, E.. (2021). Vaccine hesitancy: Causes, consequences, and the way forward. *The Lancet. Infectious Diseases*, *21*(8), e268–e276.

Dube, E., Gagnon, D., & MacDonald, N. E.. (2021). Vaccine hesitancy: Causes, consequences, and the way forward. *The Lancet. Infectious Diseases*, *21*(8), e268–e276. DOI: 10.1016/S1473-3099(20)30723-7

European Centre for Disease Prevention and Control. (2023). *Dengue cases January–December 2023*.

Gastañaduy, P. A.. (2019). Measles — United States, January 1–April 26, 2019. *Morbidity and Mortality Weekly Report*, *68*(17), 396–400. DOI: 10.15585/mmwr.mm6817e2 PMID: 31048675

Gavi, The Vaccine Alliance. (2020, September 17). *India draws lessons from polio eradication initiative*. Retrieved March 28, 2025, from https://www.gavi.org/vaccineswork/india-draws-lessons-polio-eradication-initiative

Gavi, the Vaccine Alliance. (2023). *Annual Progress Report 2023*. Retrieved March 28, 2025, from https://www.gavi.org/progress-report

Gazal, S., Sharma, N., Gazal, S., Tikoo, M., Shikha, D., Badroo, G. A., Rashid, M., & Lee, S. J. (2022). Nipah and Hendra viruses: Deadly zoonotic paramyxoviruses with the potential to cause the next pandemic. *Pathogens (Basel, Switzerland), 11*(12), 1419. DOI: 10.3390/pathogens11121419 PMID: 36558753

Ghamdi, E., MashariAayadhAlotaibi, K. N. A., MeshalBijadAlotaibi, M., Mufadhi-Alanazi, F., Asiree, A. H., Malki, S. M., & Mutairi, A. M. (2023). A comprehensive analysis of global health challenges: Public health interventions, vaccine distribution hurdles, and the socio-economic roots of health disparities. *Chelonian Research Foundation, 18*(1), 272–282.

Global Polio Eradication Initiative. (2022). *Polio eradication: Status report.* https://polioeradication.org/

Global Polio Eradication Initiative. (2023). *Global Polio Eradication Initiative: Annual Report 2022.* World Health Organization. https://polioeradication.org/resources/annual-report-2022-pre-publication-version/contentReference[oaicite:1]index=1

Greenwood, B. (2014). The contribution of vaccination to global health: Past, present and future. *Philosophical Transactions of the Royal Society of London. Series B, Biological Sciences, 369*(1645), 20130433. DOI: 10.1098/rstb.2013.0433 PMID: 24821919

Gul, S., Batool, Z., Bano, S., Aziz, A., & Tanveer, S. H. (2024). Comprehensive Details on Human Papillomavirus (HPV) Infections and Associated Malignancies: Awareness about Prophylaxis & Treatment Options Available for HPV Infections. *Bangladesh Journal of Medical Science, 23*(4), 967–983. DOI: 10.3329/bjms.v23i4.76504

Gunter, H. M.. (2021). mRNA vaccines for COVID-19: A new era of vaccinology. *Current Opinion in Immunology, 71*, 63–69. DOI: 10.1016/j.coi.2021.07.001

Haidar, G.. (2021). Booster Doses and COVID-19 Vaccine Effectiveness in High-Risk Populations. *Journal of the American Medical Association, 326*(11), 1–5.

Haidar, G., Dorritie, K., Fusco, H., Bogdanovich, T., & Samanta, P. (2021). Invasive mold infections after chimeric antigen receptor-modified T-cell therapy: A case series, review of the literature, and implications for prophylaxis. *Clinical Infectious Diseases, 73*(11), e1369–e1374. DOI: 10.1093/cid/ciaa1328 PMID: 32780792

Healy, C. M.. (2021). Safety and immunogenicity of the Novavax COVID-19 vaccine. *The New England Journal of Medicine, 384*(24), 2320–2331. DOI: 10.1056/NEJMoa2103061

Henao-Restrepo, A. M., Camacho, A., Longini, I. M., Watson, C. H., Edmunds, W. J., Egger, M., Carroll, M. W., Dean, N. E., Diatta, I., Doumbia, M., Draguez, B., Duraffour, S., Enwere, G., Grais, R., Gunther, S., Gsell, P.-S., Hossmann, S., Watle, S. V., Kondé, M. K., & Kieny, M.-P. (2017). Efficacy and effectiveness of an rVSV-vectored vaccine in preventing Ebola virus disease: Final results from the Guinea ring vaccination, open-label, cluster-randomised trial (Ebola Ça Suffit!). *Lancet*, *389*(10068), 505–518. DOI: 10.1016/S0140-6736(16)32621-6 PMID: 28017403

Johns Hopkins University. (2023). *COVID-19 dashboard*. https://coronavirus.jhu.edu/map.html

Katz, J.. (2019). Understanding vaccine hesitancy: A public health issue. *Health Affairs*, *38*(4), 587–592. DOI: 10.1377/hlthaff.2018.05248

Katz, J. M.. (2020). Oral vaccines: A practical and effective immunization strategy. *Nature Reviews. Immunology*, *20*(3), 197–210. DOI: 10.1038/s41577-019-0237-7

Kelso, J. M.. (2021). Anaphylaxis After Vaccination: Incidence, Mechanisms, and Management. *The Journal of Allergy and Clinical Immunology*, *147*(4), 1379–1389.

Li, N., Stroud, N. J., & Jamieson, K. H. (2017). Understanding the MMR vaccine controversy: An analysis of the role of journalistic tone. *Health Communication*, *32*(4), 395–402. DOI: 10.1080/10410236.2016.1138475

Matsumoto, Y.. (2021). Microneedle technology: A promising alternative for vaccine delivery. *Advanced Drug Delivery Reviews*, *178*, 113942. DOI: 10.1016/j.addr.2021.113942

Moon, S.. (2023). Overcoming vaccine inequity: Lessons from COVID-19. *Nature Medicine*, *29*(1), 15–19.

Nguyen, V. H., Huang, D. Q., Le, M. H., Jin, M., Lee, E. Y., Henry, L., Nerurkar, S. N., Ogawa, E., Thin, K. N., Teng, M. L. P., Goh, K. S., Kai, J. C. Y., Wong, C., Tan, D. J. H., Thuy, L. T. T., Hai, H., Enomoto, M., Cheung, R., & Nguyen, M. H. (2023). Global treatment rate and barriers to direct-acting antiviral therapy: A systematic review and meta-analysis of 146 studies and 1 760 352 hepatitis C virus patients. *Liver International*, *43*(6), 1195–1203. DOI: 10.1111/liv.15550 PMID: 36825358

Offit, P. A. (2005). The Cutter Incident, 50 years later. *The New England Journal of Medicine*, *352*(14), 1411–1412. DOI: 10.1056/NEJMp048180 PMID: 15814877

Oster, M. E., Shay, D. K., Su, J. R., Gee, J., Creech, C. B., Broder, K. R., Edwards, K., Soslow, J. H., Dendy, J. M., Schlaudecker, E., Lang, S. M., Barnett, E. D., Ruberg, F. L., Smith, M. J., Campbell, M. J., Lopes, R. D., Sperling, L. S., Baumblatt, J. A., Thompson, D. L., & Shimabukuro, T. T. (2022). Myocarditis cases reported after mRNA-based COVID-19 vaccination in the US from December 2020 to August 2021. *Journal of the American Medical Association, 327*(4), 331–340. DOI: 10.1001/jama.2021.24110 PMID: 35076665

Paules, C., Subbarao, K., & Fauci, A. S. (2022). Influenza: Progress and remaining challenges. *Clinical Microbiology Reviews, 35*(2), e00224–e20. DOI: 10.1128/CMR.00224-20

Polack, F. P., Thomas, S. J., Kitchin, N., Absalon, J., Gurtman, A., Lockhart, S., Perez, J. L., Pérez Marc, G., Moreira, E. D., Zerbini, C., Bailey, R., Swanson, K. A., Roychoudhury, S., Koury, K., Li, P., Kalina, W. V., Cooper, D., Frenck, R. W.Jr, Hammitt, L. L., & Gruber, W. C. (2020). Safety and efficacy of the BNT162b2 mRNA Covid-19 vaccine. *The New England Journal of Medicine, 383*(27), 2603–2615. DOI: 10.1056/NEJMoa2034577 PMID: 33301246

Polack, F. P., Thomas, S. J., Kitchin, N., Absalon, J., Gurtman, A., Lockhart, S., Perez, J. L., Pérez Marc, G., Moreira, E. D., Zerbini, C., Bailey, R., Swanson, K. A., Roychoudhury, S., Koury, K., Li, P., Kalina, W. V., Cooper, D., Frenck, R. W.Jr, Hammitt, L. L., & Gruber, W. C. (2020). Safety and efficacy of the BNT162b2 mRNA Covid-19 vaccine. *The New England Journal of Medicine, 383*(27), 2603–2615. DOI: 10.1056/NEJMoa2034577 PMID: 33301246

Prather, K. A., Wang, C. C., & Schooley, R. T. (2020). Reducing transmission of SARS-CoV-2. *Science, 368*(6498), 1422–1424. DOI: 10.1126/science.abc6197 PMID: 32461212

Ritchie, S. A., van den Hurk, A. F., Smout, M. J., Staunton, K. M., & Hoffmann, A. A. (2022). Mission accomplished? *Wolbachia* mosquito deployments eliminate dengue transmission in North Queensland, Australia. *Gates Open Research, 6*, 54. DOI: 10.12688/gatesopenres.13611.2

Sadoff, J., Gray, G., Vandebosch, A., Cárdenas, V., Shukarev, G., Grinsztejn, B., Goepfert, P. A., Truyers, C., Fennema, H., Spiessens, B., Offergeld, K., Scheper, G., Taylor, K. L., Robb, M. L., Treanor, J., Barouch, D. H., Stoddard, J., Ryser, M. F., Marovich, M. A., & Douoguih, M. (2021). Safety and efficacy of single-dose Ad26.COV2.S vaccine against Covid-19. *The New England Journal of Medicine, 384*(23), 2187–2201. DOI: 10.1056/NEJMoa2101544 PMID: 33882225

Sanders, R. W.. (2018). HIV-1 neutralizing antibodies: A comprehensive overview. *Annual Review of Immunology*, *36*, 201–232. DOI: 10.1146/annurev-immunol-042617-090014

Sarkar, J.. (2022). Needle-free vaccination: A new paradigm in immunization strategies. *Vaccine*, *40*(15), 2201–2212. DOI: 10.1016/j.vaccine.2022.02.011

Shimabukuro, T. T.. (2021). Reports of Anaphylaxis and Myocarditis After mRNA COVID-19 Vaccination. *Journal of the American Medical Association*, *325*(11), 1101–1102. DOI: 10.1001/jama.2021.1967 PMID: 33576785

Shimabukuro, T. T., Cole, M., & Su, J. R. (2021). Reports of Anaphylaxis After Receipt of mRNA COVID-19 Vaccines in the US—December 14, 2020–January 18, 2021. *Journal of the American Medical Association*, *325*(11), 1101–1102. DOI: 10.1001/jama.2021.1967 PMID: 33576785

Temereanca, A., & Ruta, S. (2023). Strategies to overcome HIV drug resistance-current and future perspectives. *Frontiers in Microbiology*, *14*, 1133407. DOI: 10.3389/fmicb.2023.1133407 PMID: 36876064

Tiwana, M. H., & Smith, J. (2024). Faith and vaccination: A scoping review of the relationships between religious beliefs and vaccine hesitancy. *BMC Public Health*, *24*(1), 1806. DOI: 10.1186/s12889-024-18873-4 PMID: 38971784

Verger, P., Botelho-Nevers, E., Garrison, A., Gagnon, D., Gagneur, A., Gagneux-Brunon, A., & Dubé, E. (2022). Vaccine hesitancy in health-care providers in Western countries: A narrative review. *Expert Review of Vaccines*, *21*(7), 909–927. DOI: 10.1080/14760584.2022.2056026 PMID: 35315308

Voysey, M.. (2021). Safety and efficacy of the ChAdOx1 nCoV-19 vaccine (AZD1222) against SARS-CoV-2: A preliminary report of a phase 3 trial. *Lancet*, *397*(10269), 99–111. DOI: 10.1016/S0140-6736(20)32661-1 PMID: 33306989

Wang, J., Peng, Y., Zhang, L., & Xu, Y. (2022). Virus-like particle vaccinology, from bench to bedside. *Cellular & Molecular Immunology*, *19*(4), 456–475. DOI: 10.1038/cmi.2014.72 PMID: 34992277

Wang, P.. (2021). Neutralizing antibody responses to SARS-CoV-2 variants. *Nature*, *593*(7857), 133–138. DOI: 10.1038/s41586-021-03398-2 PMID: 33684923

Weinreich, D. M.. (2021). mRNA COVID-19 Vaccines Efficacy and Safety. *The New England Journal of Medicine*, *384*(20), 1–15.

Weinreich, D. M., Sivapalasingam, S., Norton, T., Ali, S., Gao, H., Bhore, R., Xiao, J., Hooper, A. T., Hamilton, J. D., Musser, B. J., Rofail, D., Hussein, M., Im, J., Atmodjo, D. Y., Perry, C., Pan, C., Mahmood, A., Hosain, R., Davis, J. D., & Yancopoulos, G. D. (2021). REGEN-COV antibody combination and outcomes in outpatients with Covid-19. *The New England Journal of Medicine*, *385*(23), e81. DOI: 10.1056/NEJMoa2108163 PMID: 34587383

Wen, F.. (2021). Adverse events following immunization: Current perspectives. *Drug Safety*, *44*(7), 781–794.

Wilson, S. L., & Wiysonge, C. (2020). Social media and vaccine hesitancy. *BMJ Global Health*, *5*(10), e004206. DOI: 10.1136/bmjgh-2020-004206 PMID: 33097547

World Health Organization. Food and Agriculture Organization of the United Nations, & World Organisation for Animal Health. (2018). *Zero by 30: The global strategic plan to end human deaths from dog-mediated rabies by 2030*. https://www.who.int/publications/i/item/9789241513838

World Health Organization. (2019). *Measles*. https://www.who.int/news-room/fact-sheets/detail/measles

World Health Organization. (2020). History of smallpox vaccination. Retrieved March 28, 2025, from https://www.who.int/news-room/spotlight/history-of-vaccination/history-of-smallpox-vaccination

World Health Organization. (2021). *Ebola virus disease*. https://www.who.int/news-room/fact-sheets/detail/ebola-virus-disease

World Health Organization. (2021, April 7). *A push to vaccinate every child, everywhere, ended polio in India*. Retrieved March 28, 2025, from https://www.who.int/india/news/feature-stories/detail/a-push-to-vaccinate-every-child-everywhere-ended-polio-in-india

World Health Organization. (2022). Global progress report on HIV, viral hepatitis, and sexually transmitted infections, 2022. Retrieved from https://www.who.int/publications/i/item/9789240053779

World Health Organization. (2022). *Monkeypox*. https://www.who.int/news-room/fact-sheets/detail/monkeypox

World Health Organization. (2022). *Rabies*. Retrieved from WHO Website.

World Health Organization. (2022). *Viral diseases epidemiology*. https://www.who.int/news-room/fact-sheets

World Health Organization. (2023). *2023 WHO Global Vaccine Market Report*. Retrieved March 28, 2025, from https://www.who.int/publications/i/item/B09022

World Health Organization. (2023). *Dengue and severe dengue*. https://www.who.int/news-room/fact-sheets/detail/dengue-and-severe-dengue

World Health Organization. (2023). *Influenza*. Retrieved from WHO Website.

World Health Organization. (2024). *COVID-19 dashboard*. Retrieved from https://data.who.int/dashboards/covid19/data

World Health Organization. (2024). *COVID-19 vaccines: A guide for healthcare providers*. Retrieved from WHO Website.

World Health Organization. (2024). Global immunization efforts have saved at least 154 million lives over the past 50 years. Retrieved March 28, 2025, from https://www.who.int/news/item/24-04-2024-global-immunization-efforts-have-saved-at-least-154-million-lives-over-the-past-50-years

World Health Organization. (2024). *Influenza (seasonal)*. Retrieved from https://www.who.int/news-room/fact-sheets/detail/influenza-(seasonal)

World Health Organization. (2024). Urban health. Retrieved March 28, 2025, from https://www.who.int/news-room/fact-sheets/detail/urban-health

World Health Organization. (2025). European Region reports highest number of measles cases in more than 25 years. Retrieved March 28, 2025, from https://www.who.int/europe/news/item/13-03-2025-european-region-reports-highest-number-of-measles-cases-in-more-than-25-years---unicef--who-europe

World Health Organization. (n.d.). *COVID-19 vaccine misinformation and disinformation: An urgent public health challenge*. Retrieved March 28, 2025, from https://www.who.int/news-room/spotlight/ten-threats-to-global-health-in-2019

World Health Organization (WHO). (2022). *Challenges and solutions in global vaccine access*. Retrieved from https://www.who.int

World Health Organization (WHO). (2023). Hepatitis B Vaccines. Retrieved from https://www.who.int

Wright, P. F.. (2021). Universal influenza vaccines: Progress and challenges. *Nature Reviews. Immunology*, *21*(6), 374–390. DOI: 10.1038/s41577-020-00473-6

Yamey, G., Garcia, P. J., Hassan, F., Mao, W., McDade, K. K., Pai, M., & Zhao, F. (2023). It is not too late to achieve global COVID-19 vaccine equity. *BMJ (Clinical Research Ed.)*, *376*, e070650. DOI: 10.1136/bmj-2022-070650 PMID: 35331982

Chapter 2
Targeted Delivery and Nanotechnology:
Harnessing Nanoparticles for Precision Medicine in Viral Infection

Krishnaveni Manubolu
https://orcid.org/0000-0002-9401-1768
Narayana Pharmacy College, Nellore, Jawaharlal Nehru Technological University, Anantapur, India

Raveesha Peeriga
https://orcid.org/0000-0003-4827-6076
V.V. Institute of Pharmaceutical Sciences, India

ABSTRACT

The emergence of nanotechnology has profoundly impacted diverse fields of science and medicine, specifically in the area of targeted drug delivery systems. This proposal will investigate the revolutionary ways nanotechnology has impacted antiviral drug therapies by discussing the advantages offered by targeted delivery systems and the upcoming precision medicine. The development of a nanocarrier and a targeting mechanism is anticipated to address the current hurdles responsible for unsatisfactory antiviral therapy outcomes, including underwhelming efficacies and adverse side effects. In this paper, we will focus on reviewing present paradigms in technologies, their applications, and future considerations in developing antiviral therapy.

DOI: 10.4018/979-8-3693-9276-8.ch002

Copyright © 2025, IGI Global Scientific Publishing. Copying or distributing in print or electronic forms without written permission of IGI Global Scientific Publishing is prohibited.

1. INTRODUCTION

Different viral diseases are still on the move affecting most countries in the world and continue to exert pressure on healthcare facilities across the world. However, the modern methodologies require the further development of more advanced antiviral treatment. With the existing drugs there are so many major flaws in their effectiveness towards realisation of the goal of therapy. The effectiveness of the antiviral drugs being used today is quite moderate in achieving the expected therapeutic value. Another challenge related to drug targeting is bioavailability and interferes with most antiviral drugs to reach the site of action with low concentrations making them ineffective. This is compounded by the emergence of drug resistant forms of the viruses, which transform into a form that requires a different mode of treatment hence rendering the drugs to be of little value (Uddin et al., 2021).

Further, traditional antiviral drugs are appeared to have many side activities, that is the drug interfere with other biological targets to produce side effects and toxicity to the body. All these side effects can greatly limit dose and duration of therapy in chronic cases that may worsen treatment programs (Johnson et al., 2022). Further, solubility and stability of the antiviral agents are some other functions that define the efficacy of antiviral agents. These antiviral drugs have poor solubility intd and / or in the physiological fluids, they are also rapidly metabolized; which puts forward low therapeutic efficacy.

Another important problem in chronic viral infections such as HIV and hepatitis B is the emergence of drug-resistant virus. Extensive use of antiviral drugs for these viruses generally predisposes the emergence of resistant viral strains, which are a major challenge towards continued suppression of viral replication. It is for this reason that new methods that can reinforce the impacts of the drugs need to be discovered while minimizing on actions that the drugs are bound to develop resistance to and side-effects from. Nanotechnology has resolved some of the major strategies of antiviral drug-delivery systems through a disruptive innovation. Working at nanolevel with materials in the range of 1 to 100 nm brought a lot of unique characteristics and features to the material systems (Maus et al., 2021). There are great potentials in using nanotechnology for drug delivery in medicine; this means enhancing bio availability, increasing targeting accuracy and minimizing off target impacts and toxicity.

They include liposomes, dendrimers and polymeric nanoparticles may be prepared in a way that they releases their therapeutic agents in the infected cells or tissues only by the viral infection. On the one hand, targeted delivery would minimize delivering the nanoparticles to healthy cells; on the other hand, side effects and even systemic toxicity would also be minimized. However, the efficiency of the antiviral drugs and the therapeutic results can be further improved by coordinating the nanoparticles to interact with the viral particles or the cellular receptors (Sanna et al., 2022).

Besides all these, nanotechnology has close links to other facet, which is solubility and stability of drug. That means nanoparticles can be developed in a way that enhances the solubility of poorly soluble drugs with the objective of obtaining a desirable delivery to the site of action. Finally, the drug may also experience increased stability in association with the nanocarriers and this leads to an increase in shelf life of the drug and hence increasing its therapeutic window (Ioele et al., 2022).

It has emerged that the merging of nanotechnology with the precision approach is a new era in antiviral therapy. Precision medicine is the idea of using the distinct condition or characteristics of each patient and their disease, their genetic characteristics for example. When the knowledge based on precision medicine deploys as the branch of nanotechnology, it opens up new opportunities for combating viruses more and more effectively. Nanocarriers can be utilized to administer the correct amount of the correct drug at the correct site, in the right manner so as to achieve improved target specificity that leads to effective antiviral treatments (Majumder et al., 2019).

Many conventional approaches for antiviral drug delivery are also provoking the capacity of drawbacks such as poor tolerance and side effects caused by high dosage and recurrent applications. Some of these may defeat compliances, contribute to drugs resistance or both. Therefore, the chronic infected lifestyle gets severe adverse treatment effects on the doses for the chronic viral infections and gets new viral forms that are resistant to drugs than before making the results of treatment even worse on the patients (Song et al., 2020).

The drawbacks of the conventional style of delivering antiviral medications should therefore be a reasonable basis for thinking of new strategies which might overcome these drawbacks. The suggestion that otherwise nanotechnology could help to improve the targeting of antiviral drugs (Chakravarty & Vora, 2021); there should therefore be a possibility of eradicating most of the troubles related to conventional drug delivery systems. Nanotechnology is one of the revolutionary sectors that has come up to bring change in modern medicine. It has provided rich and innovative approaches to addressing long-standing issues related to drug delivery. In other words, nano scale design of structures, devices, and systems incorporating elements in the range between 1 and 100 nm allows scientists to develop novel sorts of materials with characteristics adapted for specific medical uses (Chakravarty & Vora, 2021). Based on modern advances in nanotechnology, new prospects are introduced into the enhancement of antiviral drugs therapy efficiency and targeting ability along with the minimization of side effects.

For example, drug nanocarriers shipped to the target infective cells or tissues with liposomes, nanoparticles or nanocapsules reduces the effect of the drugs on healthy cells (Medina-Cruz et al., 2020). This chapter describes various classes of nanocarriers employed in drug delivery for antiviral treatments and likely implications earned by nanotechnology for the prospective face of antiviral treatments. The ambitious goal of this chapter will be to convince the reader with examples of modern achievements and case histories about how nanotechnology can solve at the same time all the problems in the delivery of drugs for viral diseases at every level and contribute to the improvement of the patient's state.

Innovations in nanotechnology have offered what has been supposed to be devoid of set trends, side effects, and high toxicity that are fundamentals of the traditional antiviral drug therapy that has often formed the core of controlling viral infections. This introduction outlines the potential for nanotechnology in antiviral therapy: targeted delivery systems and development of nanocarriers, are frequently vital to the therapeutic impact and to exclude side reactions (Su & Kang, 2020).

1.1 Modern Issues Connected with Delivery of Antiviral Agents

One of the most common deficiencies in most traditional antiviral drugs is poor bioavailability that refers to the quantity and rate of absorption of the active compound or active constituent of a preparation in a suitable place in the body (Ben-Shabat et al., 2020). A very clear example with the majority of antiviral drugs is very poor solubility in aqueous media and hence does not attain adequate levels of the drug at the site of action. Also, the on-target effects which arise from the ability of drugs to affect not only the target cell or tissue, but also other cells or tissues, may lead to side effects and decreased effectiveness of the treatment regimen (Manzari et al., 2021).

Achievement of therapeutic concentrations often leads to systemic toxicity which results in overdose that also affects healthy cells and tissues (Dube et al., 2021). Such challenges explain the development of new ways to deliver the drugs as to enhance accuracy and effectiveness in antiviral medication.

1.2 Nanotechnology Revolutionization of Drug Delivery

Nanotechnology is defined as the process of controlling matter between one and 100 nanometres in size. This has developed into a wonderful technique in drug delivery system. Namely, according to the recent findings reported in the article by Chen (2013) (Gu et al., 2013), it is possible to develop absolutely new carriers for therapies, providing better delivery, stability and effectiveness of therapeutics. The nanocarriers encapsulate drugs, decreasing their degradation and delivery them progressively, which improves the therapeutic effects by decreasing side effects

(Yetisgin et al., 2020). Nanotechnology also creates the platform on which drug delivery systems can be built. These systems will release the drug or antisense oligonucleotides to the diseased cell or tissues which yield therapy extremely accurate without side-effects because off-target effect is rare (Maus et al., 2021; Yetisgin et al., 2020)

1.3 Nanocarriers: types and mechanisms

Design and function of nanocarriers are different from each other and the existing nanocarriers offer different merits in drug delivery. There are some generally recognized classes of nanocarriers; these include nanoparticles, liposomes, dendrimers and carbon nanotubes (Farah & Farah, 2019). The type mentioned above has a unique characteristic that makes it feasible to harness for optimizing drug delivery and effectiveness.

Nanoparticles: These are roundish shaped and their sizes range from 1 to 100 nanometers in dimensions. They can be shape from different materials, including metals (as in gold or silver), polymers and lipids are other examples. Because of their small size they are able to penetrate biological barriers in systems and can be concentrated selectively in certain (Cao et al., 2020; Gurunathan et al., 2020). Liposomes: These are bilayer vesicles which encapsulate hydrophilic drugs which would not ordinarily be able to pass through cell membranes. Liposomes can be controlled in a way that the liposomes have attaching targeting ligands that would enhance the specificity in infected cell receptors. standard liposome preparations were more potent and less toxic than the common forms of the drug (Suhaimi et al., 2022).

Dendrimers: These are in tree-like branched structure and have high drug load and feature release also. dendrimers had higher stability and efficacy compared to the free agents (Chis et al., 2020). Carbon nanotubes: These are cylindrical nanostructured materials that possess very high surface areas. They are therefore suitable for drug and diagnostic application delivery systems.

1.4 Pharmacological Targeting Drug Delivery Advancement

One targeted drug delivery system, therefore, may try to get a specific dose of the therapeutic agents to the diseased cells without a great impact on the normal cells. These specificities might be reached in different manners: passively targeted or actively targeted (Manzari et al., 2021). Passive Targeting: This approach builds on the ability of the nanocarriers to diffuse in the human body. For instance, EPR effect enables nanocarrier accumulation in tumor tissues resulting from the diseases' vasculature. For this reason, it can use this mechanism to reach infected tissues so that the drug delivery occurs at the site of infection (Zi et al., 2022). Active Targeting: Active targeting is achieved through the functionalization of nanocarriers with targeting agents that have affinity for mechanisms on the target cells. nanoparticles may be designed to attach to active viral surface receptors hence reducing the virus's ability to penetrate new host cells (Figueroa et al., 2021).

1.5 Assuming the Prior Directions and Conclusion

The largest accomplishment of the sector was the integration of nanotechnologies in the delivery of antiviral medications. It means that nanocarriers and targeted delivery systems give more effective drugs, side effects, and therapeutic precise than the traditional treatments (Shah et al., 2021). There are still research questions that can be pursued as follows: safety; alignment with regulatory requirements; and how the material can be engineered to work durably for mass production (Zhou et al., 2021). Further research should be aimed at creating nanocarriers, which have a higher target accumulation ability and a controlled release profile. Besides, conjugated with principles of nanotechnology, precision antiviral medicine or medicine- societal higher-order doctrine would bring the possibility of prescribing the correct medicine to the right patient, thereby on-case treatment outcomes and off- undesirable side effects (Banakar, 2021).

All in all, nanotechnology can be considered the new promising approach in the enhancement of the drug delivery with the focus on the antiviral drugs. Thus, this avenue can greatly catalyze the increase in marked therapeutic improvement and enhanced health status of the population of the entire world eliminating all the existing restrictions and conducting research into the opportunities opened by nanotechnology.

2. DRUG DELIVERY

Drug delivery can be described as one of the prominent applications of nanotechnology . Nanotechnology modifies the different scientific branches; and drug delivery is among those branches already modified. Due to the ability to control the materials at the nanometer scale, scientists have devised new ways that improve not only the speed, specificity and precision of the drug delivery systems (Sahu et al., 2021). This chapter provides an overview on different categories of nanocarriers, modes of action and the emerging trends of nanotechnology that may form the basis of future antiviral treatment.

2.1. Types of Nanocarriers

Nanocarriers are controlled and targeted delivery systems that may be designed engineered particles or structures for therapeutic agent delivery. Different categories of nanoparticles have been synthesized in different synthesis methods and sizes, and each type will meet a special requirement of drug delivery (Bhatia & Bhatia, 2016). Nanoparticles, liposomes, dendrimers and carbon nanotubes are the major forms of nanocarriers and all of them have vital roles in the development of drug delivery. Nanoparticles Nanoparticles are spherical structures described by their diameters that range from 1 to 100 nanometers (Khan et al., 2022). Depending on the application, the components of the particle may be metals, polymers or lipids. Because of such a small dimension, nanoparticles can transverse through the barriers and settle in certain tissues or cells.

Table 1. List of advanteges of nano particles

Benefits	Explanation
Enhanced Therapeutic Efficacy	Increased bioavailability results in higher drug concentrations at the target site, leading to more effective treatment outcomes.
Targeted Drug Delivery	Gold nanoparticles, for example, have specific optical properties that allow them to more effectively target and visualize infected cells (Sharma, Singh, &Yadav, 2017).
Controlled Drug Release	Nanoparticles can be engineered to release drugs in a controlled manner, ensuring sustained therapeutic effects over a longer period.
Reduced Side Effects	Controlled and targeted drug release reduces off-target effects, minimizing potential toxicity to healthy tissues.
Potential in Antiviral Therapy	Gold nanoparticle-based delivery systems have shown great promise in antiviral treatments, improving drug stability and effectiveness in combating viral infections.

From this table 1, there is a general understanding of how nanoparticles enhance the efficacy and delivery of antiviral drugs.Specific optical properties make visualization and targeting much better since they get stuck more in infected cells .ions at the target site, leading to more effective treatment outcomes.

Applications in Antiviral Therapy: The nanoparticles have shown potential in the delivery of antiviral agents through their capacity to track the cells invaded by antivirals hence enhancing therapeutic effectiveness. For example, gold nanoparticles have been applied in the delivery of antiviral drugs in such a fashion that makes them stable and more effective antiviral drugs than could be useful for curing viral diseases (Chen & Liang, 2020). Liposomes Liposomes are small, spherical, closed structures formed from one or more layers of phospholipids membrane. It means that formulations of this type can be created with the certain targeting ligands, which may increase the selectivity of the literature to target cells and, as a result, can improve the therapeutic effect.

Liposomes thus has the following advantages They show high stability and the toxicity level of the drugs would be significantly low. Liposomes encapsulation removes some challenges faced when applying drugs such as; reducing their degradation before reach effectively the body tissues (Van Tran et al., 2019). Also, it is viable to transport drugs to certain cells or tissues, and this brings a higher accuracy during the release of drugs and a decrease in the number of side effects that accompany the drug delivery.

Applications in Antiviral Therapy: There has therefore been the use of liposomal formulations that enhance delivery of drugs such as zidovudine. It has established that these liposomal formulations are more effective in their function and less net toxic than the standard drug delivery method (Chakrabarti et al., 2019). Practical properties of liposomes and encapsulation, together with the protection properties of the antiviral agents make these vesicles a possible entity that can be used in the practice of antiviral therapeutics.

Dendrimers

Dendrimers are literally known as trees and they are hyperbranched macromolecules characterized by a large number of branches. Because of this particular architecture, a dendrimer system can provide high density functionalization and drug entrapment (Nikzamir et al., 2021). In special, they can be engineered to selectively interact with certain cell receptors in order to facilitate drug delivery and outcome. The benefits associated with dendrimers are to have multiple branched shells to hold a large of drug molecules and a release profile of the drug.

Application in Antiviral Therapy: As discussed before, dendrimers increase the stability and effectiveness of the antiviral agents. An application of dendrimer based delivery system is evident in enhancing the efficacy of antiviral drugs for HIV. The therapy improves the solubility and the shelf life of drugs used in treating viral infections; therefore, one controls viral infections efficiently (Faria et al., 2021).

Carbon Nanotubes

Carbon nanotubes are elongated nanostructures that demonstrate remarkable mechanical, electrical and thermal characteristics. They possess immense electrical surface area besides being structurally versatile thus fit for use in uses like drug delivery system and diagnostics (Negri et al., 2020). The cellular level targeting could be very achievable with carbon nanotubes since they can even be functionalized, hence could provide a basis for designing various systems for drug delivery.

Carbon Nanotubes CN's are prospective in that they offer a high surface area and good mechanical properties. This means that a though come much promising about the use of carbon nanotubes, the most promising area appears to be in drug delivery. Applications in Antiviral Therapy: Carbon nanotube has been considered for its use in the delivery of antiviral active pharmaceutical ingredients and diagnostic cargoes. Self-organizing carbon nanotubes make it possible to penetrate a cell and deliver therapeutic agents directly to the site of infection, and therefore this is an important position in the development of innovative antiviral therapy.

Figure 1. Types of nanocarriers

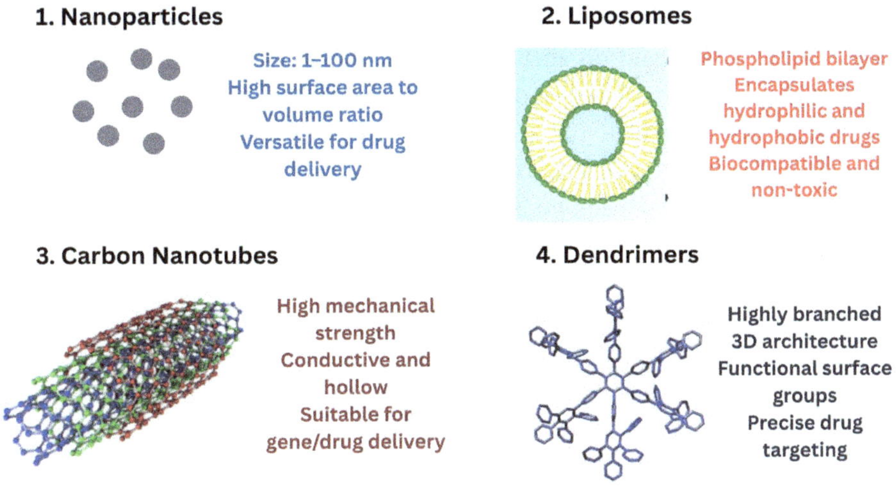

Figure 1 illustrattes and list out the characterstics of different types of nano carriers

2.2. Targeted Delivery Mechanisms

Cancer drug delivery techniques have been developed to deliver therapeutic agents to abnormal cell mass while minimizing normal cell exposure. This minimizes the side effect and increases the dose response relationship of the drug (Manubolu et al., 2023). Targeted delivery occurs through two principle routes as follows. These are passive targeting and active targeting .

Passive Targeting Passive targeting involves a spontaneous distribution of nanocarriers within the system and without the exertion of additional effort. This is mainly achieved by taking advantage of the enhanced permeability and retention effect, which leads to accumulation of nanocarriers at regions with compromised vascular barrier, for instance tumors (Attia et al., 2019).

Mechanism of EPR Effect: This is due to the observation that nanocarriers are apparently preferentially retained in tumor tissues because such tissues have structurally compromised vasculature with well developed gap junctions and an inactive lymphatic system. This configuration enables delivery of the drugs to the involved sites, which generally improves the effectiveness of the treatment (Park et al., 2019).

Applications in Antiviral Therapy: Active targeting is another approach can be used for enhance passive targeting for deliver the antiviral drugs right to the exact site of viral infection, or the infected cells. For example, the nanoparticles can be focused at the infected area and enhance the therapeutic result with the outcome of therapeutic methods in regards to antiviral actions (Shi et al., 2020). EPR effect for targeted drug delivery is one of the most important advancing towards the potential antiviral drugs. Active Targeting Active targeting is conjugation in which one or more ligands are coated on surface of the nanocarrier to whom they respond to. This would make it possible to deliver drugs only into the diseased cells and this makes the drug more efficient and with less side effects that result from off-target effectiveness (Lin et al., 2019). Mechanism of Ligand Binding: Pinpointing is based on the principle whereby ligand binds with receptors on the membrane of the target cell. Regarding the functionalization of nanocarriers which are cant play a role of recognizing and binding with suitable receptors on the surface of the cell, scientist can deliver therapeutic agents to cells (Akhter et al., 2021).

Applications in Antiviral Therapy: On the circumstances of an antiviral drugs, active targeting has been employed in the delivery of the drugs to an infected cell. For instance, nanoparticles can be linked with ligands that only have affinity to viral surface proteins. The ligand is the receptor by which the antiviral drugs are selectively transported to the target site (Cojocaru et al., 2020). This may also increase the impact of antiviral drugs while on the same note it may also decrease side effects .

Figure 2. Active and passive targetting

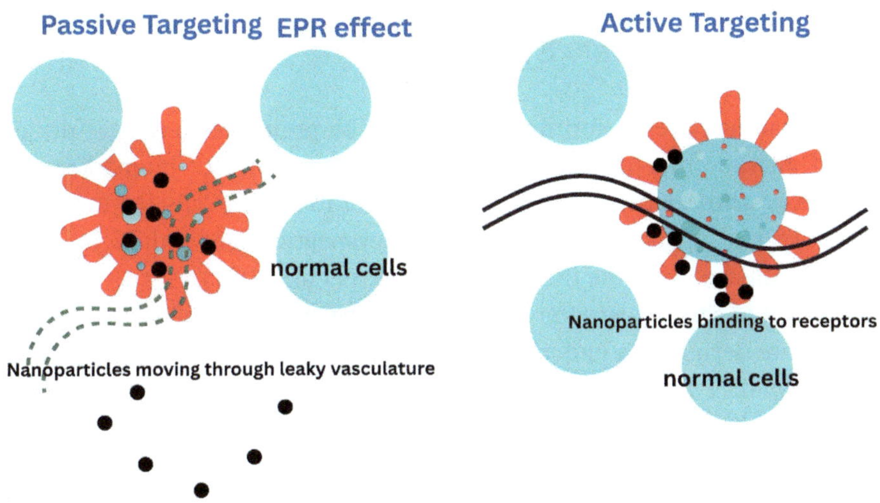

Figure 2 Schematic representations of Passive Targeting mechanism and Active Targeting mechanism.Passive Targeting: Portrays particles able to enter tumor tissues through filterable capillary leakage Depicts nanoparticles labeled with ligands that have an inherent liking for receptors.

2.3 Nanotechnology Application in Drug Delivery

Nevertheless, comparatively to all the recent achievements in the sphere of nanotechnology, there are more complicated and advanced systems of drug delivery nowadays. Among others, these are multi-stimuli smart and autonomous platforms for drug delivery such as multi-functional nanocarriers and responsive drug delivery systems and more so combination therapies.

Multifunctional Nanocarriers

Smart nanocarriers are intended to respond to alter or execute one or several tasks at the same time. In this regard, the therapeutic agents, imaging agents and the targeting ligands can be delivered by these nanocarriers. Consequently, multifunctional nanocarriers may provide the optimal solution for the disease management (Lombardo et al., 2019). Applications in Drug Delivery: Multifunctional nanocarriers can be thought of as an integral platform for drug delivery since within the vehicle of the nanoparticles, both the therapeutic and the corresponding imaging functions are integrated.

Responsive Drug Delivery Systems

Self regulating systems in drug delivery can be prompted to release drugs on account of alteration in pH, temperature and the existence of biomarkers. As a result, these system can be engineered in a manner to require only the existence of specific biological signals for the delivery of therapeutic agent with high precision and reduced impacts from the treatment (Manzari et al., 2021).

Mechanism of Stimuli-Responsive Release: The stimuli responsive drug delivery systems are designed to undergo alteration by the outside conditions or certain biological stimuli. For instance, nanoparticles may be designed to release the drug load in view of the current pH of tumor tissue or the infected cell in question. This targeted release, therefore, increases the tumour therapeutic ratio and reduces the systemic toxicity of the treatment (Souri et al., 2022).

Table 2. Applications of responsive drug delivery systems in antiviral therapy:

Applications	**Explanation**
Responsive systems	deliver antiviral agents selectively only as a response to viral indicators or other stimuli such as changes in pH or temperature.
Better Treatment Effectiveness	These systems increase the efficiency of drug administration by hitting only infected cells
Reduced systemic side effects	Targeted delivery system eliminates drug release in other areas of the body thus reducing side effects suffered from most drugs.
More Versatile Applications for Responsive Drug Delivery	Responsive drug delivery can be utilized professionally for other viral diseases
In reducing the chances of viral resistance against antiviral drugs	they can provide potential in overcoming drug resistance By directly delivering drugs to specific viral infected regions these systems may help in minimizing the risks of drug depolarization in the viral.
Enhanced Dosage Protraction	The ability to control and have the drugs released slowly in the body will increase the time of drug action

To summarize the prospect of using RDDS for improvement of antiviral therapy this table no. 2 is proposed. There is a potential to improve the antiviral therapy by using targeting drug delivery systems since the drug is released only in the presence of the viral marker or the environmental stimuli. Hence these responsive systems may have the ability to take antiviral treatments to other levels by increasing the efficiency of antiviral therapy ad decreasing side effects.

Combination Therapies

The combination therapies make use of nanocarriers in the delivery of more than one drug or therapeutic agent to the same cell, which can increase the effectiveness of the treatment because it can target features of the disease that have not been directly addressed, or issues to do with drug resistance (Shrestha et al., 2019).

Mechanism of Combination Delivery

Nanocarriers can be made in such a manner that it releases active therapeutic agents within the body for the cure of certain diseases in which the therapeutic agents function synergistically. These can address specific aspects of pleasant viral or stated pathways meant to depressed the probability of antiviral treatment in general (Lauer et al., 2020).

Applications in Antiviral Therapy: Co- delivery using nanocarriers has been proposed to deliver a combination of multiple antiviral agents or even an antiviral drug together with another therapeutic agent for optimum therapeutic purposes. It can also eradicating the mechanism of drug resistance and enhancing the outlook of treatment results all together (Delshadi et al., 2021).

3. THE ROLES OF NANOTECHNOLOGY IN ANTIVIRAL TREATMENT

They found that nanotechnology is a rather revolutionary factor within the field of antiviral therapy; new approaches have been designed to enhance modality of drug delivery, efficacy against the virus and counter the threat of resistance to the drugs. This chapter summarises how nanotechnology is now driving the treatment of viral infections by new perspectives in drug delivery systems, targeted antiviral agents and drug resistance.

3.1. Polymer Drug Delivery Systems

Nanotechnology advances in antiviral drug delivery system include alteration of pharmacokinetics and biodistribution profiles for better and safer therapy. This is important in the production of antiviral drugs and eliminating limitations resulting from standard methods of drug administration which include intramuscular, intravenous and oral (Pradhan et al., 2021).

Liposomal Formulations

Drug delivery systems in liposomal formulations represent an advanced carrier system. Liposomes describe round structures made up of lipid double layers that contain hydrophobic drugs that enhance solubility, stability, and availability (Akbarzadeh et al., 2013). This encapsulation is useful to delay any degradation process in the body system while at the same time providing controlled release primarily within the target sites. Increased solubility and stability of drugs: For instance, poor solubility antiviral drugs such as zidovudine, and oseltamivir can be encapsulated within liposomes, which increases their solubility as well as therapeutic efficacy (Chen et al., 2021). Such an improvement is useful for the drugs which have poor solubility in water, so as to be able to maintain high concentrations of the drug deliverable to the site of infection.

Prolonged Circulation and Targeted Accumulation

Liposomes promote increased circulation time of the antiviral agents in the blood and subsequently increased concentration of the drug at the site of infection. Extened release offers the best therapeutic advantages of antiviral agents through concentrated drug efficacies over the duration of longer periods (Megantara et al., 2024). For instance, liposomal formulation of oseltamivir has been said to enhance better antiviral activity and, at the same time, the dosing frequency of oseltamivir would also be reduced compared to nonliposomal oseltamivir formulation (Tang et al., 2015).

Clinical Implication: The advancement in liposome formulations has improved the effectiveness of the therapeutic approaches for several viral infections. This proves the fact from these clinical investigations, it can be recognized that liposomal formulation of zidovudine contributes in attaining much higher level of this drug concentration with the affected area of HIV infected tissues; thus providing better protection against the disease and in addition, lesser adverse effects as compared to the conventional formulation of zidovudine (Drost et al., 2021). In an even similar manner, formulation of oseltamivir using liposome increase the efficacy against the influenza infection. Hence, liposomal formulation of oseltamivir has great potential as a clinical approach more than the other existing therapy modalities (Čelková et al., 2024).

Nanoparticles in Polymers Another promising drug delivery platform is polymeric nanoparticles, which can be used to control the release of therapeutic agents and increase their efficacy (Begines et al., 2020). They are biodegradable and synthesized to have an inherent interaction with specific stimuli in their environment. Controlled release of drugs: Polymeric nanoparticles can deliver predetermined quantities of a drug at predetermined rates so that a therapeutic window of a drug is sustained for a longer duration. For instance, PLGA based nanoparticles can be designed to put out antiviral at certain times in a time span long enough to prevent the virus from getting inside the host cells, by keeping the therapeutic amounts of the drug in circulation for quite a long time (Loureiro & Pereira, 2020).

Responsive Systems to Stimuli: It is also possible to design a polymeric nanoparticle for a specific environmental stimuli, for instance pH change or biomolecules. For example, the particles of nanoparticles will discharge the drug load after arriving at the state of acid that is witnessed in tumor tissues or infected cells will assure marvellous precision of antiviral therapy. The targeted release improves drug efficiency and reduces systemic toxicity because therapeutic agents will be released only at the site of infection (Yeh et al., 2020).

Applications and Advantages: There is considerable interest in the targeted delivery of various antiviral agents based on polymeric nanoparticles; HIV and Hepatitis C have been covered in this aspect. Polymeric nanoparticles can be controlled in terms of drug release and conditions under which the drug is intended to impart therapeutic value, thereby making them valuable for creating new antiviral therapies.

3.2. Targeted Antiviral Agents

Nano technology enable the vision of treatment with antiviral agents that can selectively associate with viral particles or with infected cells; this would improve the effectiveness of the treatment without enhancing the toxicity within the Attacking Viral Surface Proteins. Of this, gold nanoparticles are a good example. Capped gold nanoparticles have the ability to stick on surface proteins of viruses including the influenza virus making them unable to penetrate host cells (Raji et al., 2020). Their relatively large surface area and versatility of the functionality that may be attained in gold nanoparticles render them a potent weapon in targeting viral proteins and enhancing antiviral properties.

Mechanism of Action: Gold nanoparticles may be functionalized with ligands or antibodies selective to viral surface proteins. Such binding may may hinder the virus from entering the host cells and can facilitate to remove the virus from the body. The studies have revealed that the gold nanoparticles show higher cytotoxicity or antiviral activity because they bind and inactivate the viral particles (Kim et al., 2020). Cell Targeting These nanoparticles can also be functionalized with surface ligands on the receptors expressed on infected cells. This idea makes the concept of having the delivery end at the site of virus replication and consequently off-target effects would also be minimized, enhancing precision in antiviral treatment (Mashel et al., 2020). Dendrimers for Cell-Specific Targeting: Dendrimers are macromolecules with an extensive branched structure that can be made to selectively bind with appropriate cell receptors. For instance, the ligands can be grafted on the dendrimer so that it hooks on the receptor of the target cell infected with HIV, which brings an antiretroviral drug delivery, enhanced plus the dose amount of the medication increased at the site of infection meaning low toxicity to the rest of the body and hence improved medical outcome (Szyk et al., 2023).

Advantages of Cellular Specific Targeting. Liposomes targeting with cellular specificity may provide more effective antiviral reagents, because they may reach the sites of viral penetration in cells. Unlike blanket therapy, targeting is less destructive to health cells, which often results in better effectiveness of antiviral medications (Nainwal, 2022).

3.3. Breaking the Back of Antidrug Resistance

Persistence of the drug resistance is one among the major threats of antiviral therapy because of its outcome results in low level efficacy and a therapy which has failed. Several solutions to the problem of drug resistance can be achieved with the help of nanotechnology: combination therapies and methodologies for sustained drug release.

Combination Therapies

Co-administration techniques involve the application of nanocarriers for the delivery of two or more antiviral compounds. Even with such a strategy, the therapy could be as effective in the sense that it attacks different phases of the viral life cycle or even different viral strains than developing an all-encompassing drug resistance strategy (Hossain et al., 2020). Mechanism of Combination Delivery: Due to their amplitude of growth and adjustability, nanocarriers can be designed to link several therapeutic components and release them cogently for improved disease control. For instance, nanoparticles may be used in encapsulating a mixture of antiviral agents targeting replication and viral entry; presenting them more holistically to the disease (Maus et al., 2021).

Benefits and Applications: Portable nanocarrier-based combination therapies could be brought forward for enhanced outcomes in managing viral diseases from optimization of antiviral treatment factors based on multiple aspects of viral infection. This also counteracts drug resistance by binding multiple viral strains or multiple modes of drug resistance and provide a chance to develop better and longer lasting antiviral drugs. The advantages of sustained-release systems include constant concentration of the antiviral drug in the body devoiding high threats of drug resistance. This method is highly applicable in chronic viral diseases particularly in illnesses that maybe treated for many years. Examples and Advantages: Current nanocarriers that can deliver antiviral drugs in a controlled manner have been designed for few antiviral drugs used in treating HIV and hepatitis. Capability of sustaining the therapeutic concentration of the drug for certain time can enhance the success of antiviral agencies as a whole and prevent the generation of new resistant form of drugs.

Eliminating Particular Forms of Resistance

Nanotechnology is even getting close to the specific mechanisms of drug resistance via nanocarriers of the autistic advocacy of antiviral drugs. It has been conceived to explain how nanoparticles could do as how the nanoparticles are engineered to avoid efflux pumps that is as often associated with drug resistant conditions (Dey et al., 2022). Overcoming Resistance Mechanism: Efflux pumps are proteins that actively transport drugs in the cells.Nanocarriers can be prepared either as efflux pump-harbouring or receptor blocking vectors, which increases the concentrations of intracellular antiviral drugs and thus increases their therapeutic coefficients. Applications and Impact: Drug resistance based nanocarrier formulations not only offer new therapeutic directions in the treatment of drug resistant viral diseases but also enhance available antiviral agents. Nanotechnology plays a crucial part in the combat against drug resistance, which increases the efficacy of existing antiviral drugs and shows the possibility of new therapeutic strategies (Gurunathan et al., 2020).

3.4. Case Studies and Examples

Recent works show that nanotechnology has the potential in creating antiviral solutions. Gold nanoparticles as antiviral therapeutic agent for the treatment of influenza virus infection. Influenza could also be treated through antiviral drugs that can be delivered by means of gold nanoparticles. Functionalised gold nanoparticles may act specifically targeting viral surface proteins in a way that we can increase efficacy of the antiviral agent and decrease viral loads.

Research Findings: Local concentrated drugs in the site of infection during the use of gold nanoparticles are real; this has been confirmed by the favorable outcome of the treatment. One example of this is the blending of antiviral agent capped gold nanoparticles wherein it was discovered that addition improves the effectiveness of the treatment outcome by facilitating targeting and neutralization of influenza viruses.

Clinical Relevance: A possible co-therapeutic agent for available treatments of flu is gold nanoparticles. This is because gold nanoparticles work through boosting the delivery of the administered drugs and targeting viral surface proteins, and so patients get efficient and target oriented antiviral drugs. Latest Zidovudine – Elemental & Liposomal Formulations & Their Roles The enhancement and optimization of the efficacy in combination with less toxic effects are proven by liposomal formulation of this antiviral drug. Liposomes enhance the stability and solubility of zidovudine, and since the concentration of the drug at the site of the infection is relatively higher (Okafor and Nnaji, 2024).

Figure 3. Applications of liposomes in antiviral drug delivery

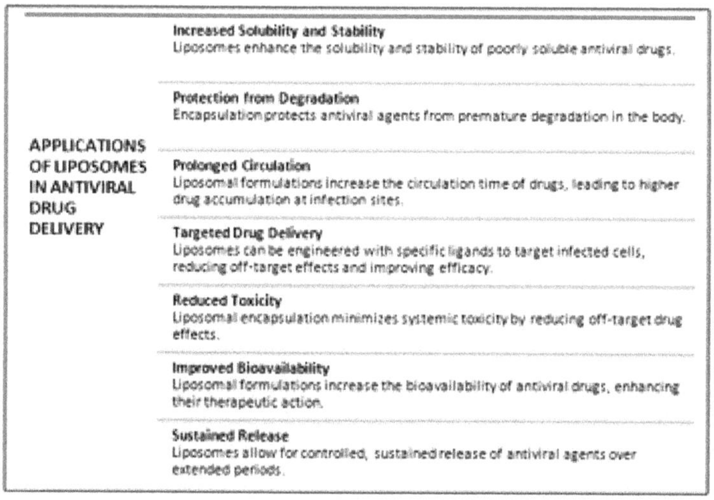

Clinical Studies: Literature review of randomised controlled clinical trials demonstrated that liposomal zidovudine offers improved efficacy and associated with less toxicity as compared to regular zidovudine. Compared to other drug formulations, Liposomes provide more protection from degradation and enhance drug bioavailability and therefore effectively check HIV infections. Advantages and Applications: Liposomal zidovudine is a major advancement in the HIV antiretroviral therapy linked to increased drug release and decreased toxicity. The aforementioned formulation improves the general effectiveness of antiviral therapy, and can be a solution to part of the presumptive deficiencies that are associated with standard zidovudine formulations. Dendrimers in HIV Therapy Dendrimers in the Management of HIV/AIDS: An overview of their application in antiretroviral drugs therapy The evolutionary path of the development of dendrimers has therefore been targeted at the improvement of drug delivery system for antiretroviral drugs in the management of HIV/AIDS (McCarthy et al., 2005). Different dendrimers are actually highly branched macromolecules and certain dendrimers are synthesized to gain access to the specific receptors present on the HIV involved cells and thus increasing its effectiveness.

Research and Findings: Among the ligand functionalized dendrimers, the targeted ligands appeared to enhance the efficiency of delivering antiretroviral drugs to infected HIV cells

Impact and Future Directions: The use of dendrimers can be foreseen in therapy against HIV in order to offer better delivery systems; thus its therapeutic effect is one of the milestones. Better overall efficiency of antiretroviral treatment results from specificity of dendrimer receptors that interact with the specific receptors of the infected cells, as well as new opportunities in managing of HIV infections appear.

4. CASE STUDIES AND APPLICATIONS

Nanotechnology is so Transformative in treatment of antiviral therapy for all the types of virus and the infections they cause. Thus, nanoparticles, liposomes, and dendrimers as some of the most studied nanotechnology based delivery system of drugs that increases targeting and drug effectiveness and counteracts resistance. The following cases and applications are used to explain how these technologies influence antiviral treatment.

4.1. Delivery: Systems Using Nanoparticles

In the present scenario, nanoparticles appear to be the backbone for promoting antiviral agents using size and surface properties for drug delivery. Here we are presenting the gold nanoparticles and polymeric nanoparticles used for the fievour of antiviral drugs delivery.

4.1.1. Applications of Gold Nanoparticles in Influenza Management

Interestingly, gold nanoparticles are currently on the spotlight because of the possible use in antiviral treatment; especially in the flu. Because of its optical and electronic characteristics, AuNPs are used in numerous biomedical sectors including drug delivery and imaging. Functionalization and Targeting: The gold nanoparticles are ligand adaptable in different types of antibodies or peptides to get the viral protein on the surface. Thus, Sharma et al. ‘s study confirmed that target antiviral drugs bonded to influenza virus more effectively when treated with anti-influenza antibodies functionalized gold nanoparticles Sharma et al. (2021). As researchers have pointed out studies indicate that the usage of oseltamivir together with AuNPs has enhanced the reaction in a therapeutic entity, reduced viral reproduction, and enhanced the response to recovery (Ghosh et al., 2022).

The processes through which this augmented delivery of antiviral agents by gold NPs is quite diverse. First, there is an important role of large area to volume ratio of AuNPs, their high drug loading ability elevates concentration of the drugs at infected area. The interaction of AuNPs with cellular membranes proves beneficial for the delivery of the antiviral drugs into cells bypassing absorption and drug distribution limitsThere is evidence which suggests that the concentrations of the drugs at site of infection is enhanced using AuNPs in comparison with conventional methods of formulation. Furthermore, the authors observed the low systemic toxicity of AuNPs. For this reason, it is believed to be a potential candidate for the development of antiviral treatment.

4.1.2. The Prospects of Polymetric Nanoparticles in HIV Treatment

Polymeric nanoparticles is another new concept in the delivery of antiviral drugs. It is made up of water soluble polymers. The payload, in addition, may be discharged in a controlled manner. Controlled Release and Targeting: Better Stability and Bio-availability: All recent studies reveal that polymeric NPs improve the stability and bioavailability of the antiretroviral agents. For instance, polymeric nanoparticles of zidovudine show higher stability and are longer circulating in the system compared to free drug forms. The improved stability will hence assure control of HIV infections and rare use of the drug .

Preclinical and Clinical Studies: Prior in vivo animal experiments have reported that polymeric nanoparticles improve the antiretroviral activity of the medicines. Trials are also going to be carried out to determine the characteristics of the drug formulations of the antiretroviral drugs using the polymeric nanoparticles for treatment of HIV. it appears that the formulations offer larger proportions of drugs for better therapeutic courses and have an indication of lesser side effects (Müller et al., 2001).

4.2. Antiviral Agents

The effects of Such antiviral agents, encapsulated in liposomes, has also been studied. It has been observed that, liposomal formulations have been one of the worthwhile and tactical ways to enhance the effectiveness and safety of antiviral agents

4.2.1. Liposomal Zidovudine Zidovudine (AZT)

Liposomal Zidovudine Zidovudine (AZT) is a potent antiretroviral drug that has attained a premier status among the widely used HIV treating medicine. A re-formulation of zidovudine in liposomes has remarkable effects on enhancing drug effectiveness and reducing toxicity. Enhanced drug solubility and stability: Liposomal encapsulation improves the solubility and stabilility of zidovudine because the issue of its low solubility in aqueous solution is minimalised. The formulation of zidovudine in liposomes exhibits better stability and circulation time compared to the conventional formulation. Consequently the stability of the drug enhances the concentration of the drug at the site of infection resulting to optimum therapeutic effect.

Liposomal zidovudine has been also found to demonstrate the ability to control HIV infection by the decrease of side effects and increasing patient's conformance.

4.2.2. Liposomal Acyclovir

Acyclovir is an antiviral agent employed mostly in herpes simplex virus disorders. Liposomal formulations of acyclovir have been prepared and these essentially aim at improving the performance of acyclovir and reduce on systemic toxicity. Due to encapsulation inside liposomes, acyclovir is not degraded and the bioavailability of this drug is enhanced; thus, is regarded as more effective treatment of herpes simplex virus infections Pandey et al. (2020). Clinical Efficacy and Safety: Clinical endeavors into liposomal acyclovir have established it's efficiency and safety when used to treat diseases caused by herpes simplex virus. As mentioned in several qualitative studies, it has been claimed that liposomal drug delivery system has a larger margin of safety than traditonal acyclovir form and reduced adverse effects and better clinical results in the patients.

4.3. Dendrimers in Antiviral Therapy

Because of their architectural characteristics, dendrimers are hyperbranched macromolecules and dendrimer can delivered the drugs to target site and accurately. Because of the larger surface area and ability to participate in functionalisation, it is very useful in antiviral therapy.

4.3.1. Dendrimers Against HIV

Some of the recent findings for applications of dendrimers reveal that dendrimers enhance the effectiveness of delivery and the antiretroviral drugs used in treating HIV/AIDS. Targeted delivery and encapsulation: The dendrimers can be designed to recognize receptors on the HIV infected cells. Thus, drug delivery becomes effective also helps to enhance the treatment result. dendrimers can incorporate antiretroviral drugs such as tenofovir and emtricitabine into the dendrimer structure thereby enhancing stability and solubility.

Figure 4. Dendrimers in Targeted Drug Delivery

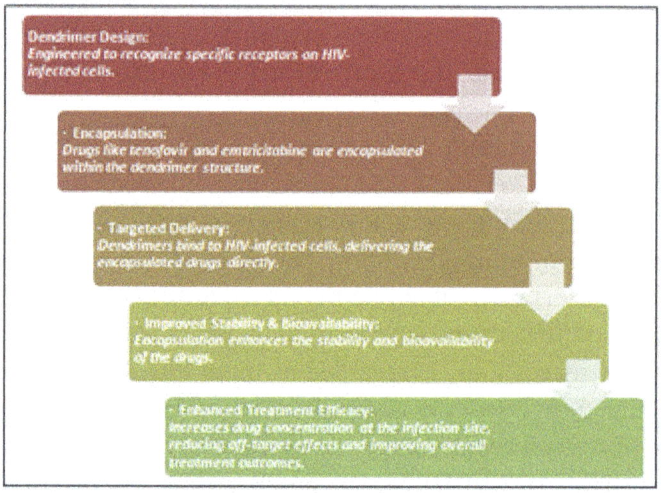

Figure 4 also shows how encapsulation and targeting occur.

Clinical Studies: From all the clinical studies, the dendrimer based formulations display enhanced HIV control with reduced side effects as compared to the conventional formulation. Of course, the fact is known that dendrimer molecules are transported or delivered by HIV infected cells to take or increase the efficiency of the drug delivery process for which a better therapeutic result is achieved.

4.3.2. Hepatitis C

Virus Dendrimers Targeting HCV infection has also been considered in dendrimers. Targeting HCV Particles: Some of the works conducted earlier have shown that dendrimers have the ability to locate HCV particles and also to improve the efficiency of transferring antiviral compounds to the affected cells.

Preclinical Findings: Based on the preclinical investigations, dendrimer mediated formulations of antiviral drugs provide promising therapeutic value and toxicity profiles to conventional antiviral schemes. The presented research shows that dendrimers have critical function in providing necessary information for developing antiviral treatment of HCV infection and for developing new antiviral therapies.

4.4. Issues and Ideas for the Future

However, there are some big barriers that need to be overcome fully harness the potential of this nanotechnology based antiviral therapy. More specifically these are: Safety and toxicity considerations, issues to do with manufacturing and standardization, and regulatory obstacles.

4.4.1. Safety and Toxicity

Poor biocompatibility and safety of nanocarriers is a key question in the application of nanotechnology in antiviral therapy. Long-term Toxicity Studies: It is considered that long-term studies are needed to study the idea of potential toxicity of nanocarriers, as well as the impact on cells that are not targeted. in order to establish first-hand knowledge on the compatibility of nanocarriers with biological systems. The determination of possible detrimental effects in nanocarriers will help in confirming that the nanocarriers are safe for use in clinical practice.

Biocompatibility would thus be paramount in the actualization of biophysical translated therapeutics developed out of nanotechnology in an operating theatre.

4.4.2. Creating Things and Making Them Generic

As important criteria for the clinical use of nanocarriers, both scalability in terms of production and standardization can be regarded as crucial. Scalability of Production The other important consideration when it comes to nanotechnology based therapies is how to develop reproducible, economic and ready for large scale production processes. It shall thus be possible to adopt large scale production methods to ensure that nanocarriers can applied in antiviral therapy. Enhancing research activities will therefore focus at how best manufacturing processes can be enhanced in order to reproduces the quality of the nanocarriers.

Standardization and Quality Control: Consequently, safety and efficacy of nanocarriers depend on formulation and its quality control. Lack of suitable standardization in characterization of nanocarriers, their formulation, and evaluation of quality would therefore be a major requirement for translation of the nanocarriers into clinical practice

5. CONCLUSION

Advances in nanotechnology in drug delivery, efficacy and the evasion of drug resistance, are markedly changing antiviral therapy. Case studies and applications using nanotechnology, including gold nanoparticles, polymeric nanoparticles, liposomal formulations, and dendrimers, show the many transformational potentialities of these technologies in the management of viral infection.

Despite this potential, much work remains to be done before the optimal expression of nanotechnology in antiviral therapy can become a reality. For example, problems include issues of safety and biocompatibility, scalability of manufacture, regulatory and ethical considerations.

Nanotechnology will increasingly play a central role as the stage of research and development continues to advance into the future of antiviral therapy. Using these technologies to address challenges and continuing exploration of their potential can help build the effectiveness of antiviral treatment and further enhance patient outcomes.

Nano-based delivery systems may serve as a potentially promising avenue through which the difficulties in designing successful antiviral drug delivery are overcome. Further research and development in the area of nanocarriers must be fostered to fully allow their potential to be realized and their effective integration into clinical use. The road ahead for innovative new antiviral therapies will continue to require inter- and multidisciplinary collaboration.

REFERENCES

Akbarzadeh, A., Rezaei-Sadabady, R., Davaran, S., Joo, S. W., Zarghami, N., Hanifehpour, Y., Samiei, M., Kouhi, M., & Nejati-Koshki, K. (2013). Liposome: Classification, preparation, and applications. *Nanoscale Research Letters*, *8*(1), 1–9. DOI: 10.1186/1556-276X-8-102 PMID: 23432972

Akhter, M. H., Beg, S., Tarique, M., Malik, A., Afaq, S., Choudhry, H., & Hosawi, S. (2021). Receptor-based targeting of engineered nanocarrier against solid tumors: Recent progress and challenges ahead. *Biochimica et Biophysica Acta (BBA)-. Biochimica et Biophysica Acta. G, General Subjects*, *1865*(2), 129777. DOI: 10.1016/j.bbagen.2020.129777 PMID: 33130062

Attia, M. F., Anton, N., Wallyn, J., Omran, Z., & Vandamme, T. F. (2019). An overview of active and passive targeting strategies to improve the nanocarriers efficiency to tumour sites. *The Journal of Pharmacy and Pharmacology*, *71*(8), 1185–1198. DOI: 10.1111/jphp.13098 PMID: 31049986

Banakar, U. V. (2021). *Pharmaceutical dissolution testing, bioavailability, and bioequivalence: Science, applications, and beyond.* John Wiley & Sons.

Begines, B., Ortiz, T., Pérez-Aranda, M., Martínez, G., Merinero, M., Argüelles-Arias, F., & Alcudia, A. (2020). Polymeric nanoparticles for drug delivery: Recent developments and future prospects. *Nanomaterials (Basel, Switzerland)*, *10*(7), 1403. DOI: 10.3390/nano10071403 PMID: 32707641

Ben-Shabat, S., Yarmolinsky, L., Porat, D., & Dahan, A. (2020). Antiviral effect of phytochemicals from medicinal plants: Applications and drug delivery strategies. *Drug Delivery and Translational Research*, *10*(2), 354–367. DOI: 10.1007/s13346-019-00691-6 PMID: 31788762

Bhatia, S., & Bhatia, S. (2016). Nanoparticles types, classification, characterization, fabrication methods and drug delivery applications. In *Natural polymer drug delivery systems: Nanoparticles, plants, and algae* (pp. 33-93). DOI: 10.1007/978-3-319-41129-3_2

Cao, J., Huang, D., & Peppas, N. A. (2020). Advanced engineered nanoparticulate platforms to address key biological barriers for delivering chemotherapeutic agents to target sites. *Advanced Drug Delivery Reviews*, *167*, 170–188. DOI: 10.1016/j.addr.2020.06.030 PMID: 32622022

Čelková, A., Búcsi, A., Klacsová, M., Fazekaš, T., Martínez, J. C., & Uhríková, D. (2024). Oseltamivir phosphate interaction with model membranes. *Biochimica et Biophysica Acta (BBA)-. Biomembranes, 1866*(3), 184273. DOI: 10.1016/j.bbamem.2024.184273 PMID: 38211646

Chakrabarti, S., Roy, P. K., Chattopadhyay, P., & Mazumder, B. (2019). Prospects of nanotechnology in brain targeting drug delivery. In *Nanotechnology* (pp. 273–314). CRC Press. DOI: 10.1201/9781351111874-10

Chakravarty, M., & Vora, A. (2021). Nanotechnology-based antiviral therapeutics. *Drug Delivery and Translational Research, 11*(3), 748–787. DOI: 10.1007/s13346-020-00818-0 PMID: 32748035

Chen, L., & Liang, J. (2020). An overview of functional nanoparticles as novel emerging antiviral therapeutic agents. *Materials Science and Engineering C, 112*, 110924. DOI: 10.1016/j.msec.2020.110924 PMID: 32409074

Chen, R., Wang, T., Song, J., Pu, D., He, D., Li, J., Yang, J., Li, K., Zhong, C., & Zhang, J. (2021). Antiviral drug delivery system for enhanced bioactivity, better metabolism and pharmacokinetic characteristics. *International Journal of Nanomedicine, 16*, 4959–4984. DOI: 10.2147/IJN.S315705 PMID: 34326637

Chis, A. A., Dobrea, C., Morgovan, C., Arseniu, A. M., Rus, L. L., Butuca, A., Juncan, A. M., Totan, M., Vonica-Tincu, A. L., Cormos, G., Muntean, A. C., Muresan, M. L., Gligor, F. G., & Frum, A. (2020). Applications and limitations of dendrimers in biomedicine. *Molecules (Basel, Switzerland), 25*(17), 3982. DOI: 10.3390/molecules25173982 PMID: 32882920

Cojocaru, F. D., Botezat, D., Gardikiotis, I., Uritu, C. M., Dodi, G., Trandafir, L., Rezus, C., Rezus, E., Tamba, B. I., & Mihai, C. T. (2020). Nanomaterials designed for antiviral drug delivery transport across biological barriers. *Pharmaceutics, 12*(2), 171. DOI: 10.3390/pharmaceutics12020171 PMID: 32085535

Delshadi, R., Bahrami, A., McClements, D. J., Moore, M. D., & Williams, L. (2021). Development of nanoparticle-delivery systems for antiviral agents: A review. *Journal of Controlled Release, 331*, 30–44. DOI: 10.1016/j.jconrel.2021.01.017 PMID: 33450319

Dey, N., Kamatchi, C., Vickram, A. S., Anbarasu, K., Thanigaivel, S., Palanivelu, J., Pugazhendhi, A., & Ponnusamy, V. K. (2022). Role of nanomaterials in deactivating multiple drug resistance efflux pumps – A review. *Environmental Research, 204*, 111968. DOI: 10.1016/j.envres.2021.111968 PMID: 34453898

Drost, M., Diamanti, E., Fuhrmann, K., Goes, A., Shams, A., Haupenthal, J., Koch, M., Hirsch, A. K., & Fuhrmann, G. (2021). Bacteriomimetic liposomes improve antibiotic activity of a novel energy-coupling factor transporter inhibitor. *Pharmaceutics, 14*(1), 4. DOI: 10.3390/pharmaceutics14010004 PMID: 35056900

Dube, T., Ghosh, A., Mishra, J., Kompella, U. B., & Panda, J. J. (2021). Repurposed drugs, molecular vaccines, immune-modulators, and nanotherapeutics to treat and prevent COVID-19 associated with SARS-CoV-2, a deadly nanovector. *Advanced Therapeutics, 4*(2), 2000172. DOI: 10.1002/adtp.202000172 PMID: 33173808

Farah, F. H., & Farah, F. H. (2019). Nanocarriers as delivery systems for therapeutics agents. *International Journal of Pharmaceutical Sciences and Research, 10*, 3487–3507.

Faria, M. J., Lopes, C. M., das Neves, J., & Lúcio, M. (2021). Lipid nanocarriers for anti-HIV therapeutics: A focus on physicochemical properties and biotechnological advances. *Pharmaceutics, 13*(8), 1294. DOI: 10.3390/pharmaceutics13081294 PMID: 34452255

Figueroa, S. M., Fleischmann, D., & Goepferich, A. (2021). Biomedical nanoparticle design: What we can learn from viruses. *Journal of Controlled Release, 329*, 552–569. DOI: 10.1016/j.jconrel.2020.09.045 PMID: 33007365

Ghosh, U., Sayef Ahammed, K., Mishra, S., & Bhaumik, A. (2022). The emerging roles of silver nanoparticles to target viral life cycle and detect viral pathogens. *Chemistry, an Asian Journal, 17*(5), e202101149. DOI: 10.1002/asia.202101149 PMID: 35020270

Gu, W., Wu, C., Chen, J., & Xiao, Y. (2013). Nanotechnology in the targeted drug delivery for bone diseases and bone regeneration. *International Journal of Nanomedicine*, •••, 2305–2317. DOI: 10.2147/IJN.S44393 PMID: 23836972

Gurunathan, S., Qasim, M., Choi, Y., Do, J. T., Park, C., Hong, K., Kim, J. H., & Song, H. (2020). Antiviral potential of nanoparticles—Can nanoparticles fight against coronaviruses? *Nanomaterials (Basel, Switzerland), 10*(9), 1645. DOI: 10.3390/nano10091645 PMID: 32825737

Hossain, M. S., Hami, I., Sawrav, M. S., Rabbi, M. F., Saha, O., Bahadur, N. M., & Rahaman, M. M. (2020). Drug repurposing for prevention and treatment of COVID-19: A clinical landscape. *Discoveries (Craiova, Romania), 8*(4), e121. DOI: 10.15190/d.2020.18 PMID: 33403227

Ioele, G., Chieffallo, M., Occhiuzzi, M. A., De Luca, M., Garofalo, A., Ragno, G., & Grande, F. (2022). Anticancer drugs: Recent strategies to improve stability profile, pharmacokinetic and pharmacodynamic properties. *Molecules (Basel, Switzerland)*, *27*(17), 5436. DOI: 10.3390/molecules27175436 PMID: 36080203

Johnson, D. B., Nebhan, C. A., Moslehi, J. J., & Balko, J. M. (2022). Immune-checkpoint inhibitors: Long-term implications of toxicity. *Nature Reviews. Clinical Oncology*, *19*(4), 254–267. DOI: 10.1038/s41571-022-00600-w PMID: 35082367

Khan, Y., Sadia, H., Ali Shah, S. Z., Khan, M. N., Shah, A. A., Ullah, N., Ullah, M. F., Bibi, H., Bafakeeh, O. T., Khedher, N. B., Eldin, S. M., Fadhl, B. M., & Khan, M. I. (2022). Classification, synthetic, and characterization approaches to nanoparticles, and their applications in various fields of nanotechnology: A review. *Catalysts*, *12*(11), 1386. DOI: 10.3390/catal12111386

Kim, J., Yeom, M., Lee, T., Kim, H. O., Na, W., Kang, A., Lim, J. W., Park, G., Park, C., Song, D., & Haam, S. (2020). Porous gold nanoparticles for attenuating infectivity of influenza A virus. *Journal of Nanobiotechnology*, *18*(1), 1. DOI: 10.1186/s12951-020-00611-8 PMID: 32209114

Lauer, S. A., Grantz, K. H., Bi, Q., Jones, F. K., Zheng, Q., Meredith, H. R., Azman, A. S., Reich, N. G., & Lessler, J. (2020). The incubation period of coronavirus disease 2019 (COVID-19) from publicly reported confirmed cases: Estimation and application. *Annals of Internal Medicine*, *172*(9), 577–582. DOI: 10.7326/M20-0504 PMID: 32150748

Lin, A., Giuliano, C. J., Palladino, A., John, K. M., Abramowicz, C., Yuan, M. L., Sausville, E. L., Lukow, D. A., Liu, L., Chait, A. R., Galluzzo, Z. C., Tucker, C., & Sheltzer, J. M. (2019). Off-target toxicity is a common mechanism of action of cancer drugs undergoing clinical trials. *Science Translational Medicine*, *11*(509), eaaw8412. DOI: 10.1126/scitranslmed.aaw8412 PMID: 31511426

Lombardo, D., Kiselev, M. A., & Caccamo, M. T. (2019). Smart nanoparticles for drug delivery application: Development of versatile nanocarrier platforms in biotechnology and nanomedicine. *Journal of Nanomaterials*, *2019*(1), 3702518. DOI: 10.1155/2019/3702518

Loureiro, J. A., & Pereira, M. C. (2020). PLGA-based drug carrier and pharmaceutical applications: The most recent advances. *Pharmaceutics*, *12*(9), 903. DOI: 10.3390/pharmaceutics12090903 PMID: 32971970

Majumder, J., Taratula, O., & Minko, T. (2019). Nanocarrier-based systems for targeted and site-specific therapeutic delivery. *Advanced Drug Delivery Reviews*, *144*, 57–77. DOI: 10.1016/j.addr.2019.07.010 PMID: 31400350

Manubolu, K., Yamini, S., Charishma, R., Kumar, N., Kishori, O., Tanusha, K., & Baba, K. H. (2023). Formulation and evaluation of phytosome and plants exhibiting anti-inflammatory activity. *Future Journal of Pharmaceuticals and Health Sciences*, *3*(1), 90–97.

Manzari, M. T., Shamay, Y., Kiguchi, H., Rosen, N., Scaltriti, M., & Heller, D. A. (2021). Targeted drug delivery strategies for precision medicines. *Nature Reviews. Materials*, *6*(4), 351–370. DOI: 10.1038/s41578-020-00269-6 PMID: 34950512

Mashel, T. V., Tarakanchikova, Y. V., Muslimov, A. R., Zyuzin, M. V., Timin, A. S., Lepik, K. V., & Fehse, B. (2020). Overcoming the delivery problem for therapeutic genome editing: Current status and perspective of non-viral methods. *Biomaterials*, *258*, 120282. DOI: 10.1016/j.biomaterials.2020.120282 PMID: 32798742

Maus, A., Strait, L., & Zhu, D. (2021). Nanoparticles as delivery vehicles for antiviral therapeutic drugs. *Engineered Regeneration*, *2*, 31–46. DOI: 10.1016/j.engreg.2021.03.001 PMID: 38620592

McCarthy, T. D., Karellas, P., Henderson, S. A., Giannis, M., O'Keefe, D. F., Heery, G., Paull, J. R., Matthews, B. R., & Holan, G. (2005). Dendrimers as drugs: Discovery and preclinical and clinical development of dendrimer-based microbicides for HIV and STI prevention. *Molecular Pharmaceutics*, *4*(4), 312–318. DOI: 10.1021/mp050023q PMID: 16053334

Medina-Cruz, D., Saleh, B., Vernet-Crua, A., Ajo, A., Roy, A. K., & Webster, T. J. (2020). Drug-delivery nanocarriers for skin wound-healing applications. In *Wound healing, tissue repair, and regeneration in diabetes* (pp. 439–488). Academic Press. DOI: 10.1016/B978-0-12-816413-6.00022-8

Megantara, S., Rusdin, A., Budiman, A., Shamsuddin, S., Mohtar, N., & Muchtaridi, M. (2024). Revolutionizing antiviral therapeutics: Unveiling innovative approaches for enhanced drug efficacy. *International Journal of Nanomedicine*, *19*, 2889–2915. DOI: 10.2147/IJN.S447721 PMID: 38525012

Müller, R. H., Jacobs, C., & Kayser, O. (2001). Nanosuspensions as particulate drug formulations in therapy: Rationale for development and what we can expect for the future. *Advanced Drug Delivery Reviews*, *47*(1), 3–19. DOI: 10.1016/S0169-409X(00)00118-6 PMID: 11251242

Nainwal, N. (2022). Treatment of respiratory viral infections through inhalation therapeutics: Challenges and opportunities. *Pulmonary Pharmacology & Therapeutics*, *77*, 102170. DOI: 10.1016/j.pupt.2022.102170 PMID: 36240985

Negri, V., Pacheco Torres, J., Calle, D., & López Larrubia, P. (2020). Carbon nanotubes in biomedicine. In *Surface-modified nanobiomaterials for electrochemical and biomedicine applications* (pp. 177-217). DOI: 10.1007/978-3-030-55502-3_6

Nikzamir, M., Hanifehpour, Y., Akbarzadeh, A., & Panahi, Y. (2021). Applications of dendrimers in nanomedicine and drug delivery: A review. *Journal of Inorganic and Organometallic Polymers and Materials*, *31*(6), 2246–2261. DOI: 10.1007/s10904-021-01925-2

Okafor, N. I., & Nnaji, N. J. (2024). Liposomes, facilitating the encapsulation and improved solubility of zidovudine antiretroviral drug. *BioNanoScience*, *3*(5), 1–8. DOI: 10.1007/s12668-024-01504-y

Pandey, M., Choudhury, H., Abdul-Aziz, A., Bhattamisra, S. K., Gorain, B., Su, J. S., Tan, C. L., Chin, W. Y., & Yip, K. Y. (2020). Advancement on sustained antiviral ocular drug delivery for herpes simplex virus keratitis: Recent update on potential investigation. *Pharmaceutics*, *13*(1), 1. DOI: 10.3390/pharmaceutics13010001 PMID: 33374925

Park, J., Choi, Y., Chang, H., Um, W., Ryu, J. H., & Kwon, I. C. (2019). Alliance with EPR effect: Combined strategies to improve the EPR effect in the tumor microenvironment. *Theranostics*, *9*(26), 8073–8090. DOI: 10.7150/thno.37198 PMID: 31754382

Pradhan, D., Biswasroy, P., Goyal, A., Ghosh, G., & Rath, G. (2021). Recent advancement in nanotechnology-based drug delivery system against viral infections. *AAPS PharmSciTech*, *22*(1), 1–9. DOI: 10.1208/s12249-020-01908-5 PMID: 33447909

Raji, V., Pal, K., Zaheer, T., Kalarikkal, N., Sabu, T., de Souza, F. G., & Asiya, S. I. (2020). Gold nanoparticles against respiratory diseases: Oncogenic and viral pathogens review. *Therapeutic Delivery*, *11*(8), 521–534. DOI: 10.4155/tde-2020-0071 PMID: 32757745

Sahu, T., Ratre, Y. K., Chauhan, S., Bhaskar, L. V., Nair, M. P., & Verma, H. K. (2021). Nanotechnology-based drug delivery system: Current strategies and emerging therapeutic potential for medical science. *Journal of Drug Delivery Science and Technology*, *63*, 102487. DOI: 10.1016/j.jddst.2021.102487

Sanna, V., Satta, S., Hsiai, T., & Sechi, M. (2022). Development of targeted nanoparticles loaded with antiviral drugs for SARS-CoV-2 inhibition. *European Journal of Medicinal Chemistry*, *231*, 114121. DOI: 10.1016/j.ejmech.2022.114121 PMID: 35114539

Shah, A., Aftab, S., Nisar, J., Ashiq, M. N., & Iftikhar, F. J. (2021). Nanocarriers for targeted drug delivery. *Journal of Drug Delivery Science and Technology*, *62*, 102426. DOI: 10.1016/j.jddst.2021.102426

Sharma, A., Mishra, R. K., Goud, K. Y., Mohamed, M. A., Kummari, S., Tiwari, S., Li, Z., Narayan, R., Stanciu, L. A., & Marty, J. L. (2021). Optical biosensors for diagnostics of infectious viral disease: A recent update. *Diagnostics (Basel)*, *11*(11), 2083. DOI: 10.3390/diagnostics11112083 PMID: 34829430

Sharma, P., & Garg, S. (2010). Pure drug and polymer-based nanotechnologies for the improved solubility, stability, bioavailability and targeting of anti-HIV drugs. *Advanced Drug Delivery Reviews*, *62*(4–5), 491–502. DOI: 10.1016/j.addr.2009.11.019 PMID: 19931328

Shi, Y., van der Meel, R., Chen, X., & Lammers, T. (2020). The EPR effect and beyond: Strategies to improve tumor targeting and cancer nanomedicine treatment efficacy. *Theranostics*, *10*(17), 7921–7924. DOI: 10.7150/thno.49577 PMID: 32685029

Shrestha, B., Tang, L., & Romero, G. (2019). Nanoparticles-mediated combination therapies for cancer treatment. *Advanced Therapeutics*, *2*(11), 1900076. DOI: 10.1002/adtp.201900076

Song, Y., Zhang, M., Yin, L., Wang, K., Zhou, Y., Zhou, M., & Lu, Y. (2020). COVID-19 treatment: Close to a cure? A rapid review of pharmacotherapies for the novel coronavirus (SARS-CoV-2). *International Journal of Antimicrobial Agents*, *56*(2), 106080. DOI: 10.1016/j.ijantimicag.2020.106080 PMID: 32634603

Souri, M., Soltani, M., Kashkooli, F. M., & Shahvandi, M. K. (2022). Engineered strategies to enhance tumor penetration of drug-loaded nanoparticles. *Journal of Controlled Release*, *341*, 227–246. DOI: 10.1016/j.jconrel.2021.11.024 PMID: 34822909

Su, S., & Kang, P. M. (2020). Recent advances in nanocarrier-assisted therapeutics delivery systems. *Pharmaceutics*, *12*(9), 837. DOI: 10.3390/pharmaceutics12090837 PMID: 32882875

Suhaimi, N. A., Ahmad, S., Husna, S. M., Sarmiento, M. E., Acosta, A., Norazmi, M. N., Ibrahim, J., Mohamud, R., & Kadir, R. (2022). Application of liposomes in the treatment of infectious diseases. *Life Sciences*, *305*, 120734. DOI: 10.1016/j.lfs.2022.120734 PMID: 35760094

Szyk, P., Czarczynska-Goslinska, B., Mlynarczyk, D. T., Ślusarska, B., Kocki, T., Ziegler-Borowska, M., & Goslinski, T. (2023). Polymer-based nanoparticles as drug delivery systems for purines of established importance in medicine. *Nanomaterials (Basel, Switzerland), 13*(19), 2647. DOI: 10.3390/nano13192647 PMID: 37836288

Tang, Y., Zhang, H., Lu, X., Jiang, L., Xi, X., Liu, J., & Zhu, J. (2015). Development and evaluation of a dry powder formulation of liposome-encapsulated oseltamivir phosphate for inhalation. *Drug Delivery, 22*(5), 608–618. DOI: 10.3109/10717544.2013.863526 PMID: 24299495

Uddin, T. M., Chakraborty, A. J., Khusro, A., Zidan, B. R., Mitra, S., Emran, T. B., Dhama, K., Ripon, M. K., Gajdács, M., Sahibzada, M. U., Hossain, M. J., & Koirala, N. (2021). Antibiotic resistance in microbes: History, mechanisms, therapeutic strategies and future prospects. *Journal of Infection and Public Health, 14*(12), 1750–1766. DOI: 10.1016/j.jiph.2021.10.020 PMID: 34756812

Van Tran, V., Moon, J. Y., & Lee, Y. C. (2019). Liposomes for delivery of antioxidants in cosmeceuticals: Challenges and development strategies. *Journal of Controlled Release, 300*, 114–140. DOI: 10.1016/j.jconrel.2019.03.003 PMID: 30853528

Yeh, Y. C., Huang, T. H., Yang, S. C., Chen, C. C., & Fang, J. Y. (2020). Nano-based drug delivery or targeting to eradicate bacteria for infection mitigation: A review of recent advances. *Frontiers in Chemistry, 8*, 286. DOI: 10.3389/fchem.2020.00286 PMID: 32391321

Yetisgin, A. A., Cetinel, S., Zuvin, M., Kosar, A., & Kutlu, O. (2020, May 8). Therapeutic nanoparticles and their targeted delivery applications. *Molecules (Basel, Switzerland), 25*(9), 2193. DOI: 10.3390/molecules25092193 PMID: 32397080

Yetisgin, A. A., Cetinel, S., Zuvin, M., Kosar, A., & Kutlu, O. (2020). Therapeutic nanoparticles and their targeted delivery applications. *Molecules (Basel, Switzerland), 25*(9), 2193. DOI: 10.3390/molecules25092193 PMID: 32397080

Zhou, Y., Fletcher, N. F., Zhang, N., Hassan, J., & Gilchrist, M. D. (2021). Enhancement of antiviral effect of plastic film against SARS-CoV-2: Combining nanomaterials and nanopatterns with scalability for mass manufacturing. *Nano Letters, 21*(24), 10149–10156. DOI: 10.1021/acs.nanolett.1c02266 PMID: 34881894

Zi, Y., Yang, K., He, J., Wu, Z., Liu, J., & Zhang, W. (2022). Strategies to enhance drug delivery to solid tumors by harnessing the EPR effects and alternative targeting mechanisms. *Advanced Drug Delivery Reviews, 188*, 114449. DOI: 10.1016/j.addr.2022.114449 PMID: 35835353

Chapter 3
Personalized Antiviral Therapy TailoringTreatments to Genetic Environmental and Lifestyle Factors

Reyed M. Reyed
https://orcid.org/0000-0001-5027-3478
Genetic Engineering and Biotechnology Research Institute (GEBRI), City of Scientific Research and Technological Applications (SRTA-City), Alexandria, Egypt

Pranav Kumar Prabhakar
https://orcid.org/0000-0001-8130-1822
Research and Development Cell, Parul University, India

ABSTRACT

This chapter aims to discuss the individualized approach to antiviral therapy targeting genetic, environmental, and lifestyle issues. It talks about how pharmacogenomics, AI, and computational modeling help identify better drug targets and reduce the odds of resistance. These comprise genetic screening in personalized medicine, biomarkers, and clinical trials in therapeutics. It also examines infection control, the axis between the gut microbiome and the host, and probiotics as novel biotherapeutics. Positioned under the umbrella of a one-health concept, it entails concierge wellness nutrition and precision medicine to optimize therapeutic interventions, combat antiviral resistance, and advance patient satisfaction. The prospects provided are intended to open new avenues in the composition of antiviral drugs.

DOI: 10.4018/979-8-3693-9276-8.ch003

Copyright © 2025, IGI Global Scientific Publishing. Copying or distributing in print or electronic forms without written permission of IGI Global Scientific Publishing is prohibited.

1 INTRODUCTION

This area of antiviral treatment has been transitioning from the traditional model to a targeted and effective individual approach. Today's antiviral treatments are primarily pharmacological and functional, such as brand-name antibiotics, which do not distinguish between genetic or phenotypic variability or between patients. This not only reduces therapeutic effects but also leads to the development of side effects or the emergence of drug resistance (Duan et al., 2024). Personalized, rational antiviral treatment appears to be one of the most promising directions for employing genomic, bioinformatics, and molecular biology approaches to predict the efficacy of treatment for different individuals (Su et al., 2024). This strategy integrates data regarding the genetic makeup of every patient exhibiting features of particular viral infections with the characteristics of the infection to ensure efficient and effective treatment with minimal side effects.

Many people mention one of the most promising techniques in this context: the application of the CRISPR–Cas9 system. This cutting-edge technology involves the delicate process of editing viral DNA, which could open the door to eradicating infections caused by some of the most difficult-to-eradicate strains of viruses (Hussen et al., 2024). Furthermore, other advanced techniques, such as pharmacogenomics, viral genotyping, and immune-modulating therapies, are integrated into personalized treatments to increase treatment efficacy and reduce the occurrence of drug-resistant states. These factors are crucial in managing both chronic and acute viral illnesses, such as AIDS (Frericks et al., 2025), hepatitis B virus (HBV) (Lee et al., 2024), hepatitis C virus (HCV) (Fasano et al., 2024), influenza (Lane et al., 2014), human metapneumovirus (hMPV) (WHO 2025) and the new invasive virus 2019 novel coronavirus, also known as severe acute respiratory syndrome coronavirus 2 (SARS-CoV-2) (Li et al., 2025).

However, several challenges arise regarding the concept of targeted antiviral therapy for viral diseases. The following are some drawbacks related to genetic screening: the genetic testing procedure is technical and therefore relatively time-consuming, there are ethical challenges concerning the privacy of information, and the costs of personalizing treatment are relatively high. Importantly, however, access to these therapies for patients, especially those from developing countries, should be made reasonably available. Most of these issues are best resolved systematically and cannot be addressed by a single discipline: technology fixes, policy, and public health approaches are needed.

Probiotics, prebiotics and synbiotics represent a cutting-edge concept in pharmacology that enhances a specific and targeted strategy, deviating from the empirical, chemical, and biological control of viral diseases in general. This chapter thus presents theoretical frameworks, scientific developments, and applications of this new perspective of individualized therapy and offers the reader an up-to-date account of how the approach is transforming the treatment of viral diseases. The chapter begins with the concept of personalized antiviral therapy and a comparative evaluation of the differences with conventional treatment strategies. This highlights the deficiencies of conventional methods: the absence of reflection of interpatient variability and the continuously evolving threat of drug resistance. The processes of personalization of genetic, phenotypic, and clinical characteristics are described and correlated to subsequent sections of complementary instruments.

A brief discussion of concepts and devices is then conducted thoroughly. For example, CRISPR-Cas9 is commonly used to edit viral DNA and can be utilized to treat viral diseases such as AIDS, hepatitis, and HPV. Pharmacogenomics and viral genotyping are two key technologies for understanding patient outcomes related to certain drugs and selecting suitable antiviral molecules. Advanced techniques in artificial intelligence and machine learning facilitate the identification of these drugs and increase the quality of treatment regimens, whereas nanotechnology, especially delivery systems such as nanovaccines, increases therapeutic value and reduces side effects.

In this chapter, approaches to antiviral therapy are emphasized, as well as the role of nutrition, where multivitamins and minerals such as vitamin C, D, zinc, and omega-3 fatty acids are beneficial in enhancing immune strength. However, the report addresses the impact of probiotics, prebiotics, synbiotics and the gut microbiota, particularly on the systemic immune system, and how they may improve antiviral results. Moreover, physical activity, stress, and environmental factors, including nutrition, influence the immune response and therapeutic outcomes.

As a model for personalized antiviral therapy, case studies providing concrete considerations concerning clients with HIV, hepatitis, seasonal flu, and COVID-19 are discussed in detail. The use of information technology in genomics and pharmacogenomics is examined, specifically issues that are most susceptible to drugs and efforts to combat drug resistance. Concrete examples demonstrating how it is possible to improve a patient's state through the application of genetic and clinical data will also be discussed.

The chapter also discusses challenges surrounding the adoption of personalized antiviral therapies. These include the high cost of screening for genetic disorders in contrast to the necessary treatment services, the issue of personal and equitable access to genetic data and information, and the technology, as well as the physical infrastructure needed to support adoption, especially in the developing world. Finally, advancements in antiviral therapy through interdisciplinary approaches combining genomics, nutrition, artificial intelligence, and biotherapeutics are outlined in the last section of the chapter. Some potential ways to mitigate these risks involve resolving the cost issue, advocating for favorable laws, and ensuring worldwide access to effective treatments.

This chapter describes a shift in antiviral medicine aimed at the further development of various technologies, including CRISPR, pharmacogenomics, and immune-focused treatment interventions, that transform antiviral treatment into a more personalized model, such as personalized antiviral therapy. Through a discussion of scientific fundamentals, ethical considerations, and practical applications, the findings highlight the opportunities that personalized antiviral therapy presents for enhancing treatment outcomes and establishing a new paradigm in combating viral diseases.

2 OVERVIEW OF PERSONALIZED ANTIVIRAL THERAPY

This is one of the actively developing trends in today's practice of medicine, which means that the method of treatment is highly specific and depends on the genetic, phenotypic, and clinical characteristics of patients. This approach is better than the norm that involves administering a single medication to all people because the antiviral effects differ significantly across individuals. On the basis of current technologies, such as genomics, bioinformatics, and patient-tailored diagnostics, one can determine the genetic characteristics, immunogenicity, and particular virus to develop particular treatment trials to achieve maximal efficacy with fewer side effects (Hoang, et al., 2024). The umbrella concept of this approach is reminiscent of the concept of increasing specificity to yield better results and limit drug resistance in personalized antiviral therapy (Table 1). Patient-specific antiviral therapy is best applicable for chronic viral diseases, exemplified by HIV, hepatitis B, and hepatitis C, which are caused by viral mutations, issues of drug resistance or variations in the immune system response. Like some patients with chronic diseases need customized management of chronic renal diseases, acute viral-like flu and new pandemics, such as COVID-19, could also ensure that high-risk patients receive more effective and safer therapies (Jose-Abrego et al., 2023). The idea of combining genetic differences and viruses in the therapy plan makes the concept of personalized therapy a novel concept for antiviral therapy.

Table 1. Significantly expressed genes and their important functions in personalized antiviral therapy. Table showing important terms and their meanings connected to personalized antiviral therapy, emphasizing the cutting-edge features of this developing area

Term	Definition	Significance
Tailored Antiviral Treatment	The treatment plan consisting of the individualized therapy considering a patient genotype, phenotype, and clinical characteristics (Su et al.,2024). The incorporation of genetic differences and viruses in the treatment approach makes personalized therapy a new approach for antiviral treatment.	**Enhances appropriate therapeutic intervention and the least side effects thus improving the result of the patients.**
Pharmacogenomics	**The study of how genes affect a person's response to drugs, allowing for tailored medication regimens (Zhou et al.,2024).**	Enhances the precision of antiviral therapies by predicting individual responses to specific antiviral agents.
Viral Genotyping	Assessing viral strains to uncover mutations associated with therapeutic resistance (Chakraborty et al., 2025).	**Crucial for choosing the right treatments when facing drug-resistant viral strains.**
Identification of biological markers	**The detection of biomarkers that predict a patient's response to antiviral therapy (Motomura et al.,2023).**	Makes it easier to choose the right treatment plans based on how patients are expected to respond.
Focused therapeutic intervention	A method that utilizes virus-specific features or host genetics to create personalized therapies (Parambi et al., 2022).	**Enhances the effectiveness of treatment by targeting the unique characteristics of the infection or the patient.**
Personalized medical plans	**Personalized treatment plans that combine genetic, environmental, and clinical factors unique to the patient (Stefanicka-Wojtas et al., 2023).**	Increases treatment compliance and reduces the chances of side effects from medications.
Drug resistance assessment	Analyzing resistance patterns in viruses to refine drug selection and achieve better therapeutic results (Muteeb et al., 2023).	**Enables healthcare providers to select the most effective therapies, decreasing the risk of treatment failure.**
Integration of immune-targeting treatments	**To improve the host's antiviral defenses, antiviral medications are combined with immune-modulating therapies (Costa et al., 2023).**	Supports antiviral treatments by enhancing the immune response to fight viral infections.

continued on following page

Table 1. Continued

Term	Definition	Significance
Precision targeting for susceptible populations	Approaches aimed at recognizing and managing individuals who are at higher risk for severe viral infections (Shyr et al., 2021).	**Makes sure that at-risk groups get personalized and effective antiviral treatments.**
Personalized public health strategies	**Using personalized medicine concepts on a community scale to handle and prevent viral outbreaks (Dopazo et al., 2023).**	Seeks to enhance community health results by tackling the unique needs and risks in different populations.

3 NOVEL THERAPEUTIC AGENTS IN PERSONALIZED ANTIVIRAL THERAPY: FOCUS ON CRISPR-CAS9 TECHNOLOGY

Currently, antiviral therapy approaches are in transition, and personalized medicine is becoming the focus of interest in the field. All of these have been spearheaded by the cutting-edge tool called CRISPR Cas9, which enables the targeting and altering of viral genomes. This approach not only improves the efficacy of antiviral therapies but also paves the way for the development of symptomatic approaches for chronic viral diseases, which are difficult to treat.

Knowledge About the CRISPR–Cas9 System

CRISPR–Cas9 is the present-day cutting-edge gene editing tool, whereby one can modify sections of an individual's genome with precision. Originating from bacteria for the purpose of defending against virus attacks, CRISPR-Cas9 has been patented for several uses, especially in the field of medicine. In antiviral therapy, this technology holds great potential for eradicating dormant viral leagues, or "sanctuaries," harboring viruses but not actively producing them. This feature is most important for chronic infections where the virus can remain in the organism even after treatments that are usually applied. The CRISPR-Cas9 (Chakraborty et al., 2025) system comprises two essential components: Cas9, which functions as the molecular scalpel to cut the DNA viral genome, and RNA (gRNA), which helps Cas9 identify the precise viral DNA sequence that has been targeted for alteration. This directed approach allows for the targeted modification of viral genomes, potentially leading to the eradication of integrated viral DNA.

Figure 1. Provides a summary of the key aspects of CRISPR-Cas9 technology (Domingues et al.,2024), including its components, mechanism, potential applications, benefits, limitations, and future prospects. This highlights the importance of CRISPR–Cas9 technology in the context of individualized antiviral treatment."

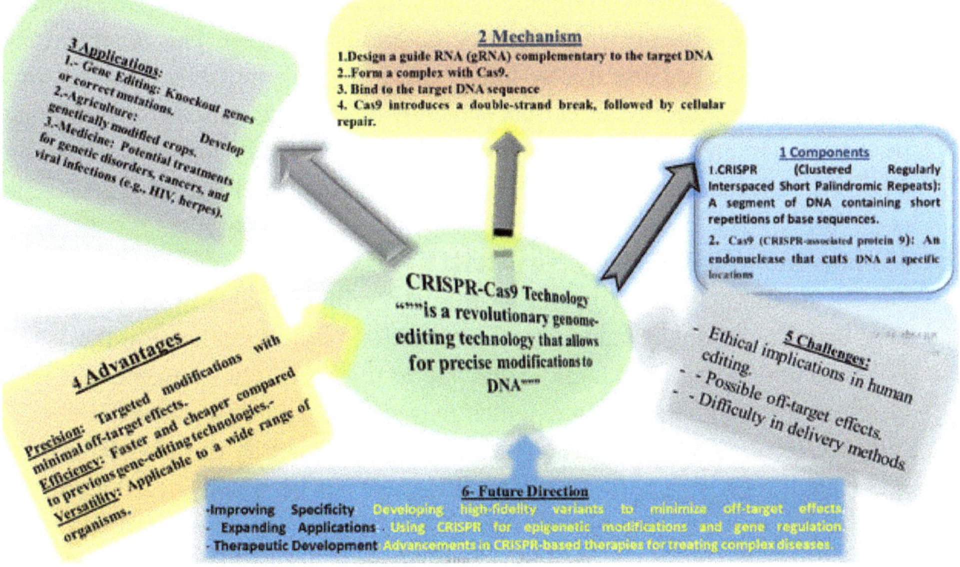

Applications of CRISPR-Cas9 in Antiviral Therapy

CRISPR-Cas9 has been considered for antiviral therapy on the basis of target-dependent and off-target mechanisms (Fatima et al., 2025; Najafi et al.,2022). The benefits of CRISPR-Cas9 have made it possible to develop new strategies for antiviral therapy due to its flexibility, and viral genomes can be targeted with CRISPR-Cas9. The above approach has shown promise for different viral infections, both in the management of diseases and in the permanent eradication of viruses. Several key applications of CRISPR-Cas9 in antiviral therapy are described below:

1. **Hepatitis B virus (HBV)** is a chronic liver infection that commonly causes chronic liver disease, cirrhosis, and liver cancer. Previously used therapies for chronic HBV have focused on reducing viremia via antiviral agents; however, no cure has been identified, and lifelong treatment is needed. CRISPR-Cas9 provides a novel strategy by directly controlling and eradicating **HBV covalently closed circular DNA (cccDNA)**, which is located in the nucleus of infected hepatocytes. This form of viral DNA is a stable reservoir, and its existence is always a major challenge to current treatments because the HIV virus is never

completely cleared (Kumar, et al.,2024). An effective example of the application of CRISPR-Cas9 was its ability to separate and delete HBV DNA from infected cells and reduce the viral load. Alterations made with CRISPR-Cas9-targeted specific regions of the HBV genome in animal models, demonstrated in pre-clinical studies, render it a potentially functional cure by lowering the number of copies of the virus present in liver cells (Yao et al.,2024). This strategy may be more effective because it is no longer just trying to control the virus; it aims to eradicate the cause of infection, which may save the patient from lifelong treatment. The next challenge is how to optimize delivery methods to deliver CRISPR components to liver cells without side effects (Zhang, et al 2024; Cai et al.,2019).

2. **Human immunodeficiency virus (HIV)** HIV remains one of the biggest current threats to human health worldwide, with available ART treatments now being powerful at managing the condition but not eradicating the virus. HIV acquires a universal code for its site in the host's chromosome and then develops dormant viral storage, which reawakes at ART cessation (Chou, et al 2024). One of the challenges has therefore been the ability of the HIV virus to form latent reservoirs; however, CRISPR-Cas9 has been considered a tool for attacking these reservoirs (Hussein et al., 2023). It has been engineered into guide RNAs that have the ability to bind selectively to integrated HIV proviral DNA. Upon binding Cas9, the nuclease targets the HIV genome for cleavage, and the signals for replication or reactivation of the virus become ineffective. Research conducted on animal models has demonstrated that this method significantly decreases the viral load and possibly eradicates the viral load. In addition, there have also been recent developments in dual-targeting strategies, where CRISPR-Cas9 can be incorporated with other gene editing or immunotherapy strategies. These strategies target not only the viral genome but also the immune system to eliminate infected cells. While it remains unproven, considerable potential exists with this new technology—the prospect of finding a cure for HIV has remained elusive for several decades (Zahedipour, et al.,2024)
3. **Human papilloma virus (HPV)** is a normal sexually transmitted virus known to be associated with several forms of cancer, including cervical cancer. Currently, cancer treatments that target HPV include surgery, chemotherapy, and radiation, none of which are aimed at the genome of a virus. HPV is known to form its DNA in host cells, which in effect turns normal cells into cancerous cells. Such direct penetration into infected cells to eliminate oncogenic HPV genomes represents the strength of CRISPR–Cas9 technology (Zhen, et al., 2023). Experimental work conducted on cervical cancer cell lines has shown that the CRISPR–Cas9 system can cleave HPV DNA selectively in regions encoding the E6 and E7 oncogenes that contribute to the unlimited growth of infected cells (Dust et al.

2022). The deletion of these oncogenes causes a decrease in viral load and stops the development of cancer. In addition, CRISPR-Cas9 has potential applications for suppressing the formation of HPV-related cancers. According to the information given by the developers of this technology, it could be used to target HPV at an early stage of illness, which might help prevent the development of cancer in high-risk patients. In addition, both the antiviral and anticancer uses of CRISPR-Cas9 are believed to have implications for the development of novel therapeutic interventions for HPV-associated carcinomas) Weiet al.,2022).

4. **Zika virus** The virus is most commonly transmitted by Aedes mosquitoes and came to the center stage in 2016, with the World Health Organization characterizing it as a result of microcephaly, a condition where babies are born abnormally small-headed. Previous methods include community-level vector control and public health measures (Nyathi, et al.2024); however, the CRISPR–Cas9 system offers the potential of a genetically engineered solution to the Zika virus via modified mosquitoes. Recently, by applying the CRISPR–Cas9 methodology, mosquitoes have been engineered such that they cannot be infected by the Zika virus or are incapable of reproducing (Kefi, et al.,2024; Li et al., 2019). One of them is that it is possible to interfere with segments in a mosquito's genome in which the virus needs to reproduce and continue to exist. As a result, genetically modified mosquitoes are not as effective at transmitting the virus to humans. Another technique involves deploying CRISPR to insert gene traders that spread these virus resistance characteristics through wild mosquito populations within a very short time (Zhang, et al 2024). In laboratory environments, these genetically altered mosquitoes have a high rate of reduction in the spread of the Zika virus. This approach has not yet been used in the field of eradication; however, it is a promising way to control not only Zika but also other mosquito-borne viruses, such as dengue and chikungunya (Naidoo, & Oliver, 2024; Ogaugwu, et al., 2019).
5. **Human metapneumovirus (hMPV)** Bronchial viral particles called hMPV may trigger deadly infections of the lower airway tract, particularly in children and those with a weakened immune system. Both therapeutic and diagnostic applications of CRISPR-Cas9 have shown great potential:

Diagnostics: When combined with RT-RPA, CRISPR-Cas12a enhances diagnostic capabilities and increases the rapid, specific, and powerful early identification of hMPV (Tian et al., 2021).

Therapeutics: CRISPR interference (CRISPRi) is a new approach for analyzing hMPV-host interactions. For example, the regulation of the editosome ADAR1 gene, which edits hMPV genomes, has provided insights into the replication kinetics of the virus (van den Hoogen et al., 2022). Furthermore, the ability of CRISPR-Cas9 to disrupt essential hMPV genes could prevent viral replication and reduce disease manifestations. Given how CRISPR-Cas9 was utilized in the management of hMPV, a recent and acute viral threat, this technology can therefore be regarded as capable of responding to emerging viral threats. When combined with conventional antiviral drugs or immunomodulatory strategies, the inclusion of the CRISPR-Cas9 system appears to improve treatment outcomes and minimize the development of resistance.

Thus, CRISPR-Cas9 represents a revolutionary concept in antiviral treatment (Table 2), addressing long-term viruses such as HBV and HIV, as well as acute viruses such as Zika and hMPV. Its features in genome editing and diagnostics make it a critical platform for future therapeutic techniques, with the potential to revolutionize global healthcare and molecular medicine. Further studies and clinical work are necessary for this agent to fully realize its potential in combating viral diseases.

The table compares standard antiviral therapies to novel CRISPR/Cas9 techniques, emphasizing significant variations in accuracy, flexibility, precision, and long-term effects. CRISPR/Cas9 represents a paradigm shift, offering a more specialized, efficient, and adaptable approach to combating viral illnesses while addressing issues such as distribution, adverse outcomes, and ethical dilemmas.

Table 2. Comparison of standard antiviral treatments and CRISPR/Cas9 approaches

Category	Traditional Antiviral Treatments	CRISPR/Cas9 Approaches
Precision Targeting	**Broad-spectrum approaches targeting common viral features, leading to side effects and harmful consequences. (Wang et al .,2024)**	Employs gRNA for precise targeting of viral genomes, inhibiting replication with minimal off-target effects.) Chen, et al ., 2025)
Resistance Development	High likelihood of resistance development due to viral mutations (e.g., HIV, influenza). (Wang H. 2024).	**Adaptable to new mutations, allowing rapid response to evolving viruses like SARS-CoV-2. (Cheng, et al 2024)**
Specificity	**Nonspecific targeting affecting both infected and normal cells, reducing immunity and causing adverse impacts. (Bhargavi, & Subbian, 2024).**	Targets viral or host factors with high precision, sparing healthy cells and minimizing side effects. (Chehelgerdi, et al., 2024)
Personalization	Limited scope for individualized treatments based on genetic variations. (Zhou et al.,2024)	**Tailors therapies to host and viral genotypes, enhancing efficacy and reducing risks. (Macarrón Palacios et al.,2024)**
Versatility Across Viruses	Restricted to specific viruses, lacking flexibility across different viral families. (Jones, J et al 2024)	Effective against both DNA and RNA viruses, including HBV and Zika, making it a versatile tool.) Bahrulolum, et al., 2024)
Side Effects	**Potentially significant due to the nonspecific nature of treatments. (Tian, et al ., 2023)**	Reduced side effects due to pre cise targeting and minimal impact on healthy cells. (Yang, et al.,2024)

continued on following page

Table 2. Continued

Category	Traditional Antiviral Treatments	CRISPR/Cas9 Approaches
Long-lasting Effects	**Requires repeated drug administration for chronic infections.** (Shyr, **et al ., 2021**)	**Introduces heritable genome changes, offering potential cures for chronic infections. (Zhang et al., 2024)**
Efficiency	**Moderate efficiency with potential for unintended effects on nontarget sites.**) Mahajan, **et al.,2021)**	High efficiency with minimal off-target effects, ensuring accurate outcomes. (Chehelgerdi, et al .,2024)
Delivery Systems	Relies on systemic administration methods with varying success. (Žigrayová, et al., 2023)	**Faces challenges in delivering components effectively; viral vectors and nanoparticles are potential solutions (Chehelgerdi, et al.,2024) .**
Off-target Risks	**Not applicable.**	Improvements in guide RNA design reduce but do not eliminate off-target risks. (Choi, et al .,2024)
Regulatory and Ethical Issues	Not significant for most traditional treatments.	**Raises concerns about germline modifications and regulatory compliance for individualized therapies. (Hadi, et at., 2024)**

KEY DIFFERENCES BETWEEN CRISPR-CAS9 AND CRISPR-CAS12: MECHANISMS, APPLICATIONS, AND EFFICIENCY

Table 3 summarizes the differences between CRISPR-Cas9 and CRISPR-Cas12 on the basis of various key characteristics. This comparison helps to understand their distinct mechanisms, applications, and efficiencies.

Table 3. Comparison between CRISPR-Cas9 and CRISPR-Cas12. This table highlights the differences between CRISPR-Cas9 and CRISPR-Cas12, showing that while CRISPR-Cas9 excels in gene editing and modifications, CRISPR-Cas12 is more advantageous for diagnostic applications because of its higher precision, sensitivity, and flexibility in identifying specific DNA sequences.

Feature	CRISPR-Cas9	CRISPR-Cas12
Target Mechanism	Binds to double-stranded DNA (dsDNA), requires a specific PAM sequence (e.g., NGG)	Cuts both dsDNA and ssDNA, more flexible PAM sequence (e.g., TTTN or TTTV), and shows collateral cleavage activity
Protein Structure	Single protein, uses a single guide RNA (sgRNA)	Multidomain protein, requires only PAM sequence for interaction, capable of collateral cleavage
Applications	Gene knockout, insertion, and replacement in genetic research and agriculture	Primarily used for diagnostics, including virus detection (e.g., COVID-19), and genome editing
Precision and Side Effects	Higher likelihood of off-target effects, requires careful guide RNA design	More specific with fewer off-target effects, better suited for diagnostics
Flexibility and Efficiency	General-purpose for genetic manipulations, limited by PAM sequence	More flexible targeting and efficient in diagnostics with rapid DNA cleavage
Medical Diagnosis and Treatment	Used in gene therapy, genetic research, and chronic genetic diseases	Used in diagnostic applications due to high sensitivity, low cost, and quick pathogen detection

4 FUTURE PERSPECTIVES AND CHALLENGES

In 2025, Jiang et al. reported on the development of the first CRISPR/dCas9 hybrid SERS biosensor, which the authors named dCasSERS, which is portable via smartphones. The presented paradigm addresses several critical concerns in early pathogen identification, including low sensitivity, instability, and extended analysis durations. By utilizing LAMP for DNA amplification and CRISPR/dCas9 for specific recognition, the system assembles gold nanoparticles to enhance SERS signals, enabling highly sensitive detection of *S. aureus, P. aeruginosa*, and *E. coli* O157:H7 at 1 CFU/mL in under 50 minutes. This technology further highlights the possibility of other applications for CRISPR systems beyond diagnostics, including viral identification and targeted medicine on the basis of the patient's individual genetic makeup.

However, there are significant hindrances to the clinical application and therapeutic use of the CRISPR-Cas9 system for viral diseases. The challenges include the delivery of CRISPR components to infected cells, off-target effects, and ethical and regulatory issues. As evidenced by vector control techniques, CRISPR has the ability to help prevent epidemics, which is a critical factor in managing infectious diseases and precision medicine. To optimize the deliverability of this project in the future, forthcoming work should focus on improving the delivery of CRISPR components, especially through the use of viral vectors or nanoparticles, thus targeting infected cells without affecting normal cells (Model et al., 2022).

Furthermore, enhancing Cas9 variants with high fidelity is essential for increasing safety and specificity, according to a study by Dubey and Mostafavi 2023. In terms of applications, CRISPR has great potential in the development of antiviral therapy as a new approach for the treatment of viral diseases, especially those that have not yet been effectively treated. When combined with other healthcare systems, such as immunotherapies and pharmacological treatments, the potential for targeted delivery via CRISPR is strengthened and can transform medicine when implemented in individual practice.

ADVANTAGES OF CRISPR-CAS9 IN ANTIVIRAL THERAPY

It has been demonstrated that CRISPR-Cas9 has a high affinity for viral agents, and owing to this high specificity, this tool presents unique opportunities in antiviral therapy. By far, its chief strength is its high specificity, which significantly reduces potential off-target effects. These controls make it possible to modify viral DNA or RNA while avoiding harm to the rest of the host cells. This feature is especially useful in diseases that take a long time to manifest, such as HIV or even hepatitis, where the target is purely the virus, not healthy cells. These strategies increase the potential of antiviral treatments and the safety of their use.

CRISPR-Cas9 holds another revolutionary promise in terms of obtaining functional cures for chronic viral infections. For that reason, generic antiviral drugs designed to slow or stop viral replication are quite distinguishable from genes excised from a host organism's genome: CRISPR-Cas9 aims to excise them permanently. This could dramatically change the approach to diseases such as HIV that integrate into the host's genome and for which current remedies are ineffective. The advantage of CRISPR-Cas9 is that it directly removes the viral genetic material and creates the possibility of curing the disease rather than just controlling the virus.

Another factor that increases the attraction of the CRISPR–Cas9 system as a platform for antiviral solutions is its flexibility. The opportunity to edit the gRNA sequences allows the targeting of a vast number of existing viruses, including newly emerging ones or those with a high mutation rate, such as influenza and coronavirus. This versatility makes CRISPR-Cas9 an effective system for addressing both present and future viral threats while offering the possibility of improvement over time, highlighting the need for better and innovative antiviral therapies.

Table 4 summarizes the main strengths of CRISPR-Cas9-based antiviral therapy.

Table 4. Advantages of using CRISPR-Cas9 for antiviral therapy. The chart highlights the primary advantages of CRISPR-Cas9 in the treatment of viral infections. By minimizing damage to healthy cells, precision targeting ensures that treatments are accurately directed. By potentially removing the virus from the host's DNA, CRISPR may provide more than mere viral suppression, as evidenced by the potential for functional cures. Owing to its adaptability, CRISPR can effectively target a wide array of viruses by modifying their gRNA sequence.

Advantages of CRISPR-Cas9 in Antiviral Therapy	**Description**	**References**
1. Precision Targeting	CRISPR-Cas9 has high specificity and reduces off-target effects, hence increasing the safety of treatments. It is very useful in the treatment of slow-growing infections, which require targeting the virus without damaging other cells of the body.	Sinclair et al., (2023)
2. Potential for Functional Cures	Notably, CRISPR-Cas9 can ideally pull out the virus from the host DNA someday in the future, providing a novel way to cure, or more precisely, a functional cure as opposed to mere suppression of viral replication, in contrast to all current forms of antiviral drugs that inhibit viral replication.	Baddeley et al., (2023)
3. Adaptability	Due to the flexibility of the CRISPR–Cas9 system, it is very easy to change the gRNA sequence to target any of the viruses below. For this reason, it is a very suitable platform for future antiviral therapy innovations.	Dimitrievska et al., (2024)

CHALLENGES AND SOLUTIONS FOR PERSONALIZING CRISPR-CAS9-BASED ANTIVIRAL THERAPY

The main challenges associated with applying CRISPR technology to antiviral therapy are presented in Table (5) below, along with several potential solutions. Although CRISPR holds enormous promise as a treatment for viral and genetic disorders, its widespread use is fraught with difficulties related to cost, implementation, and ethical dilemmas. To accelerate the development and application of this technology in both individual and group therapy, the table outlines the challenges and potential solutions.

Table 5. Issues and suggestions for enhancing the economic role of CRISPR-Cas9 in antiviral treatment. Key issues surrounding the implementation of CRISPR-Cas9 in antiviral therapy, particularly in personalized medicine, are outlined in this table. Additionally, it presents suggested solutions to these challenges with the aim of reducing costs and improving the feasibility of CRISPR-based therapies in medical settings. To make CRISPR technology more efficient and accessible for patients, solutions focus on optimizing the development process, simplifying delivery methods, and addressing ethical and legal concerns.

Challenge	Description	Proposed Solution
High Initial Development Costs	It is extremely expensive to initiate CRISPR for certain viruses since the methods and required tools are often only accessible through costly infrastructure and require clinical trials to test the therapy.	If public health organizations, academic institutions, and pharmaceutical companies work in concert, they should be able to reduce the costs of research and development. Additionally, an emphasis on ensuring that efforts to scale up the use of CRISPR therapies are cost-effective will lead to steady and continuous cost reduction in the long run.
Delivery Systems & Manufacturing Costs	Some factors include the synthesis of components such as Cas9 proteins and gRNAs, which can be costly; furthermore, the delivery of components to target cells is not cheap.	This can be improved through better biomanufacturing and optimization of the manufacturing of CRISPR components, as well as enhancements in cheaper and more efficient delivery systems (viral vectors or lipid nanoparticles). If these systems are optimized for the mass production of these treatments, the costs would additionally decrease.
Some Barriers to Personalized Cancer Treatment	CRISPR technology offers potential for antiviral therapy tailored to individual patients; however, this opportunity is currently limited due to high costs, especially for targeted actions based on genetic tests.	By combining it with diagnostic methods and AI-assisted individualized medicine platforms, treatments could be developed and brought to market more quickly and affordably. A more extensive overall healthcare system incorporating CRISPR may also contribute to streamlining healthcare processes, thus reducing the costs associated with these therapies, making them more accessible.
Regulatory and Ethical Considerations	The regulatory actions and ethical codes governing the development of CRISPR therapy that must be applied in clinics may restrain the pace of adopting CRISPR-derived antiviral therapies.	With proper planning and ethical agreements regarding related applications of CRISPR, better coordination on protocols, regulation of such techniques, and approval of therapeutic applications can be significantly enhanced, thereby lowering overall costs by managing trials needed for addressing various virus classes.

In general, the CRISPR–Cas9 technique is a rapidly changing conventional antiviral therapy that provides information on how a virus's genome can be targeted and altered (Koujah et al., 2019). This unique strategy is a potentially exciting line of research with respect to chronic viral illnesses and could provide functional cures. However, further work will be needed to address the current issues associated with this technology to enhance its utilization in normal clinical practices. Moving further in the journey of antiviral treatment, the use of CRISPR-Cas9 has demonstrated the benefits of inventions that can enhance patients' lives and help combat viral diseases worldwide (Zahedipour, et al.,2024).

5 IMPORTANCE OF INDIVIDUALIZED TREATMENTS IN ANTIVIRAL DRUG DEVELOPMENT

The shift toward individualized drug therapy and individualized medicine in antiviral treatment thus stems mainly from the problems that conventional approaches have following their inability to consider interpatient variability. Host genetic factors, together with the immune response and genetic variation of the virus, are key determinants of variability in antiviral drug efficacy (Bottino, et al.,2023). Some of these factors affect the therapeutic outcome of the drug, whereas others affect the safety of the drug, that is, the toxicity and reactions that may be elicited by the drug. For example, in HIV treatment, drug resistance mutations may reduce the effectiveness of first-line antiretroviral drugs, thereby resulting in poor outcomes (Kumar et al., 2020). New developments in pharmacogenomics, the study of how genetic variations and viral proteomics, or how virus proteins operate, have allowed scientists to identify the necessary markers and proteoforms to address how patients respond to antiviral treatments. Such knowledge helps in the development of specific protein drugs either to attack such markers or suit each patient's individual requirements (Su et al., 2022). For example, therapeutic protein drugs used to treat SARS-CoV-2 necessitate versatility among drugs because they are expected to target different viral strains. The integration of proteoformics into antiviral treatment makes it possible to target specific proteoforms within viral strains, making treatment much more specific. Moreover, the molecular design of therapeutic protein drugs has been transformed through the application of artificial intelligence (AI) and machine learning. These technologies are based on detailed, individual patient models that integrate genomic and proteomic information to choose the most appropriate therapies (Elste, et al.,2024). It not only enhances the efficacy of therapy but also eliminates the spiral of trial and error in drug prescription for large-scale targeted prevention in health risk assessment-identified high-risk populations (Bolarin et al., 2021).

CASE STUDIES ON PERSONALIZED ANTIVIRAL THERAPY

Antiviral medications that target the patient individually and perform phenotypic and genotypic analyses of the virus are the principles of this new approach. The following is a set of case studies of specific viral infections that can be treated by personalized antiviral therapy. For instance,

In Patient 1, patients with HIV drug resistance benefitted from a differential treatment plan designed after a pharmacogenomic test. This approach facilitated the pindowning of antiretroviral therapy, resulting in better results, a rare emergence of drug resistance, and reduced side effects. In accordance with Lakhman et al.'s (2021) observations, these findings support the possibility of the use of pharmacogenomics in enhancing HIV care, given that drug resistance is a real consideration. Specific to therapeutic responses, in **Case 2** of SARS-CoV-2 (COVID-19), a proteoform-based drug design optimized for distinct viral strains was shown to improve outcomes. This strategy, which was aimed at the differentiation of the structural and functional characteristics of viral proteoforms, led to better recovery and minimal side effects, according to Haas et al. (2021). This study also aimed to assess the importance of individualized antiviral treatments owing to the constant emergence of viral mutations, especially during pandemics such as COVID-19.

In **Case 3** patients suffering from hepatitis C, the appropriate therapy was delivered after a combination of proteomics for the individual and the viral genotype analysis was provided. O' Kane et al. (2023) revealed that this more individualized therapy approach shortened the treatment time and lowered the risks associated with adverse effects, leading to enhanced patient results. This case demonstrates the importance of combining viral genotyping with other patient information to increase the precision of viral treatments.

The example of **Case 4**, where AI-supported medication choice of antiviral drugs for influenza (H1N1) was proposed depending on the patient's genes, is also relevant to the topic of individualization. Compared with conventional treatments, AI models have been applied to predict the best antiviral therapies, which results in a decrease in viral impact and an increased rate of recovery (Li et al., 2024). This case illustrates the increasing importance of artificial intelligence to improve antiviral treatment by processing more intricate information about the patient and the virus.

Similarly, in **Patient 5,** hepatitis B patients were reaped from the pharmacogenomic and proteomic analyses since the results indicated drug resistance. This strategy increased the therapeutic outcome in patients with specific genetic changes and guaranteed better lifelong antiviral control (Geng et al., 2024). This case also demonstrated that although hepatitis B is a chronic viral infection, the therapeutic response varies with mutation; thus, a patient-specific long-term treatment plan is needed. For cytomegalovirus (CMV) (Sudevan, et al.,2024; Khan et al., 2019), **case 6** demonstrated the application of machine learning for identifying specific factors of the immune response of a given patient to enhance drug treatments. This was followed by targeted immune modulation therapy that enhances the potentiation of the immune response, as reported by Wang et al. (2023) in immunocompromised individuals. The application of machine learning in this regard is not just how computational solutions are valuable in enhancing patient-specific antiviral treatment. In **case 7**, a strategy in which the high-risk population was targeted in accordance with the overall health risk assessment during the Ebola virus outbreak as well as the individual doses of drugs contributed to the increase in survival rates. High-risk populations such as HCWs and other susceptible populations benefit from this targeted approach to therapeutic manipulations of their risk profiles (Hoffmann Dahl, et al., 2024). Finally, **Case 8** discussed the Zika virus by applying precision public health and therefore treatment by examining the genomics of the virus (Woodson, et al.,2024). These interventions contributed to the prevention and management of disease transmission and enhanced relevant patients' experiences in deficient areas, as supported by da Costa Castilho et al. (2024). With reference to the details of this case, it is possible to speak about the relationship between the personalization of interventions and both personal and population health. Similarly, **in Patient 9**, experimental treatment with CRISPR-Cas12 via targeted therapy with human metapneumovirus (hMPV) yielded better results. Patient genotyping for hMPV and phenotyping confirmed the inability of the virus to respond to conventional anti-viral medications. By employing CRISPR-Cas12, researchers were able to control the virus by selectively cleaving viral genes related to replication and observed a decrease in viral copy number and time to heal. According to Xuan et al. (2024), CRISPR-Cas12 is considered effective for rapid diagnostics and targeted genome editing, which is very important in this study. The individual patient care model, which incorporates gene modification with an antiviral regimen, diminishes side effects and offers a specific readjustment method targeting a specific viral strain and the patient's unique genetics. This case strengthens the need for the utilization of precision medicine in enhancing the treatment of patients with acute viral infections, such as hMPV, in immunocompromised individuals.

6. CURRENT CHALLENGES IN STANDARD ANTIVIRAL THERAPY AND PERSONALIZED APPROACHES

Few antiviral drugs have been used to treat viral infections effectively; nevertheless, drastic drawbacks are observed in the conventional treatment paradigm. One of them seems to be the development of resistance to drugs, which is especially promptly evident in such viruses as HIV, influenza, and hepatitis C. This requires either the formulation of new drugs that would work in combination with one another or even new classes of drugs, which is challenging in the long-term management of the patient (Matthew et al., 2021). Furthermore, conventional antiviral treatments often employ monodirectional drugs, also known as body-attacking drugs, as most of them are designed to combat various strains of the virus or many types of viruses. This may sound like something positive) Kausar al., 2021), but, in fact, in the overwhelming majority of cases, it leads to nonspecific effects, that is, the drug affects the virus and normal cells. For example, nucleoside and nucleotide analogs in many HIV and hepatitis medications cause side effects such as mitochondrial toxicity owing to the ability of the compound to inhibit host enzymes. Similarly, polymerase inhibitors such as remdesivir and favipiravir used during the COVID-19 outbreak (Hegazy et al.,2023) work differently in different patient categories; their results are strongly dependent on the viral load, immune status, and genetic factors (**Bauer et al.,2024**). Another very serious issue associated with antiviral treatment is the question of when exactly the given chemical substance will need to be taken. If therapy is initiated late, for example, in acute viral infections, it is suboptimal because the virus has already replicated significantly. For example, in the treatment of influenza, neuraminidase inhibitors such as oseltamivir are less effective if they are administered beyond 48 hours after the onset of symptoms, underscoring the cytocratic nature of early diagnosis and intervention (Ikematsu, et al., 2024). Owing to the future perspective of personalized medicine, pharmacological interventions aimed at the inhibition of viral replication continue to experience a call for more selective and specific antiviral approaches. For example, entry inhibitors and integrase inhibitors provide effective possibilities for HIV treatment since they target some particular stages of the virus's life cycle and do not influence other processes, which reduces the likelihood of developing drug resistance (Ch Chou et al.,2024). Moreover, the use of interferons in the context of virus treatments, such as hepatitis or SARS-CoV-2 infection, has been discussed, with a focus on immune-driven therapies and immunotherapy. The adoption of an AI-based drug discovery approach also enhances the determination of patient-sorted treatments, such as protease inhibitors, for treating virus-resistant strains, as proposed by Resop et al. (2020).

7. OUTSTANDING REAL-LIFE RESEARCH FINDINGS ON ANTIVIRAL THERAPY RESISTANCE AND PERSONALIZED THERAPEUTICS.

Many of these cases demonstrate how different forms of individual plans have emerged as solutions to common shortfalls in the use of standardized antiviral drugs and therapies for specific forms of resistant infections.

Patient 1 received combination therapies involving entry inhibitors as well as protease inhibitors. These individualized treatments for complex cases result in enhanced viral load suppression in clients with multiple drug-resistant mutations, making personalized treatment regimens complementary to the treatment of clients with complicated HIV cases (Clipman et al., 2022).

In **Patient 2 (H1N1),** neuraminidase inhibitors were also effective at treating the condition unless they were administered within the first 48 hours of symptom onset. This case therefore supports the need to start antiviral treatment early enough to maximize efficacy (Heneghan, et al.,2016; Louie, et al., 2012).

The **third case** was related to hepatitis C, in which nucleoside analogs were incorporated alongside protease inhibitors in regimens for patients with resistant strains. The results revealed that treatment was shortened and that adverse effects were minimized, which highlights that any specific combination of these drugs could enhance the results (Sallam, et al 2024; Lawitz et al., 2013).

For **Patient 4,** who utilized remdesivir for COVID-19 treatment, immune profiling was used to consider patient variance in viral loads and the immune response. Although the results have been mixed, the study captures the effectiveness of integrating data to personalize interventions according to individual patient characteristics (Godwin et al., 2024).

For hepatitis B, **Patient 5** demonstrated that the use of nucleotide analogs with interferons led to enhanced long-term viral suppression and immune response modulators in patients with certain genetic backgrounds, which supports the generalization of prospective antiviral approaches (He et al 2024; Hagiwara, et al., 2015). **In case 6**, an example of artificial intelligence in therapy management for cytomegalovirus (CMV) was presented, with the use of integrase inhibitors to overcome drug resistance. This is due to a more individualized approach to treatment that allows for optimal dosing of treatment regimens, especially those at risk of being immunocompromised, and the derivation of accurate formulas for antiviral therapy (Wang et al. (2023)).

8. OBJECTIVES AND SCOPE OF PERSONALIZED APPROACHES IN ANTIVIRAL THERAPY

The goal of personalized antiviral therapy is to prescribe treatment regimens tailored to the individual patient's genetic, immunologic, and virologic profiles. By focusing on these specific features, this new approach attempts to maximize therapeutic effectiveness while minimizing adverse effects and minimizing the risk of developing drug resistance. In contrast, traditional therapies commonly rely on a unified treatment strategy that ignores the variability of patient response (von Delft et al., 2023). To realize these objectives, a variety of advanced strategies are being investigated. They identified key methodologies for predicting drug efficacy, including biomarkers, pharmacogenomic analyses to determine patient susceptibility to adverse reactions, and therapies that specifically target specific viral strains or proteins. Precise editing of viral genomes with CRISPR-Cas9 technology has recently been highlighted as a means for treating persistent infections that are difficult to treat, such as hepatitis B (Najafi et al., 2022). The landscape of personalized antiviral approaches is broad, embracing a range of cutting-edge technologies and frameworks. Among these methods, proteomics and viral genomics are useful for identifying specific biomarkers that can help predict how patients respond to antiviral agents. In addition, nanovaccines have become a sophisticated delivery mechanism for vaccine technology, allowing antigens to be directed to target cells with specificity superior to that of off-target effects while avoiding associated side effects (Bezbaruah, et al. 2022).

Furthermore, individualized antiviral approaches are expected to improve patient outcomes and decrease aggregate health care costs by averting unwanted procedures and hospitalizations associated with antiviral toxicity and poor therapeutic outcomes. With the ongoing changes in reformed healthcare systems, an increasing number of value-based approaches seem to focus on the merits of personalized treatment options, particularly within the context of chronic and recurrent viral diseases (Khoo, et al., 2021).

• **Artificial intelligence (AI) and machine learning (ML)** have been adopted as cutting-edge technologies that have led to breakthrough achievements in the field of healthcare, particularly with respect to treatment outcomes, by introducing the principle of individualization. Learning from large structured datasets derived primarily from clinical trials, medical IT solutions, also known as electronic medical records (EMRs), make it possible to identify patterns, molecular markers, and other traits that determine the efficacy of specific treatments for given patients. It is now possible to utilize AI to better inform practitioners on how best to treat individual patients while minimizing toxicity (Maleki, & Forouzanfar, 2024).

Adaptive Intelligence in COVID-19 Management and Treatment

The management of COVID-19 has illustrated how crucial AI is in battling viral diseases. For example, patients' electronic medical records have been analyzed to develop an AI-based model to forecast the future health status of patients and potentially at-risk populations. Experts' opinions and published medical research have demonstrated how AI can be used to categorize COVID-19 patients on the basis of the immune response. One example is the comparison of models involving early cytokine assessment for severe COVID-19 cases, leading to distinct inflammation patterns. This discovery resulted in the use of immunomodulators such as tocilizumab, improving survival rates among severely affected patients (Xu, et al., 2022).

A second area of clinical application of AI during the pandemic was the identification of repurposable drugs. Scientists have used an ML approach based on interactions between viral proteins and known antiviral compounds to predict the effectiveness of COVID-19 treatments. Expanding AI techniques revealed that remdesivir was an effective treatment candidate, a finding later confirmed by clinical studies (Winkler, D. A. 2024). All these AI-based methodologies greatly helped save time in developing therapeutic protocols during the crisis.

Artificial Intelligence in Chronic Disease Management

In addition to COVID-19, the application of AI has been utilized to manage other viral infections. In the treatment of hepatitis C, AI systems have been employed to examine factors such as genetics and demographics to predict patients' responses to antiviral regimens. They tailor treatments to achieve the best combination of therapy, thereby reducing drug resistance rates and treatment failure (Ashique, et al., 2022).

Future prospects

The applicative value of AI in RT–PCR extends further to therapeutic settings by enhancing dataset analysis. Genomic, proteomic, and clinical datasets can be utilized with AI to predict viral evolution and guide the creation and deployment of vaccines and therapies (Gawande, et al.,2025). For example, current efforts are being made to apply AI to combat human metapneumovirus (hMPV), which has mutated into resistant strains of treatment, by selecting optimal treatment areas and improving patient care experiences in the post-SARS-CoV-2 era (Bakkers, et al.,2024).

RECENT ADVANCES AND FUTURE DIRECTIONS

- Medicine, with its complex individualized approach and new technologies, has expanded the spectrum of therapeutic interventions for chronic and acute viral diseases. The application of nanovaccines in combination with artificial intelligence analytics has introduced new opportunities in approaches for individualized therapy as well as suggested ways for practical, efficient, and suitable preventative measures for high-risk groups (Tian et al., 2022).
- Case studies of treatments have shown that personalized therapy works better than standard treatments do, especially in issues such as drug resistance. For example, in HIV and hepatitis C treatment and prevention, individual approaches have enhanced the best drug regimens adjusted according to patient characteristics, minimizing the likelihood of resistance and enhancing the durability of viral control (Van Hout et al., 2021).
- There are a number of advancements in the field, clearly demonstrating the areas of viral disease treatment that may benefit from personalized antiviral therapy:

These studies indicate that tailor-made treatment models can improve therapeutic results compared with conventional antiviral treatment methods.

- In **Case Study 1**, drug-resistant HIV patients used genomic information to receive individualized antiretroviral therapy. This results in better suppression of the viral load and a low incidence of drug-resistant mutations, making this treatment approach more effective in individuals with multipart mutations (Al-Omairi et al., 2019).
- As noted in **Case Study 2** of COVID-19, which was enhanced by AI, individualized treatment measures were effective in enhancing the recovery period and shortening the hospital stay. With the help of AI-driven algorithms for providing optimal treatment regimens for patients, healthcare practitioners have been able to fully personalize therapies according to the disease progression of patients, thereby improving outcomes (Bekbolatova et al., 2024).
- In hepatitis C (**Case study 3**), an integrated assessment of proteomics and viral sequences allowed tailoring of therapy and resulted in shorter exposure to treatment and fewer side effects. Because of the unique genotypes of the virus and the response of specific patients, treatment plans have been implemented more effectively, demonstrating that chronic viral infections can be controlled better through the concept of personalized medicine (Schiano et al., 2024).
- **Case Study 4** described how the technology of data mining algorithms can be applied for the recommendation of antiviral agents in patients with influenza (H1N1). These algorithms enabled a shorter recovery period and milder symptoms of the illness, which made the therapy more accurate than nonadvanced methods (Hung et al., 2023). Hepatitis B (**case study 5**) identified the use of genetic screening to target and adapt antiviral treatment to patients. Owing to this individualized therapy concerning specific gene markers, treatment outcomes and sustained antiviral suppression were recorded (Shah, et al., 2021).
- In **Case Study 6**, which involved treating immunocompromised patients with cytomegalovirus (CMV), the application of individual immune response management enhanced treatment. Prospectively identifying patients and individualizing treatment according to the immune response allowed the improvement of therapeutic effects in a vulnerable group of patients to be optimized (Wang et al. (2023)).
- **Case Study 7** also involved targeted, individualized, and risk-based treatment options initiated for the Ebola virus in the treatment of HCWs and high-risk exposure groups. Such measures improved mortality rates, especially among vulnerable persons, highlighting the value of personalized care in high-risk viral contingencies (Hoffmann Dahl, et al., 2024).
- In managing the Zika virus outbreak (**Case Study 8**), the organization emphasized population interventions and was equipped with accurate interventions contributing to the successful control of the virus and better patient condi-

tions in the affected areas. These findings indicate that specific public health interventions can prevent viral spread in stressed populations (Noorbakhsh et al. (2019)).

- As witnessed by the appearance of SARS-CoV-2 variants (**Case Study 9**), variations in therapeutic targets have made treatments more efficient and less toxic to individuals with distinct strains of the virus. This case study aims to understand the role of developing treatment plans that consider the genetic variation of viruses, especially during the COVID-19 pandemic (Haas et al. 2021).
- Last but not least, in **Case Study 10**, the application of customized procedures for HPV vaccination, on the basis of risk factors, made it possible to increase immune responses and reduce the risk of HPV-associated cancer. This case demonstrated the importance of a targeted vaccination approach for wasting virus-related malignancies (Bolarin et al., 2021).
- As a final point, antiviral treatment tailored to the patient is a revolutionary concept that departs from a previously used, nonspecific approach with antiviral medications. Because of the specific genetic, virological, and immunological profile of the patient, better outcomes are guaranteed than those of mass-produced drugs because of their safety and efficiency as well as lower costs. With the emergence of technologies such as CRISPR-Cas9, nanovaccines, and artificial intelligence, it is foreseeable that every advancement can be an important contribution to the development of antiviral therapy beyond the horizon of traditional therapies and stand as the beginning of customized medicine (Maserat. 2022).

10- ROLE OF GENETIC FACTORS IN ANTIVIRAL THERAPY

Pharmacogenomics: enhancing drug efficacy

Pharmacogenomics is a relatively new science with a focus on the impact of genetic differences on patients' drug reactions. Pharmacogenomics, as a subdisciplinary approach to genomics combined with pharmacology, provides essential information about how genetic polymorphisms affect the pharmacokinetics and pharmacodynamics of antiviral drugs (Qahwaji et al., 2024). Knowledge of these genetic properties is crucial for enhancing antiviral medication dosing, especially as the incidence of viral infections grows worldwide. Studies conducted in the past few years have explained in detail how certain genetic differences may significantly affect the efficacy of antiviral drugs. For example, the CYP2B6 gene is involved in the metabolism of a range of antiviral drugs, including those used to treat hepatitis

C. Different allelic expression of this gene can greatly alter the half-life of drugs and influence their efficacy and toxicity profiles (Langmia et al., 2021). This insight underscores the need for a genetic test before beginning antiviral therapy so that healthcare providers can offer safer and more efficient treatment procedures. Case Analysis of Hepatitis C Treatment: For hepatitis C patients, pharmacogenomic testing has shown success within the health field in terms of adapting treatment and fine-tuning drug administration. DAAs in hepatitis C patients have replaced older regimens as a standard of care, but the results vary among patients. Failure to treat can be diagnosed genetically through polymorphisms that are distinguishable on the basis of the cause of treatment failure and hence the development of individual treatment options on the basis of an organism's genetic makeup (Lange et al.,2024; Asselah et al.,2018).

SPECIFIC EXAMPLES OF PHARMACOGENOMICS IN HEPATITIS C PATIENTS INCLUDE THE FOLLOWING:

- **Sofosbuvir and Genotype Variants**: Hepatitis C genotype 3: IL28B variation is related to lower rates of response to therapy in patients. Patients with some forms of the IL28B gene undergo lower effective treatment outcomes in the case of HCV genotype 3. Given these genetic factors, clinicians can tailor therapy and, in many cases, involve multiple types of DAAs to increase patients' chances of achieving a sustained virologic response (SVR) (Mohamed Abdelnajid et al., 2023).
- **Ribavirin Dosing**: The gene encoding the enzyme ribavirin kinase RTP4 has been proposed to be associated with interindividual differences in ribavirin metabolism. By using genetic testing, clinicians can adjust the profile of ribavirin according to the metabolic capacity of the body, thus minimizing cases of toxicity and enhancing therapy (Ali, et al.,2024; Borden et al., 2010).
- **NS5A resistance-associated variants (RAVs):** NS5A RAV resistance testing is useful when choosing the best combination therapy for patients with hepatitis C. For example, patients with certain mutations in the NS5A domain need different antiviral drugs than do patients with no mutations in the NS5A domain, hence enhancing treatment outcomes and decreasing the number of cases of virologic failure (Tsai, et al., 2021).

Pharmacogenomics has a significant role in the evolution of an exact antiviral treatment for individual genetic characteristics, increasing patient prognosis. Several examples illustrate that pharmacogenomics can be effectively implemented to individualize antiviral therapy for different viral diseases.

- In **Case Study 1**, with an IL28B variant, patients with hepatitis C received a personalized regimen of a direct-acting antiviral (DAA). These consequences predispose patients to significantly improved SVR rates in line with the fruition of individual genetic factors through personalized treatments, especially in genotype 3 infections (Salum et al., 2019).
- Regarding **HIV in Case Study 2**, the CCR5 Delta 32 mutation had beneficial effects on the response to CCR5 antagonist therapy. The use of pharmacogenomic knowledge of the impacts of HIV therapy on P450s has helped healthcare practitioners increase treatment outcomes in genotyped individuals prone to genetically sensed adverse therapy effects, which works well in developing the direction of HIV therapy (Mohamed et al., 2022).
- In **Case Study 3**, patients with NAT2 polymorphisms involved in hepatitis B received adjustments to the dosage of lamivudine. This change reduces the toxicity of drugs while increasing their effectiveness, further demonstrating that individualized dosing in accordance with a patient's genes can dramatically improve outcomes (B, K., Appaji, et al.,2024; Zhang, et al., 2018).
- A specific case from **Case Study 4** was on influenza treatment, whereby, depending on the HLA-B variants, patients were given specific antiviral selection. This pharmacogenomic approach reduced the recovery time and proved the future of genetically driven therapy in recovery from viral illnesses (Biswas et al., 2022).
- In relation to Cytomegalovirus in **Case Study 5**, extended alleles of MTHFR led to patients with variant genotypes being given tailored ganciclovir dosages, resulting in lower tolerance of toxicity by patients, in addition to improved total scores. This shows that people have unique genetic variations in the metabolism of drugs and beliefs about their effectiveness (Parkhitko et al., 2022).
- **In Case Study 6**, the finding of E6/E7 oncogenes in HPV patients suggested the principal vaccination strategy. The act of following such an approach led to a greater immunogenic reaction, thus highlighting the importance of using genetic and viral markers while designing preventive measures (Pal et al., 2020).

Pharmacogenomic testing has been found to help clinicians individualize antiviral therapies, thereby improving both patient outcomes and safety. This integration not only provides greater definitions for treatments but also leads to a better way of handling viral infections (Qahwaji et al., 2024).

11. IDENTIFYING GENETIC VARIATIONS IN VIRAL AND HUMAN HOSTS

Screening for genetic polymorphisms in the virus and its host is central to enhancing the directed administration of antiviral agents. Mutations in the viral genome can cause massive differences in the ability of a virus to neutralize antiviral drugs. For example, certain polymorphisms affecting germline DNA can be associated with resistance to NRTIs, which is fundamental to HIV treatment (Udeze et al., 2020). Additionally, the dynamics of evolution play a critical role in the COVID-19 situation, concerning both the efficacy of vaccines and monoclonal antibodies targeting SARS-CoV-2 spike proteins; new mutations in the spike protein appeared in 2021 (Chen et al., 2022). Sequencing of both the viral and host genomes allows for a clinically useful determination of the optimum antiviral regimen in a particular patient. More recent advances in next-generation sequencing (NGS) have made it possible to characterize genetic changes rapidly in both pathogens and patients. This approach seems particularly useful in diseases where host-directed therapy related to hepatitis C might be used; specific mutations in NS5A that inform responses to direct-acting antiviral therapies have been explored (Queret al., 2022) (Table 6). In addition, genetic differences in the human host, such as gene polymorphisms governing immune responses, can be applied to enhance the course of therapy. For example, polymorphic HLA genes can affect the clinical outcomes of antiviral agents by affecting T-cell reactions, which are important for subsequent immunity against infections (**Crux** et al., 2017).

Table 6. Summarizes genetic variations in viral hosts and their implications for antiviral therapy.

Virus	**Genetic Variation**	**Impact on Treatment**	**Reference**
HIV-1	Mutations in reverse transcriptase	Resistance to NRTIs	(Saehlee, et al., 2024; Das et al. 2013)
Hepatitis C	NS5A mutations	Predicts response to direct-acting antivirals	Lashgarian, et al. (2021)
Influenza A	HA gene mutations	Influences vaccine efficacy	Cao et al. (2021)
SARS-CoV-2	Spike protein mutations	Impacts efficacy of monoclonal antibodies	Chen et al., 2022
Hepatitis B	Precore and core mutations	Influence on interferon treatment response	Ahmed et al. (2023)
Zika Virus	NS5 mutations	Affects vaccine development and efficacy	Wang et al. (2022)
Human Papillomavirus	E6 and E7 oncogene variations	Alters response to therapeutic vaccination	Dust et al. (2022)

As indicated in Table 6, genetic differences in viral hosts have differential effects on antiviral therapy, which is important with respect to the formulation of treatment approaches for viral infections. This highlights the importance of specific genetic alterations in various viruses and their implications for prevention and treatment strategies, as described in the table. These findings emphasize how these mutations affect treatment responses, medication resistance, and vaccination efficacy, thereby highlighting the value of genetic analysis in virology and personalized medicine research.

Case studies: Impact of genetics on antiviral drug sensitivity

Some examples of the use of pharmacogenomics in antiviral therapy are outlined below. For example, one study investigated the effect of genetic make-up on the efficacy of lamivudine in HIV-positive patients. Compared with the absence of such mutations, specific resistance mutations in the reverse transcriptase gene M184V were associated with reduced responses to lamivudine therapy. Such a result indicates the profound need to perform genetic testing before beginning therapy to achieve the best results (Jennings, et al., 2022). One of the discussed case studies addresses the treatment of hepatitis C with the use of sofosbuvir and velpatasvir. The outcomes of patients designated with NS5A-RAS versus those without this marker showed that NS5A-RAS patients had dramatically different patterns. The existence of these mutations leads to decreased SVR rates, which shows that knowledge of genetic traits can greatly help patients select their treatment (Howe et al., 2023). Furthermore, in research on hepatitis B, the reactions of patients with particular precore and core mutations to interferon treatment are dissimilar. These mutations require individual therapy to improve the clinical outcomes of patients due to therapy obligations (Chen et al.,2024; Wang et al., 2005). Similarly, NS5 mutations could impact vaccine effectiveness for the Zika virus, suggesting that there should be a personalized approach with respect to vaccine creation (Howe et al., 2023). These points help healthcare providers improve antiviral strategies, thereby improving patient success rates owing to considerations of specific viral and human genomic variations.

12 GENETIC SCREENING AND PRECISION ANTIVIRAL THERAPY: BRIDGING GENOMICS AND TREATMENT OUTCOMES

Genetic screening in the context of antiviral therapy has led to a paradigm shift owing to improvements in the effectiveness and differentiation of treatments that actually result in better prognoses for patients. As NGS (Marcinkowska-Swojak, et al.,2024), PCR (Wang, et al. 2025) and microarray analyses have unfolded (Katsiki, et al.,2024), it is possible to address genetic markers contributing to drug responses, thus identifying better treatment regimens. The current article identifies emerging technologies in genetic screening. DNA sequencing by synthesis (SBS) (McCombie et al., 2019) and next-generation sequencing (NGS) technology (Satam, et al., 2023) remain the key solutions to genomic analyses with versatile viral genotyping ability. This approach enables the identification of traditional and novel genetic factors that can determine antiviral drug efficacy in patients. Illumina and PacBio (Rayamajhi,2022) are the two primary platforms where Illumina is capable of providing more accuracy, whereas PacBio is designed for long-read sequencing of complex genomes. This integrated approach is especially useful in chronic infection treatment, as genetic analysis is the key to effective therapy (Rayamajhi,2022). PCR is useful for fast detection of such mutations in viral quits, as drug resistance, in particular, is consistently high. For example, in the treatment of HIV, the use of PCR can identify lesions such as M184V, which make a patient resistant to lamivudine, thereby facilitating changes in treatment (Ávila-Ríos et al., 2020). Although PCR offers faster as well as specific mutation identification, NGS is more versatile, enabling simultaneous identification of several varieties and presenting a more diverse analysis. The most important uses of microarray analysis include the ability to identify several genetic variations at once, making it useful for large-scale testing. Despite these drawbacks, the global gene expression patterns predicted by microarray analysis are valuable, especially in the study of genetic factors affecting antiviral drug responses in vaccines (Rao et al., 2019). These technologies are backed by bioinformatics, which enables the analysis of genome-level big data to identify mechanisms for predicting drug responses. The machine learning algorithms used in the bioinformatics field improve these models, increasing their accuracy in determining the best treatment options given a patient's genetic makeup.

Comparative Insights into Genetic Screening Technologies

- NGS is extremely sensitive in that it can identify known and novel mutations; it is particularly useful in diseases characterized by chronic viral infection (e.g., hepatitis C), where a detailed understanding of the genomic context is necessary (Di Stefano et al., 2023).

- PCR is more specific and better at producing quick results in acute clinical scenarios, whereas it genetically targets known mutations (Ávila-Ríos et al., 2020).
- Microarrays are less expensive and generally useful for screening multiple genes but can overlook other novel variants that NGS can detect (Rao et al., 2019).

For example, NGS is more comprehensive in the study of hepatitis C resistance markers; PCR is most crucial in the prompt handling of acute HIV drug resistance. This microarray analysis is used for purposes such as HPV vaccination strategies where more than a single gene is considered with respect to the vaccine.

The role of pharmacogenomics in the treatment of viral ailments In addition, therapeutic decision-making and genetic antiviral drug options are equally important to genetic screening. Hepatitis C has been used as a major example with respect to this argument since screening for the polymorphism IL28B is able to predict the response of a patient to interferon-based therapies, thus assisting clinicians in finding better treatment strategies (Mohamed Abdelnajid et al., 2023).

Likewise, pharmacogenomic databases are in demand for HIV treatment today. Such databases of drug resistance profiles associated with genetic mutations assist in choosing the right antiretroviral therapy for patients, thus reducing the predominant emblematic approach of trial and error (Yu et al., 2021).

This has transformed into valuable drug regimens fundamental to the model of precision medicine, hence enhancing inhabitants' health status and decreasing troublesome effects.

For example, prediction for genomics in hepatitis C patients includes decision-making for choosing direct-acting antivirals on the basis of patients' gene characteristics to increase treatment efficacy without side effects.

Genetic Variations and Their Impact on Antiviral Therapy

The following section outlines several viral infections, the associated genetic variations, and the effects of these variations on treatment outcomes.

- **HIV-1:** These changes make resistance to this class of drugs easier, as they alter reverse transcriptase (Das et al. 2013).
- **Hepatitis C:** Mutations in NS5A make it possible to estimate patients' responses to direct-acting antivirals to provide more individualized treatment (Kane et al. 2023).

- **Influenza A**: HA gene mutations affect the effectiveness of vaccines and have become the key element of public health programs (Cao et al. 2021).
- **SARS-CoV-2**: Spike protein mutations influence monoclonal antibody treatments, and new mutations must be constantly monitored (Shen et al., 2024).
- **Hepatitis B**: Genetic testing is critical for antiviral therapy in patients with hepatitis C and should be used to optimize interferon therapy on the basis of mutations present, such as precore and core mutations (Lazarevic et al.2024; Wang et al., 2005).
- **Zika virus**: The presence of NS5 mutations hampers vaccine development; therefore, genetic screening is relevant in vaccine research (Wang et al. 2022).
- **Human papillomavirus (HPV):** Therapeutic vaccines are highly dependent on E6 and E7 oncogene variations; therefore, the prescription of vaccines needs to be individualized (Dust et al. 2022).

Future Directions: A Historic and Scientific Examination of the Concept of Personalized Antiviral Treatment Recent breakthroughs in genetic testing are promising for taking antiviral therapy to a new generation of personalized medicine. Since PCR, NGS, and microarray methods are increasingly utilized in clinical settings, it is possible to prescribe antiviral drugs on the basis of patients' genetic characteristics to increase efficiency and minimize adverse reactions. In the future, the variety and specificity of interventions, including bioinformatics and advanced machine learning, which influence the analysis of the interindividual variability in genetic factors and viral infection, will expand as the field advances. The ultimate goal is to utilize these innovations to provide more effective, patient-specific antiviral therapies, transforming how viral infections are treated and managed in the coming years.

14. APPLICATIONS AND MECHANISMS OF ANTIVIRAL DRUG SELECTION

The general overview of antiviral therapy is changing dramatically, and one can speak of the dominance of precision medicine at this stage. This shift is based on increased genetic testing ability and a better understanding of how several drugs act, allowing doctors to design specific antiviral treatments for a given person. However, most importantly, such a personalized approach not only advances therapeutic processes but also dramatically alters the quality of life of patients.

Antiviral Drug Classes and Their Mechanisms of Action

It is essential to know which class of antiviral agent is used for treatment and how exactly it works. In the next subsection, we describe major classes of antiviral drugs and how these agents work at the molecular level to assist clinicians in the optimization of patient care.

1. The integer inhibitors **dolutegravir and bictegravir** are among the integrase inhibitors that are used in HIV management. These agents act specifically to develop the HIV integrase enzyme, which integrates viral DNA into the host cell's DNA, which is a vital stage in the replication process of HIV. Integrase inhibitors act to block this integration process, thus denying that the virus has a chance to establish a chronic infection. This class of drugs is particularly beneficial because of their low resistance, which simply means that the chances of the virus mutating and acquiring cellular changes are low (Elliott et al., 2020). Furthermore, integrase inhibitors have moderate toxicity profiles, which makes them a mainstay in current HIV treatment and paves the way for tremendous advancements in viral suppression and patient outcomes.
2. The protease inhibitors **lopinavir and ritonavir** are important drugs for treating a number of viral infections, including HIV and hepatitis C. These agents act by irreversibly binding to viral proteases, which are essential for the cleavage of polyproteins into functional proteins required for the development of new viruses. This process makes it impossible to manufacture dangerous orders of new viruses, thereby reducing the quantity of virus produced in the body (Talha et al., 2024). However, the effectiveness of protease inhibitors is somewhat limited by certain obstacles, such as drug–drug interactions, which require careful evaluation of patients with this type of therapy plan.
3. **Nucleoside and nucleotide reverse transcriptase inhibitors are** abbreviated NRTIs. These NRTIs, such as tenofovir and emtricitabine, constitute the basis of retroviral infection treatment, especially for HIV. These drugs are structurally similar to natural nucleotides and are incorporated into the growing DNA strand during the reverse transcription process; thus, they act as chain terminators. This mechanism stops viral DNA synthesis and thus blocks viral replication (Udeze et al., 2020). Although they are widely used in HIV therapy and have been shown to improve patient outcomes, the development of resistance due to mutations in the reverse transcriptase gene underscores the importance of genetic screening to personalize therapy.

In conclusion, such an understanding of classic antiviral drugs and their classification is crucial for improving treatment efficacy and patient outcomes. The challenge of incorporating genetic screening into the selection of antiviral drugs is quite remarkable; it forms a strong and sound factorial platform that elucidates the differences in response to drugs by different individuals. Our data suggest that as the field of personalized medicine progresses further, featuring the consistent creation of more elaborate predictive models and comparative databases, clinicians would be able to address the challenges presented by viral infections properly. To this end, genetic research places early emphasis on understanding patients as well as establishing proactive and data-focused strategies that make it possible to deliver the most effective treatment to patients depending on their genetic make-up, thereby advancing the possibilities of medical practice and overall health.

Summary Of Antiviral Drug Classes And Mechanisms Of Action

Table 7. Classes of Antiretroviral Drugs and Their Mechanisms of Action, Integrase Strand Transfer Inhibitors (INSTIs), Protease Inhibitors (PIs), Nucleoside Reverse Transcriptase Inhibitors (NRTIs), Nontucleoside Reverse Transcriptase Inhibitors (NNRTIs), C-C Chemokine Receptor Type 5 (CCR5).

Drug Class	Example Drugs	Mechanism of Action	Reference
Integrase Inhibitors (INSTIs)	Dolutegravir, Bictegravir, Raltegravir, Elvitegravir	Inhibit viral integrase, preventing the integration of viral DNA into the host genome	Smith et al. (2021)
Protease Inhibitors (PIs)	Lopinavir, Ritonavir, Darunavir, Atazanavir	Inhibit viral proteases, which disrupts the cleavage of viral polyproteins, hindering maturation of the virus	Foo et al. (2022)
Nucleoside Reverse Transcriptase Inhibitors (NRTIs)	Tenofovir, Emtricitabine, Zidovudine, Lamivudine	Act as nucleoside analogs, interfering with reverse transcription and halting viral replication	Feng et al. (2022)
Non-Nucleoside Reverse Transcriptase Inhibitors (NNRTIs)	Efavirenz, Nevirapine, Etravirine, Rilpivirine	Bind to reverse transcriptase at a site distinct from NRTIs, altering enzyme conformation and preventing viral RNA replication	Namasivayam et al. (2019)
Fusion Inhibitors	Enfuvirtide	Block the fusion of viral and host cell membranes, preventing viral entry into cells	Düzgüneş, et al. (2021)
Entry Inhibitors	Maraviroc	Block CCR5 receptors on host cells, preventing the virus from attaching and entering	Faivre, et al. (2024)

Major types of antiretroviral medications are categorized in this table, along with examples and detailed mechanisms of action. These medications focus on several phases of the HIV replication cycle, including viral maturation, integration, reverse transcription, and host cell entry, underscoring their role in comprehensive HIV treatment plans.

The provided table can be considered a helpful resource for tracing how various antiviral drugs function and the effects they have on viral infection processes. Each class of antiviral drugs targets different parts of the virus or halts its function in some way, making them important contributors to antiviral therapies.

Case Studies in Antiviral Drug Selection

Table 8 below presents various case studies that emphasize the significant influence of genetic screening on the selection of antiviral drugs.

Table 8. Genetic features and their applications in disease management. This table emphasizes the importance of genetic traits in guiding disease-focused applications, including treatment options, vaccine development, and the improvement of antiviral therapies. It includes examples of extensive databases used for genetic analysis and citations for further research.

Disease	Genetic Feature	Application	Comprehensive Database	Reference
Chronic Hepatitis C	IL28B Polymorphisms	Predicts the likelihood of response to pegylated interferon and ribavirin	N/A (Genetic markers associated with treatment outcomes)	Medina, et al . (2024) Berger et al. (2012)
HIV	Mutations in Reverse Transcriptase	Informs the selection of appropriate antiretroviral therapy	HIV Resistance Database	Clipman et al. (2023)
Hepatitis B	HBV Genotype Variants	Guides decisions regarding antiviral therapy options	N/A (Genotypes correlated with treatment response)	Varghese et al. (2024)
Influenza	HA Gene Mutations	Affects vaccine formulation and efficacy	Global Influenza Surveillance and Response System	Maqsood et al. (2024)
SARS-CoV-2	Spike Protein Mutations	Predicts vaccine efficacy and antibody response	GISAID (Global Initiative on Sharing Avian Influenza Data)	Shen et al., 2024
Tuberculosis (TB)	Mutations in rpoB Gene	Guides the selection of rifampicin-based therapies	Tuberculosis Drug Resistance Mutation Database	Zeng et al. (2021)

continued on following page

Table 8. Continued

Disease	Genetic Feature	Application	Comprehensive Database	Reference
Herpes Simplex Virus (HSV)	Thymidine Kinase Gene Mutations	Informs the choice of antiviral agents, such as acyclovir	N/A (Genetic markers associated with antiviral susceptibility)	Hyun, et al .,(2023); Nguyen et al ., (2024)
Ebola Virus	Variations in VP40 Gene	Assists in selecting appropriate treatment regimens	N/A (Genetic features linked to treatment outcomes)	Landeras-Bueno et al. (2021)
Zika Virus	NS5 Gene Mutations	Influences vaccine development strategies	N/A (Genetic markers related to viral pathogenesis)	Estévez-Herrera et al. (2021)

15. ENVIRONMENTAL AND LIFESTYLE FACTORS AFFECTING ANTIVIRAL TREATMENTS

Impact of Environmental Exposure on Immune Function Pollution, Climate, and Disease Susceptibility

The harmful effects of pollution and certain climatic changes significantly impact the immune system, reducing the body's ability to fight diseases and diminishing the effectiveness of antiviral medications. Air pollution, particularly fine particles such as PM2.5, has been shown to increase the incidence of respiratory infections, such as influenza and COVID-19, and exacerbate existing viral conditions. Research has demonstrated that these airborne impurities damage lung tissue and impair immune cells, weakening the body's defenses and increasing the susceptibility of individuals to infections. This compromised immune function hinders the efficacy of antiviral treatments, as the body's ability to respond to therapy is diminished by ongoing exposure to environmental pollutants (Monoson et al., 2023). The interplay between pollution and climate change creates a greater burden on health, especially for populations in highly polluted areas, where respiratory and viral infections are more prevalent and severe because of a weakened immune system. This growing body of evidence underscores the critical need to address environmental factors to enhance public health outcomes. (Burbank, 2023). Research that has been conducted with samples from populations living in urban areas has clearly shown that there is a strong relationship between episodes of air pollution and the admission of patients suffering from respiratory viral infections. In one study, increased air pollution was linked to increased hospital admissions resulting from respiratory diseases, thus emphasizing the harm that environmental pollutants cause to health

(Monoson et al., 2023). Moreover, owing to global warming and changes in climate circumstances, new features of viral transmission have been observed; changes in temperature and humidity may affect vector-borne diseases and increase the risk of outbreaks of viruses such as Zika or dengue (Mojahed et al., 2022). In addition, the conditions of the environment can cause the expression of viral illnesses and even aggravate them. Some of the external variables include relative humidity, temperature, and time of the year since the stability of the virus and the mechanism of transmission vary with these environmental factors. For example, temperatures below fifty and low levels of humidity may increase the transmission of flu, which is more stable in colder, dry climates (Peci et al., 2019). Moreover, some conditions of the environment can create a background for the appearance of new viral strains. Zoonotic transmission is a process by which a virus is transmitted from animals to humans and is commonly attributed to climate change, such as deforestation. These changes in markets increase interactions between humans and animals, likely leading to virus mutations and the emergence of viruses such as Ebola as well as different coronaviruses (Ellwanger et al., 2021).

Table 9. Summary of the environmental factors affecting immune function.

Environmental Factor	Impact on Immune Function	Associated Viral Infections
Air Pollution (PM2.5)	Impairs lung function and immune response	Influenza, COVID-19
Climate Change	Alters disease transmission patterns	Zika, Dengue
Pesticides	Disrupts endocrine function and immune response	West Nile Virus
Heavy Metals	Impairs immune response	Respiratory viruses

This table offers a comprehensive summary of the numerous environmental elements that might affect our immune system and the viral infections associated with them. It emphasizes significant issues such as air pollution (especially PM2.5), climate change, pesticides, and heavy metals. The chart lists the ways in which each of these factors might compromise immune function, including problems such as decreased lung capacity, abnormalities in hormone control, and a generalized weakening of the immunological response. Furthermore, it highlights certain viral infections that may develop as a result of these environmental factors. Understanding how these factors can increase the risk of disease and influence public health is made easier with the aid of this knowledge.

LIFESTYLE FACTORS AND THE ANTIVIRAL RESPONSE

Role of Key Nutrients, Physical Activity, and Disease Resistance

Nutrition and physical activity, as part of lifestyle, determine the state of immunity in response to viral infections. The immune system is a combination of factors in a network that need many nutrients for proper function. Eating a diet that is rich in vitamins and minerals helps the immune system work effectively, as does consuming foods rich in antioxidants. These nutrients not only modulate immune function but also augment the anticancer and anti-infective activities of the body **(Banerjee, et al., 2025)**.

1. **Vitamin C**: This hydrophilic vitamin is well known for its ability to counteract the oxidative effects on immune cells. Owing to their ability to recognize pathogens, macrophages and T cells are two types of immune cells with heightened functions in this process. According to Reyed & Prabhakar. In 2025, adequate consumption of vitamin C has a positive effect on the treatment of viral diseases and the shortening of the illness period as well as the body's immune system.
2. **Vitamin D**: Commonly known as the "unshine vitamin," vitamin D helps to moderate immune reactions. essential for the release of immune cells, such as T cells or macrophages, which are immune cells that can hunt for pathogens. Persons with adequate levels of vitamin D are likely to record improved results in antiviral therapy, specifically with hepatitis C. Thus, this study revealed that vitamin D could increase the effectiveness of antiviral agents and promote patient outcomes (Sizar et al., 2024; Reyed, & Prabhakar. 2025).
3. **Zinc**: This important micronutrient is essential for the immune system. In addition, zinc plays a vital role in the development and function of T cells, which are crucial in the immune response. It has also been demonstrated to have mild anti-inflammatory effects that may regulate inflammatory reactions within the body. The necessity of enforcing a diverse diet and consuming rich zinc-containing foods has been proven to be connected to a decrease in the severity of viral infections and may lead to quicker recovery (Reyed, & Prabhakar. 2025)
4. **Omega-3 fatty acids** are considered to have anti-inflammatory properties. They have crucial functions in regulating the immune response and may help the body clear viral infections. Omega-3 fatty acids may modulate the synthesis of related molecules called eicosanoids, which are involved in the inflammatory process. Research shows that the consumption of omega-3 fatty acids may increase viral immunity and decrease the number of respiratory infections (Reyed, & Prabhakar. 2025).

EFFECTS OF NUTRITION STATUS ON ANTIVIRAL THERAPY:

Previous research has revealed the essential role that nutritional status markers play in assessing the effectiveness of antiviral therapy.

Effective antiviral treatment: A recent systematic review and meta-analysis of clinical trials revealed that patients with better nutritional status, including vitamin D, had better outcomes with antiviral treatments. For example, patients with hepatitis C who were receiving treatment with normal vitamin D had a better prognosis than did those with a deficiency of this vitamin. On the basis of the results of this study, nutritional supplementation, especially vitamin D supplementation, may complement the results of antiviral drugs (Eini. et al., 2023).

These findings suggest that the human body has a complex association with nutrition and immunity, particularly with respect to the antiviral immune response; thus, balanced diets are needed. Taking vitamins, minerals, and omega-3 fatty acids can enhance immunity and may be beneficial for viral infections and treatments (Reyed, & Prabhakar. 2025). Subsequent studies should therefore endeavor to establish profitable outcomes when targeted nutrient supplements are used in combination with conventional antiviral medicines to increase the health of populations with vulnerabilities to recurrent viral infections. Therefore, it is crucial to implement nutritional interventions alongside population and patient management programs to strengthen the adult body's immune system, especially for high-risk groups (Alam et al., 2021).

Stress and immune function: Implications for antiviral therapy

One of the effects of chronic stress on the immune system is that it weakens the ability of the human body to fight viral infections. Chronic stress is associated with misregulation of the immune system and elevated inflammation; this can compromise a patient's adaptation and responsiveness to antiviral medicines (Morey et al., 2015). Some related studies have shown that people who endure chronic stress become sick more often and have worse reactions to antiviral medicines. For example, a cross-sectional study on the link between psychological stress and immunological responses revealed that higher stress levels correspond to reduced concentrations of antiviral cytokines and higher viral loads in flu patients (Mays et al., 2012).

Case Study: Exploring the interaction of stress and influencing outcomes in a cross-sectional study among HCWs during the influenza season, the authors noted that HCWs with high perceived stress developed more influenza illness reports with severe symptoms than did HCWs with low perceived stress. Hence, this study highlights the need to focus on the effects of stress on the effectiveness of antiviral treatment (Zhang et al., 2022). The social determinants of health population characteristics, such as education and access to antiviral health care and social and economic conditions, also play a large role in shaping people's health and their ability to react to antiviral remedies. People who come from less privileged backgrounds in contact with primary care are likely to experience numerous barriers to accessing care to contribute to delayed identification and treatment, which adversely impacts health (Braveman et al., 2014). In addition, vulnerable populations can be predisposed to diseases by the way in which they live, including diet and exercise, which influence the immune system. For example, those who live in places with poor access to high-quality foods, such as fresh fruits and vegetables, have poor immune-boosting nutrient intake since they live in food deserts (Suardi et al., 2021).

The impact of social determinants of health on antiviral therapy outcomes

Thus, social determinants of health work primarily on how health outcomes are impacted and the success of antiviral medications. For example, a patient's socio-economic status determines the level of care they can receive. Poor people experience hurdles that cause them to seek treatment much later than those in better-off positions and therefore tend to experience less favorable health (Braveman et al., 2014). This delayed access may lead to worsening of infections, making antiviral therapy a challenging proposition. The fourth one is education and knowledge, or actually health literacy, as one of the key determinants of health. As a rule, people with a higher level of education have more comprehensive knowledge about the treatment process, and consequently, they have better chances of following antiviral therapy regimens and better self-management. However, poor health literacy may result in confusion and the unsuitable handling of antiviral treatments (Braveman et al., 2014). It is also equally important that such food is available for use in improving health conditions among people. A lack of proper nutritional factors can weaken the immune system and cause people to be less protected against viruses and less sensitive to medications (Suardi et al., 2021). The healthy nutrient status of the body increases the efficiency of antiviral treatments by strengthening the body's immune system against viruses. As a result, stable housing remains crucial for sound psychological and physiological health. Housing insecure people who live with chronic stress may have weakened immunity and may therefore be less responsive to antiviral interventions (Braveman et al., 2014). To this end, a stable living environment is an important dimension that improves one's health and enables effective treatment. Therefore, understanding the interrelationships between such factors and antiviral therapy indicates that more efforts are needed to address them and make interventions more effective and equitable.

A CLOSER LOOK AT ENVIRONMENTAL VARIABLES AND PUBLIC HEALTH FACTORS: LESSONS LEARNED FROM CASE RESEARCH ON VIRAL CONSEQUENCES

The studies as a whole paint a picture of how environmental, dietary, lifestyle, and psychological conditions affect disease progression and the development of diseases, immune system responses, and the occurrence of infectious diseases. External factors, including air pollution, climate change, and pesticide treatments, are key in determining the degree of disease spread and severity. Pollution indicators, particularly when air quality is above normal levels, are associated with high rates of hospitalization and mortality. Many people suffer from nutrient deficiencies, including vitamin D deficiency, and have adverse lifestyle indicators, such as residing in food deserts, which weaken the immune system and make individuals vulnerable to infections. Moreover, stress plays a key organizational role by increasing the risk of infection and disease severity, particularly among healthcare workers; thus, stress management and mental health promotion are needed. Additionally, climate change and environmental disturbances, such as deforestation and urban sprawl, create new zoonotic diseases; therefore, there is a need to adopt global preventive interventions to mitigate these effects. Taken together, these results emphasize the conclusion that environmental, social, and health factors form an interdependent system and that their interaction defines the state of public health.

Table 10. **Environmental and** *lifestyle factors influencing viral infections and immunity.*

Case Study	Description	Key Findings	Reference
Impact of Air Pollution on COVID-19 Severity	Analysis of COVID-19 cases in Italy relative to air pollution levels.	**Higher levels of PM2.5 correlated with increased hospitalization and mortality rates from COVID-19.**	Hernandez Carballo (2022)
Chronic Stress and Influenza Outcomes	**Study examining healthcare workers' stress levels during flu season.**	Workers with high stress had a significantly higher incidence of influenza infection and more severe symptoms.	Luo et al. (2020)
Nutritional Status and Hepatitis C Treatment	Meta-analysis on vitamin D levels in patients undergoing hepatitis C treatment.	**Adequate vitamin D levels correlated with better treatment outcomes for hepatitis C, suggesting the need for supplementation.**	Anagnostis et al., 2024 Villar et al. (2013)
Environmental Factors and Zika Virus Transmission	**Study investigating the impact of climate change on Zika virus outbreaks in Brazil**	Higher temperatures and increased rainfall were associated with greater Zika transmission rates.	Li et al., (2019)

continued on following page

Table 10. Continued

Case Study	Description	Key Findings	Reference
Pesticides and West Nile Virus	Investigation of the relationship between pesticide exposure and West Nile Virus infection rates.	**Areas with higher pesticide use exhibited increased rates of West Nile Virus infections, indicating a potential link.**	Petersen et al. (2013)
Heavy Metals and Respiratory Viral Infections	**Analysis of the effects of heavy metal exposure on respiratory health and viral infections.**	Increased exposure to heavy metals was linked to a higher incidence of respiratory viral infections, impairing immune function.	Zhang et al. (2024)
Food Deserts and Immunity	Study assessing the relationship between living in food deserts and immune health outcomes.	**Individuals in food deserts had lower nutritional intake and higher rates of infections, indicating compromised immune function.**	Duncan et al. (2020)
Climate Change and Emerging Infectious Diseases	Review of the influence of environmental changes on the emergence of new viral pathogens.	Environmental changes, including deforestation and urbanization, were linked to increased emergence of zoonotic diseases.	Meyer et al. (2022)

This table presents significant case studies that examine various factors influencing viral infections, treatment outcomes, and immune responses. These findings illustrate the complex relationships among stress, nutrition, and environmental factors and their impact on community health, particularly concerning viral diseases. The results offer valuable insights for guiding future research, health policy, and therapeutic strategies.

16. BIOTHERAPEUTICS AND INFECTION CONTROL

The development of antiviral therapy has been revolutionary, characterized by a significant prevalence of the concept of a targeted approach. The most developed branch within this field is biotherapeutics, which were further developed in the form of probiotics, interdisciplinary approaches, and position-focused solutions. These methods are gradually being appreciated for their applicability to infection prevention and for the efficient handling of viral disorders. This article further analyzes the application of these sophisticated techniques in line with cases, in addition to trends showing their effectiveness in enhancing patients' postoperative conditions and immune system responses.

Probiotics as Novel Therapeutics

Probiotics are known to function in the gut microbiota, but their ability to manage viral infections is a relatively new concept. These findings indicate that live ingredients, when taken at the right doses, can be used to improve health, enhance the immune system, and prevent viral replication. In antiviral therapy, their utilization is the most promising, as scientific data show that certain types of probiotics enhance immune status and act as supplements in viruses.

New studies have started to reveal new health benefits of probiotics that extend beyond gut health, especially in the case of viral infections. Table 11 below presents an outline of several significant earlier case studies of how select probiotic strains have been employed to alter various immune substations to bolster antiviral interventions.

Table 11. Probiotic Interventions in Viral Infections: A Summary of Clinical Studies

Pathogen	Probiotic Strain	Findings	Study
Respiratory Syncytial Virus (RSV)	*Lactobacillus plantarum*	This study found that *Lactobacillus plantarum* reduced viral replication and improved immune responses in pediatric patients with RSV.	Ji et al. (2021)
Hepatitis C	*Lactobacillus rhamnosus*	The probiotic helped modulate gut microbiota, improved liver function, and enhanced the antiviral response when used in combination with standard therapies.	Maslennikov et al. (2023)
Influenza Virus	*Bifidobacterium animalis*	Among elderly patients, the probiotic significantly reduced flu severity and shortened the duration of illness.	Wang et al. (2022)
COVID-19	*Saccharomyces boulardii*	The use of *Saccharomyces boulardii* in a controlled study reduced COVID-19 symptoms and shortened hospital stays.	Tian, et al. (2023)

These examples indicate increased acceptance of the use of probiotics in the adjunct treatment of viral diseases as research has increased. Of special interest is the ability of these molecules to modulate immune responses and enhance antiviral activity, which suggests many potential uses for further therapeutic interventions. The different and combined ways through which probiotics trigger antiviral effects are described below. They include the regulation of immune reactions, strengthening of the barrier action of epithelial cells, and the synthesis of antimicrobial agents.

PROBIOTIC MECHANISMS IN INFECTION CONTROL

It is essential to link probiotic interventions to potential mechanisms that help control infection among adult and pediatric patients admitted to the hospital. Notably, the approaches through which probiotics exert their antiviral effects are multiple and intricate. Among these mechanisms, immune modulation, promotion of the epithelial barrier, and the formation of antimicrobial peptides are some examples. Lactobacillus and Bifidobacterium species are associated with antiviral cytokines, natural killer (NK) cell formation activity, and macrophage stimulation capacity. Furthermore, probiotics can help limit the growth of these diseases through pathogen exclusion by occupying nutrient receptors on the gut and respiratory tract for themselves (Tian, et al. 2023).

PREBIOTICS IN VIRAL TREATMENT: THEIR MODULATION OF THE GUT MICROBIOTA AND HOST IMMUNE SYSTEM ENHANCEMENT

Prebiotics are gradually being recognized as effective treatment strategies for viral diseases, especially those caused by drug-resistant viruses. Prebiotics are specific and unique compounds in food that are not absorbed in the small intestine but selectively promote the immunomodulation of beneficial gut bacteria. They include bacteria, fungi, and viruses that live in symbiosis with ten trillion human-associated microbiota, serving as optimal immune system regulators. First, the gut–lung axis and gut–brain axis are components of systemic immunology; second, prebiotics can target these axes (Deehan, et al., 2024).

The types of prebiotics that have been found to promote friendly bacteria such as Bifidobacterium and Lactobacillus, which are known for their immunomodulatory properties, include inulin, fructo-oligosaccharides (FOSs), and galacto-oligosaccharides (GOSs). These bacteria participate in the production of short-chain fatty acids (SCFAs), with butyrate being one of the most vital, as it strengthens the barrier function of the intestinal tract, suppresses the inflammatory process, and enhances the immune system's capacity to combat infections (Obayomi, et al.,2024). This phenomenon is particularly important in viral diseases, including COVID-19 (Vakili, et al.,2024) and influenza (Wang et al., 2024), where immune modulation is crucial for reducing disease severity and viral replication.

Giovanetti et al. (2024) reported that prebiotic intake increases the count of Bifidobacteria and consequently improves immune responses to respiratory viruses. In addition to positively regulating the gut microbiota, prebiotics bolster the body's ability to combat viral diseases, thereby reducing the viral load and recovery time while increasing the rate of recovery.

SYNERGISTIC EFFECTS OF PREBIOTICS AND PROBIOTICS IN FIGHTING VIRAL INFECTIONS: A SYNBIOTIC APPROACH

This study also revealed that immune responses and the ability to fight viral infections are enhanced when prebiotics are taken with probiotics. This combination of prebiotics and probiotics is referred to as a synbiotic, where the synergistic effect produces optimal results for the human body and system (Amrouche, et al., 2024).

Probiotics have favorable effects on the skin microbiome, promote the activity of immune cells, and suppress viral replication. Akram et al. (2024) mentioned synbiotics in relation to COVID-19 treatment, particularly their effects on patients with severe respiratory symptoms. Maintaining a healthy microbiota, stimulating immune cells, and controlling viral replication are essential roles of synbiotics. Akram's research tested the impacts of synbiotics, with a focus on COVID-19 patients with severe respiratory manifestations.

Published work has indicated that both dietary prebiotics (FOSs) and probiotics (Lactobacillus) improve the immune response and disease symptoms. Furthermore, the consumption of Lactobacillus species increased the levels of interferon and antiviral cytokines necessary for treating viral diseases. In human patients with hMPV infection, which causes both upper and lower respiratory tract infections (Strauss et al. 2023), synbiotics facilitate a numerical reduction in viral shedding and improve lung function compared with the preinfection state, providing full support for concepts linked to the gut microbial community, its effects on respiratory health, and immune system modulation (Malik, et al., 2021).

Similarly, in cytomegalovirus (CMV) synthesis, the majority of CMV patients who received synbiotics showed an improvement in immune status and reduced viral titers, in contrast to the groups that did not receive synbiotics. These synbiotics achieve this goal by increasing T-cell responses and enhancing mucosal immunity, which protects against viral infection and replication (Przemska-Kosicka, et al., 2016).

THE STRENGTHS OF SYNBIOTICS IN THE INDUCTION OF AUTOPHAGY AND VIRAL DEFENSE

It took a very long time even to discover that an indigenous process of the cell, referred to as autophagy, is crucial in breaking down cellular pathogens such as viruses. The role of autophagy is defensive; it enables cells to disintegrate other structures, such as viral particles, thus limiting the replication of the virus.

Current studies reveal that the use of synbiotics may lead to the activation of autophagy, thereby playing another protective role against viral infections. How prebiotics and probiotics influence the gut microbiota has previously been reported to alter autophagy pathways through short-chain fatty acids (SCFAs), such as butyrate (Tsui, et al.,2024).

Probiotics and prebiotics combined with their cooperative functions, including synbiotics, are effective courses of action against drug-resistant viral infections such as COVID-19, CMV, and influenza. In total, the natural compounds described herein have broad-spectrum strategies to regulate the gut microbiome, enhance immune cells, and induce autophagy for immunity against viruses. Furthermore, synbiotic-induced autophagy represents another defense mechanism in the human body that may help eradicate viral agents and reduce the severity of viral infections. However, these observations emphasize the need to elucidate the gut microbiome, immune system modulation, and autophagy for therapeutic intervention against viral diseases and for improving human health in the context of resistant viral strains.

17 BACTERIOPHAGES AS POWERFUL BIOLOGICAL PROBIOTICS

Another functional group of biotherapeutic agents is the use of bacteriophages, commonly known as phages (García-Cruz et al., 2023), for controlling infections. Bacteriophages are viruses that target and destroy bacterial cells and are hence potential agents for controlling bacteria-related diseases. Unlike other types of probiotics, which include friendly bacteria, phages attack only pathogenic bacteria without negatively affecting helpful bacteria, which makes eliminating only unwanted bacteria that cause infections possible. Recent studies have revealed that phages can be produced to target coinfected bacteria, which are often observed in viral diseases, and that these bacteria worsen disease severity. In respiratory viral infections, for example, patients develop secondary bacterial infections that slow the recovery process and increase morbidity. Phage therapy is another way by which such bacterial pathogens can be degraded without exacerbating the high level of antibiotic resistance prevalent in current medical practice (Cui et al., 2024).

Case Studies on Bacteriophages in Antiviral Contexts

The use of bacteriophages in combination with antiviral drugs has been considered by the authors of several recent publications. For example, in patients with COVID-19 who are particularly at risk of developing bacterial pneumonia while on antimicrobial therapy, phages have been used to control multidrug-resistant bacteria that would otherwise worsen the underlying infection. An actual case report conducted in 2024 by Cui et al. demonstrated the effectiveness of clinical bacteriophage application in managing bacterial pneumonia in COVID-19 patients. Phage therapy accelerated the recovery rate and lowered the need for broad-spectrum antibiotics; this finding also indicates that bacteriophages may be useful for addressing coinfections related to viral diseases. Likewise, in influenza infections, where complications from secondary bacterial infection, such as *Streptococcus pneumoniae,* can often be fatal, phage therapy has been proven to effectively counter bacteria while leaving the viral aspect of the infection intact (Mueller Brown et al., 2022). The ability to selectively target bacterial profiles within a complex microbiome is important because bacteriophages can be an effective addition to multifaceted microbial control approaches. Another potential use of bacteriophages is in treating, for instance, cystic fibrosis, where chronic bacterial and viral infections exist in a mutually interacting manner. Phage therapy has been used to drive down bacterial loads in the lungs, thereby enhancing respiratory status and comprehensive management of viral illnesses that precipitate the condition (Ling et al., 2023).

Prospects of bacteriophage therapy

The increasing concern for phage therapy as one of the biotherapeutic intervention approaches is hinged on its specificity to hazardous bacteria, particularly in situations where antibiotic intervention is ineffective. In the context of antiviral therapy, the use of bacteriophages provides a dual-action approach that focuses on the elimination of bacterial pathogens in conjunction with the enhancement of immunity against viral illnesses (Lin et al., 2017). In future developments, as technology toward phage therapy improves and given that synthetic biology technology enables specific phages to be developed, the possibility of using bacteriophages in conjunction with probiotics and antiviral drugs will create the necessary layers of defense against infections (Lin et al., 2017).

Future Therapy Conclusions

The incorporation of probiotics in antiviral treatment, together with the recently discovered possibility of bacteriophages, has been oriented toward the inclusion of broader concepts for managing infection. More studies related to the gut–lung axis of interaction, the selective targeting of the gut immune system, and the effects of bacteriophages in the regulation of the microbiome indicate that these biotherapeutic agents will likely target not only GI infection but also respiratory and systemic viral infections in the future (Wang et al 2022). Moreover, bacteriophages may also significantly contribute to moderate the severity of viral disease by controlling subsequent bacterial infections, increasing rates of cure, and possibly minimizing the occurrence of viral proliferation by strengthening host immune responses. The combination of probiotics and bacteriophages, on the other hand, may be complementary, promoting improvements in the immune system and preventing microbial dysbiosis during antiviral therapy (Lin et al., 2017).

For that reason, therapeutics, including the use of probiotics and bacteriophages, should be considered promising avenues in antiviral approaches. By incorporating them into antiviral therapies, they could increase patient recovery from viral infections and decrease the severity of viral and bacterial infections (Cui et al., 2024). The use of probiotics, which stimulate the body's immune functions, and bacteriophages, which selectively kill bacteria causing diseases, may change the outlook of infections in the future. With more research in this area, it could be argued that these biotherapeutics are an important addition to individualized antiviral treatments and could indeed revolutionize the ways of differentiating and combating infections. Further studies on such therapies will remain important for further unlocking and expanding the potential of such treatments within the future of antivirals (Mazziotta et al 2023).

Possible Directions and Trends in Antiviral Treatment

Currently, antiviral therapy is at the brink of a revolution in terms of technological innovation and improved knowledge of viral infections. Through genetic determination and new approaches to cures, the trend toward targeted virus infections is now shifting. The idea is to develop individualized therapies that improve the health status of patients and, vice versa, lessen the global disease virus toll.

Another important aspect of highly tailored antiviral treatment approaches, which are rapidly evolving now, has been linked to the use of artificial intelligence and machine learning. These technologies are reshaping the forward-prognosis of treatment results on the basis of data on patients' genotypes as well as their responses to prior treatments. For example, such AI models for hepatitis C treatment have allowed the best set of regimens for 95% of patients to be disclosed, whereas standard practices can affect 75% of patients (Liu, et al., 2022). This enables clinicians to predict outcomes more rapidly and accurately, thus decreasing the amount of time patients spend on nonresponsive treatment.

Another remarkable trend is that compounds with multiple targets have received increasing attention, as they damage different aspects of viral infections. Other classes of antiviral drugs that have been previously developed include protease inhibitors, which usually act in phases within the life cycle of a virus. On the other hand, new oriented antiviral therapy strategies explore the synchronization of blocking viral entrance, replication, and assembly to avoid resistance (Kausar, et al., 2021).

The map of antiviral drugs is changing owing to the advantages of the CRISPR–Cas9 system in gene modification, which enables the targeting of the viral genome. This capability is highly relevant to chronic infections such as hepatitis B and HIV infections. New investigations have shown that the delivered commercial CRISPR–Cas9 system can efficiently erase the integrated HBV DNA from infected cells, opening the possibility of a functional cure (Najafi et al., 2022).

In addition, there is an emerging class of biological-based therapies, such as probiotics and immunomodulators. For example, the addition of probiotics to conventional hepatitis C therapies increases liver health and decreases the amount of virus present in the body (Russo, et al., 2022). Antiviral personalized therapy is more effective because it addresses the patient's characteristics, including genetics, history, and way of living, which is important for chronic viral diseases such as hepatitis C. Improving genetic screening allows the targeting of specific genetic markers and achieves higher sustained virologic response (SVR) rates (Essa et al., 2021). Furthermore, the engagement of patients in the decision-making process of their payments enhances compliance with laid-down procedures, thus enhancing the result. New case descriptions prove the efficacy of individual antiviral management, which has a high SVR rate as a result of highly individualized pharmacogenomics based on genetic tests and individual analytical models (Qahwaji et al., 2024)). In summary, the concept of individualized antiviral care is a new model for approaching further infections. Through the combination of CRISPR-Cas9, AI models, and bioderived therapies, clinicians can improve the results of treatments. Thus, the concept of the individual approach opens the prospect for better shifting the therapeutic approach area and, consequently, provides hope for patients' improved quality of life in their fight against viral diseases (Domingues et al.,2024).

CONCLUSION

For that reason, differentiated antiviral treatment represents a new step in the management of viral diseases as a result of the development of genomic, proteomic, and bioinformatics approaches. This strategy frequently takes into account personal genetic and phenotypic variations along with overall virus characteristics, aiming to use these factors to achieve optimal treatment outcomes. Techniques such as CRISPR-Cas9, pharmacogenomics, viral genotyping, and AI-supported metrics have demonstrated revolutionary potential in enhancing the accuracy and efficacy of treatment, preventing side effects, and overcoming drug resistance. This approach not only redefines antiviral treatment but also suggests that chronic diseases, such as HIV, hepatitis, and severe acute respiratory syndrome coronavirus 2 (SARS-CoV-2), may be potentially treatable to achieve functional cures.

Nutrition and probiotics represent additional components within this framework, supporting the immune system and coordinating with specific antiviral treatment methods. Moderation of vitamins, minerals, omega-3 fatty acids, and antioxidants is crucial for sustaining immunity according to established nutritional guidelines. For example, vitamin D regulates immune functions and enhances antiviral activity, whereas zinc and omega-3 fatty acids possess anti-inflammatory and immunomodulatory properties. The use of probiotics effectively alters the gut microbial composition, thereby increasing systemic immunity and increasing the body's capacity to manage inflammation and treat. The incorporation of these interventions best embodies a biologic and patient-oriented treatment paradigm that aims not only to neutralize the virus directly but also to strengthen the foundation on which it thrives.

However, several issues hinder the large-scale development of personalized antiviral therapy. High costs, ethical concerns, and the technical challenges associated with genetic screening and delivery complicate usability, especially in low-income communities. The way in which these innovative treatments are disseminated requires collaborative diplomacy across policy, public health, and technology. Therefore, eliminating these barriers is essential to promote a comprehensive and personalized approach to antiviral therapy.

In the future, one can imagine only the potential of combining technologies such as machine learning, nanotechnology, and immune-modulating therapy with individualized antiviral approaches. These advancements could improve treatment delivery, bolster the immune system, and tailor antiviral therapies to unprecedented levels of accuracy. Additionally, a new standard for antiviral treatments may be established by integrating nutritional interventions and the optimized use of probiotics within this sophisticated framework. Further research and development should focus on delivery method selection, the associated costs of these therapies, and the ethical considerations required to make these treatments available to everyone.

The future of personalized antiviral medications will likely be characterized by a collaborative approach involving science, technology, and holistic health strategies. Such innovations can be effectively incorporated into plans for combatting viral diseases, shifting approaches toward more humane and accurate means of addressing health issues and benefiting not only individual patients but also the broader global community and humanity as a whole.

REFERENCES

Akram, N., Saeed, F., Afzaal, M., Shah, Y. A., Qamar, A., Faisal, Z., Ghani, S., Ateeq, H., Akhtar, M. N., Tufail, T., Hussain, M., Asghar, A., Rasheed, A., & Jbawi, E. A. (2023). Gut microbiota and synbiotic foods: Unveiling the relationship in COVID-19 perspective. *Food Science & Nutrition, 11*(3), 1166–1177. DOI: 10.1002/fsn3.3162 PMID: 36911846

Al-Omairi, O., Elgalib, A., & Al Kindi, H. (2019). HIV Drug Resistance among Patients Failing Therapy at a Tertiary Center in Oman: A Case Record Review. *Oman Medical Journal, 34*(6), 490–495. DOI: 10.5001/omj.2019.91 PMID: 31745412

Alam, S., Bhuiyan, F. R., Emon, T. H., & Hasan, M. (2021). Prospects of nutritional interventions in the care of COVID-19 patients. *Heliyon, 7*(2), e06285. DOI: 10.1016/j.heliyon.2021.e06285 PMID: 33615017

Ali, R., Awadalla, E., Amin, Y., Fouad, S. S., Ahmed, M. A.-E. B., Hassan, M. H., Abdel-Kahaar, E., & Abdel-Aziz, R. H. (2024). The deleterious effects of sofosbuvir and ribavirin (antiviral drugs against hepatitis C virus) on different body systems in male albino rats regarding reproductive, hematological, biochemical, hepatic, and renal profiles and histopathological changes. *Scientific Reports, 14*(1), 5682. DOI: 10.1038/s41598-024-55950-5 PMID: 38453980

Amrouche, T., Lammi, S., & Drider, D. (2024). Probiotics and prebiotics intervention in respiratory and digestive infections linked to COVID-19. *Probiotics and Antimicrobial Proteins.* Advance online publication. DOI: 10.1007/s12602-024-10404-2 PMID: 39614066

Anagnostis, P., Livadas, S., Goulis, D. G., Rees, M., & Lambrinoudaki, I. (2023). Vitamin D, Menopausal Health and COVID-19: Critical Appraisal of Current Data. *Journal of Clinical Medicine, 12*(3), 916. DOI: 10.3390/jcm12030916 PMID: 36769563

Ashique, S., Mishra, N., Mohanto, S., Garg, A., Taghizadeh-Hesary, F., Gowda, B. H. J., & Chellappan, D. K. (2024). Application of artificial intelligence (AI) to control COVID-19 pandemic: Current status and future prospects. *Heliyon, 10*(4), e25754. DOI: 10.1016/j.heliyon.2024.e25754 PMID: 38370192

Asselah, T., Marcellin, P., & Schinazi, R. F. (2018). Treatment of hepatitis C virus infection with direct-acting antiviral agents: 100% cure?. *Liver international: official journal of the International Association for the Study of the Liver, 38 Suppl 1*(Suppl 1), 7–13. https://doi.org/DOI: 10.1111/liv.13673

Ávila-Ríos, S., Parkin, N., Swanstrom, R., Paredes, R., Shafer, R., Ji, H., & Kantor, R. (2020). Next-Generation Sequencing for HIV Drug Resistance Testing: Laboratory, Clinical, and Implementation Considerations. *Viruses, 12*(6), 617. DOI: 10.3390/v12060617 PMID: 32516949

Appaji, R., Thomas, L., & Baral, T., N, S., Chaithra, Saravu, K., ... & Rao, M. (2024). Characteristics of isoniazid-induced psychosis: A systematic review of case reports and case series. *European Journal of Clinical Pharmacology, 80*(11), 1725–1740.

Baddeley, H. J. E., & Isalan, M. (2021). The Application of CRISPR/Cas Systems for Antiviral Therapy. *Frontiers in Genome Editing, 3*, 745559. DOI: 10.3389/fgeed.2021.745559 PMID: 34723245

Bahrulolum, H., Tarrahimofrad, H., Rouzbahani, F. N., Nooraei, S., Sameh, M. M., Hajizade, A., & Ahmadian, G. (2023). Potential of CRISPR/Cas system as emerging tools in the detection of viral hepatitis infection. *Virology Journal, 20*(1), 91. DOI: 10.1186/s12985-023-02048-5 PMID: 37158910

Bakkers, M. J. G., Ritschel, T., Tiemessen, M., Dijkman, J., Zuffianò, A. A., Yu, X., van Overveld, D., Le, L., Voorzaat, R., van Haaren, M. M., de Man, M., Tamara, S., van der Fits, L., Zahn, R., Juraszek, J., & Langedijk, J. P. M. (2024). Efficacious human metapneumovirus vaccine based on AI-guided engineering of a closed prefusion trimer. *Nature Communications, 15*(1), 6270. DOI: 10.1038/s41467-024-50659-5 PMID: 39054318

Bauer, A. N., Williams, J. F., Pokhrel, L. R., Garcia, S., Majumdar, N., Eells, J. B., Cook, P. P., & Akula, S. M. (2024). Evaluating Molecular Mechanism of Viral Inhibition of Aerosolized Smart Nano-Enabled Antiviral Therapeutic (SNAT) on SARS-CoV-2-Infected Hamsters. *Toxics, 12*(7), 495. DOI: 10.3390/toxics12070495 PMID: 39058147

Bekbolatova, M., Mayer, J., Ong, C. W., & Toma, M. (2024). Transformative Potential of AI in Healthcare: Definitions, Applications, and Navigating the Ethical Landscape and Public Perspectives. *Healthcare (Basel), 12*(2), 125. DOI: 10.3390/healthcare12020125 PMID: 38255014

Berger, C. T., & Kim, A. Y. (2012). IL28B polymorphisms as a pretreatment predictor of response to HCV treatment. *Infectious Disease Clinics of North America, 26*(4), 863–877. DOI: 10.1016/j.idc.2012.08.010 PMID: 23083820

Bezbaruah, R., Chavda, V. P., Nongrang, L., Alom, S., Deka, K., Kalita, T., Ali, F., Bhattacharjee, B., & Vora, L. (2022). Nanoparticle-Based Delivery Systems for Vaccines. *Vaccines, 10*(11), 1946. DOI: 10.3390/vaccines10111946 PMID: 36423041

Bhargavi, G., & Subbian, S. (2024). The causes and consequences of trained immunity in myeloid cells. *Frontiers in Immunology*, *15*, 1365127. DOI: 10.3389/fimmu.2024.1365127 PMID: 38665915

Biswas, M., Sawajan, N., Rungrotmongkol, T., Sanachai, K., Ershadian, M., & Sukasem, C. (2022). Pharmacogenetics and Precision Medicine Approaches for the Improvement of COVID-19 Therapies. *Frontiers in Pharmacology*, *13*, 835136. DOI: 10.3389/fphar.2022.835136 PMID: 35250581

Boilesen, D. R., Nielsen, K. N., & Holst, P. J. (2021). Novel Antigenic Targets of HPV Therapeutic Vaccines. *Vaccines*, *9*(11), 1262. DOI: 10.3390/vaccines9111262 PMID: 34835193

Bolarin, J. A., Oluwatoyosi, M. A., Orege, J. I., Ayeni, E. A., Ibrahim, Y. A., Adeyemi, S. B., Tiamiyu, B. B., Gbadegesin, L. A., Akinyemi, T. O., Odoh, C. K., Umeobi, H. I., & Adeoye, A. B. (2021). Therapeutic drugs for SARS-CoV-2 treatment: Current state and perspective. *International Immunopharmacology*, *90*, 107228. DOI: 10.1016/j.intimp.2020.107228 PMID: 33302035

Borden, K. L., & Culjkovic-Kraljacic, B. (2010). Ribavirin as an anticancer therapy: Acute myeloid leukemia and beyond? *Leukemia & Lymphoma*, *51*(10), 1805–1815. DOI: 10.3109/10428194.2010.496506 PMID: 20629523

Bottino, P., Pastrone, L., Curtoni, A., Bondi, A., Sidoti, F., Zanotto, E., Cavallo, R., Solidoro, P., & Costa, C. (2023). Antiviral Approach to Cytomegalovirus Infection: An Overview of Conventional and Novel Strategies. *Microorganisms*, *11*(10), 2372. DOI: 10.3390/microorganisms11102372 PMID: 37894030

Braveman, P., & Gottlieb, L. (2014). The social determinants of health: it is time to consider the causes of the causes. *Public health reports (Washington, D.C.: 1974), 129 Suppl 2*(Suppl 2), 19–31. DOI: 10.1177/00333549141291S206

Burbank, A. J. (2023). Risk Factors for Respiratory Viral Infections: A Spotlight on Climate Change and Air Pollution. *Journal of Asthma and Allergy*, *16*, 183–194. DOI: 10.2147/JAA.S364845 PMID: 36721739

Cai, B., Chang, S., Tian, Y., & Zhen, S. (2023, May). CRISPR/Cas9 for hepatitis B virus infection treatment. *Immunity, Inflammation and Disease*, *11*(5), e866. DOI: 10.1002/iid3.866 PMID: 37249290

Cao, L., Zhao, S., Lou, J., Zheng, H., Chan, R. W. Y., Chong, M. K. C., Chen, Z., Chan, P. K. S., Zee, B. C. Y., & Wang, M. H. (2021). Differential Influence of Age on the Relationship between Genetic Mismatch and A(H1N1)pdm09 Vaccine Effectiveness. *Viruses*, *13*(4), 619. DOI: 10.3390/v13040619 PMID: 33916601

Chakraborty, S. S., Ray Dutta, J., Ganesan, R., & Minary, P. (2025). The Evolution of Nucleic Acid-Based Diagnosis Methods from the (pre)CRISPR to CRISPR era and the Associated Machine/Deep Learning Approaches in Relevant RNA Design. *Methods in Molecular Biology (Clifton, N.J.)*, *2847*, 241–300. DOI: 10.1007/978-1-0716-4079-1_17 PMID: 39312149

Chehelgerdi, M., Chehelgerdi, M., Khorramian-Ghahfarokhi, M., Shafieizadeh, M., Mahmoudi, E., Eskandari, F., Rashidi, M., Arshi, A., & Mokhtari-Farsani, A. (2024). Comprehensive review of CRISPR-based gene editing: Mechanisms, challenges, and applications in cancer therapy. *Molecular Cancer*, *23*(9), 9. Advance online publication. DOI: 10.1186/s12943-023-01925-5 PMID: 38195537

Chen, K. K., Tsung-Ning Huang, D., & Huang, L. M. (2022). SARS-CoV-2 variants - Evolution, spike protein, and vaccines. *Biomedical Journal*, *45*(4), 573–579. DOI: 10.1016/j.bj.2022.04.006 PMID: 35526825

Chen, T., Li, S., Deng, D., Zhang, W., Zhang, J., & Shen, Z. (2024). Key role of interferon regulatory factor 1 (IRF-1) in regulating liver disease: progress and outlook. IRF-1 : . *Journal of Zhejiang University. Science. B.*, *25*(6), 451–470. DOI: 10.1631/jzus.B2300159 PMID: 38910492

Chen, X., Zheng, M., Lin, S., Huang, M., Chen, S., & Chen, S. (2025). The application of CRISPR/Cas9–based genome-wide screening to disease research. *Molecular and Cellular Probes*, *79*, 102004. DOI: 10.1016/j.mcp.2024.102004 PMID: 39709065

Cheng, H., Deng, H., Ma, D., Gao, M., Zhou, Z., Li, H., Liu, S., & Teng, T. (2024). Insight into the natural regulatory mechanisms and clinical applications of the CRISPR–Cas system. *Heliyon*, *10*(20), e39538. DOI: 10.1016/j.heliyon.2024.e39538 PMID: 39502233

Choi, W., Cha, S., & Kim, K. (2024). Navigating the CRISPR/Cas Landscape for Enhanced Diagnosis and Treatment of Wilson's Disease. *Cells*, *13*(14), 1214. DOI: 10.3390/cells13141214 PMID: 39056796

Chou, T. C., Maggirwar, N. S., & Marsden, M. D. (2024). HIV Persistence, Latency, and Cure Approaches: Where Are We Now? *Viruses*, *16*(7), 1163. DOI: 10.3390/v16071163 PMID: 39066325

Clipman, S. J., Solomon, S. S., Srikrishnan, A. K., McFall, A. M., Gomathi, S., Saravanan, S., Anand, S., Vasudevan, C. K., Kumar, M. S., Celentano, D. D., Mehta, S. H., & Lucas, G. M. (2022). Antiretroviral Drug Resistance in HIV Sequences From People Who Inject Drugs and Men Who Have Sex With Men Across 21 Cities in India. *Open Forum Infectious Diseases*, *9*(10), ofac481. Advance online publication. DOI: 10.1093/ofid/ofac481 PMID: 36225747

Costa, B., & Vale, N. (2023, January 3). Modulating Immune Response in Viral Infection for Quantitative Forecasts of Drug Efficacy. *Pharmaceutics*, *15*(1), 167. DOI: 10.3390/pharmaceutics15010167 PMID: 36678799

Crux, N. B., & Elahi, S. (2017). Human Leukocyte Antigen (HLA) and Immune Regulation: How Do Classical and Non-Classical HLA Alleles Modulate Immune Response to Human Immunodeficiency Virus and Hepatitis C Virus Infections? *Frontiers in Immunology*, *8*, 832. DOI: 10.3389/fimmu.2017.00832 PMID: 28769934

Cui, L., Watanabe, S., Miyanaga, K., Kiga, K., Sasahara, T., Aiba, Y., Tan, X. E., Veeranarayanan, S., Thitiananpakorn, K., Nguyen, H. M., & Wannigama, D. L. (2024). A Comprehensive Review on Phage Therapy and Phage-Based Drug Development. *Antibiotics (Basel, Switzerland)*, *13*(9), 870. DOI: 10.3390/antibiotics13090870 PMID: 39335043

da Costa Castilho, M., de Filippis, A., Machado, L., de Lima Calvanti, T., Lima, M., Fonseca, V., Giovanetti, M., Docena, C., Neto, A. M., Bôtto-Menezes, C. H. A., Kara, E. O., de La Barrera, R., Modjarrad, K., Giozza, S. P., Pereira, G. F., Alcantara, L. C. J., Broutet, N. J. N., Calvet, G. A., Wallau, G. L., & Franca, R. (2024). Evidence of Zika Virus Reinfection by Genome Diversity and Antibody Response Analysis, Brazil. *Emerging Infectious Diseases*, *30*(2), 310–320. DOI: 10.3201/eid3002.230122 PMID: 38270216

Das, K., & Arnold, E. (2013). HIV-1 reverse transcriptase and antiviral drug resistance. Part 1. *Current Opinion in Virology*, *3*(2), 111–118. DOI: 10.1016/j.coviro.2013.03.012 PMID: 23602471

Deehan, E. C., Al Antwan, S., Witwer, R. S., Guerra, P., John, T., & Monheit, L. (2024). Revisiting the concepts of prebiotic and prebiotic effect in light of scientific and regulatory progress—A consensus paper from the Global Prebiotic Association. *Advances in Nutrition*, *15*(12), 100329. DOI: 10.1016/j.advnut.2024.100329 PMID: 39481540

Di Stefano, M., Ismail, M. H., Leitner, T., Faleo, G., Alwazzeh, M. J., Mbisa, J. L., Fiore, J. R., & Santantonio, T. A. (2023). A novel candidate hepatitis C virus genotype 4 subtype identified by next generation sequencing full-genome characterization in a patient from Saudi Arabia. *Frontiers in Microbiology*, *14*, 1285367. DOI: 10.3389/fmicb.2023.1285367 PMID: 38029191

Dimitrievska, M., Bansal, D., Vitale, M., Strouboulis, J., Miccio, A., Nicolaides, K. H., El Hoss, S., Shangaris, P., & Jacków-Malinowska, J. (2024). Revolutionizing healing: Gene editing's breakthrough against sickle cell disease. *Blood Reviews*, *65*, 101185. DOI: 10.1016/j.blre.2024.101185 PMID: 38493007

Domingues, W., Folgosi, V. Â., Sanabani, S. S., Leite, P. D.Junior, Assone, T., & Casseb, J. (2024). Novel approaches for HTLV-1 therapy: Innovative applications of CRISPR-Cas9. *Revista do Instituto de Medicina Tropical de São Paulo*, *66*, e48. DOI: 10.1590/s1678-9946202466048 PMID: 39194140

Dopazo, J., Maya-Miles, D., García, F., Lorusso, N., Calleja, M. Á., Pareja, M. J., López-Miranda, J., Rodríguez-Baño, J., Padillo, J., Túnez, I., & Romero-Gómez, M. (2021, May 26). Implementing Personalized Medicine in COVID-19 in Andalusia: An Opportunity to Transform the Healthcare System. *Journal of Personalized Medicine*, *11*(6), 475. DOI: 10.3390/jpm11060475 PMID: 34073493

Duan, X. P., Qin, B. D., Jiao, X. D., Liu, K., Wang, Z., & Zang, Y. S. (2024). New clinical trial design in precision medicine: Discovery, development and direction. *Signal Transduction and Targeted Therapy*, *9*(1), 57. DOI: 10.1038/s41392-024-01760-0 PMID: 38438349

Dubey, A. K., & Mostafavi, E. (2023). Biomaterials-mediated CRISPR/Cas9 delivery: Recent challenges and opportunities in gene therapy. *Frontiers in Chemistry*, *11*, 1259435. DOI: 10.3389/fchem.2023.1259435 PMID: 37841202

Dust, K., Carpenter, M., Chen, J. C., Grant, C., McCorrister, S., Westmacott, G. R., & Severini, A. (2022). Human Papillomavirus 16 E6 and E7 Oncoproteins Alter the Abundance of Proteins Associated with DNA Damage Response, Immune Signaling and Epidermal Differentiation. *Viruses*, *14*(8), 1764. DOI: 10.3390/v14081764 PMID: 36016386

Düzgüneş, N., Fernandez-Fuentes, N., & Konopka, K. (2021). Inhibition of Viral Membrane Fusion by Peptides and Approaches to Peptide Design. *Pathogens (Basel, Switzerland)*, *10*(12), 1599. DOI: 10.3390/pathogens10121599 PMID: 34959554

Eini, P., Hatami, M., Eini, P., & Poorolajal, J. (2023). Evaluation and Comparison of Serum Vitamin D Levels between Injection Drug Users with and without Chronic Hepatitis C Infection in Hamadan-Iran. *Maedica*, *18*(3), 437–441. DOI: 10.26574/maedica.2023.18.3.437 PMID: 38023763

Elliott, J. L., & Kutluay, S. B. (2020). Going beyond Integration: The Emerging Role of HIV-1 Integrase in Virion Morphogenesis. *Viruses*, *12*(9), 1005. DOI: 10.3390/v12091005 PMID: 32916894

Ellwanger, J. H., & Chies, J. A. B. (2021). Zoonotic spillover: Understanding basic aspects for better prevention. *Genetics and Molecular Biology*, *44*(1, Suppl 1), e20200355. DOI: 10.1590/1678-4685-gmb-2020-0355 PMID: 34096963

Elste, J., Saini, A., Mejia-Alvarez, R., Mejía, A., Millán-Pacheco, C., Swanson-Mungerson, M., & Tiwari, V. (2024). Significance of Artificial Intelligence in the Study of Virus–Host Cell Interactions. *Biomolecules*, *14*(8), 911. DOI: 10.3390/biom14080911 PMID: 39199298

Essa, S., Siddique, I., Saad, M., & Raghupathy, R. (2021). Modulation of Production of Th1/Th2 Cytokines in Peripheral Blood Mononuclear Cells and Neutrophils by Hepatitis C Virus Infection in Chronically Infected Patients. *Pathogens (Basel, Switzerland)*, *10*(11), 1519. DOI: 10.3390/pathogens10111519 PMID: 34832674

Estévez-Herrera, J., Pérez-Yanes, S., Cabrera-Rodríguez, R., Márquez-Arce, D., Trujillo-González, R., Machado, J. D., Madrid, R., & Valenzuela-Fernández, A. (2021). Zika Virus Pathogenesis: A Battle for Immune Evasion. *Vaccines*, *9*(3), 294. DOI: 10.3390/vaccines9030294 PMID: 33810028

Faivre, N., Verollet, C., & Dumas, F. (2024). The chemokine receptor CCR5: Multifaceted hook for HIV-1. *Retrovirology*, *21*(2), 2. Advance online publication. DOI: 10.1186/s12977-024-00634-1 PMID: 38263120

Fasano, M., Ieva, F., Ciarallo, M., Caccianotti, B., & Santantonio, T. A. (2024). Acute Hepatitis C: Current Status and Future Perspectives. *Viruses*, *16*(11), 1739. DOI: 10.3390/v16111739 PMID: 39599853

Fatima, H., Singh, D., Muhammad, H., Acharya, S., & Aziz, M. A. (2025). Improving the use of CRISPR/Cas9 gene editing machinery as a cancer therapeutic tool with the help of nanomedicine. 3 Biotech, 15(1), 17. https://doi.org/DOI: 10.1007/s13205-024-04186-1

Feng, J. Y., Du Pont, V., Babusis, D., Gordon, C. J., Tchesnokov, E. P., Perry, J. K., Duong, V., Vijjapurapu, A., Zhao, X., Chan, J., Cohen, C., Juneja, K., Cihlar, T., Götte, M., & Bilello, J. P. (2022). The Nucleoside/Nucleotide Analogs Tenofovir and Emtricitabine Are Inactive against SARS-CoV-2. *Molecules (Basel, Switzerland)*, *27*(13), 4212. DOI: 10.3390/molecules27134212 PMID: 35807457

Foo, C. S., Abdelnabi, R., Kaptein, S. J. F., Zhang, X., Ter Horst, S., Mols, R., Delang, L., Rocha-Pereira, J., Coelmont, L., Leyssen, P., Dallmeier, K., Vergote, V., Heylen, E., Vangeel, L., Chatterjee, A. K., Annaert, P. P., Augustijns, P. F., De Jonghe, S., Jochmans, D., & Neyts, J. (2022). HIV protease inhibitors Nelfinavir and Lopinavir/Ritonavir markedly improve lung pathology in SARS-CoV-2-infected Syrian hamsters despite lack of an antiviral effect. *Antiviral Research*, *202*, 105311. DOI: 10.1016/j.antiviral.2022.105311 PMID: 35390430

Frericks, N., Klöhn, M., Lange, F., Pottkämper, L., Carpentier, A., & Steinmann, E. (2025). Host-targeting antivirals for chronic viral infections of the liver. *Antiviral Research, 234*, 106062. DOI: 10.1016/j.antiviral.2024.106062 PMID: 39716667

García-Cruz, J. C., Huelgas-Méndez, D., Jiménez-Zúñiga, J. S., Rebollar-Juárez, X., Hernández-Garnica, M., Fernández-Presas, A. M., Husain, F. M., Alenazy, R., Alqasmi, M., Albalawi, T., Alam, P., & García-Contreras, R. (2023). Myriad applications of bacteriophages beyond phage therapy. *PeerJ, 11*, e15272. DOI: 10.7717/peerj.15272 PMID: 37101788

Gawande, M. S., Zade, N., Kumar, P., Gundewar, S., Weerarathna, I. N., & Verma, P. (2025). The role of artificial intelligence in pandemic responses: From epidemiological modeling to vaccine development. *Molecular Biomedicine, 6*(1), 1. DOI: 10.1186/s43556-024-00238-3 PMID: 39747786

Geng, N., Ma, L., Jin, Y., Lu, J., Zheng, Y., Wang, J., Wang, X., & Chen, X. (2024). Prediction Model for the Clearance of Hepatitis B Surface Antigen in Patients with Chronic Hepatitis B before Interferon Therapy: A Prospective Case–Control Study. *Diagnostics (Basel), 14*(1), 118. DOI: 10.3390/diagnostics14010118 PMID: 38201427

Giovanetti, M., Pannella, G., Altomare, A., Rocchi, G., Guarino, M., Ciccozzi, M., Riva, E., & Gherardi, G. (2024). Exploring the Interplay between COVID-19 and Gut Health: The Potential Role of Prebiotics and Probiotics in Immune Support. *Viruses, 16*(3), 370. DOI: 10.3390/v16030370 PMID: 38543736

Godwin, P. O., Polsonetti, B., Caron, M. F., & Oppelt, T. F. (2024). Remdesivir for the Treatment of COVID-19: A Narrative Review. *Infectious Diseases and Therapy, 13*(1), 1–19. DOI: 10.1007/s40121-023-00900-3 PMID: 38193988

Haas, P., Muralidharan, M., Krogan, N. J., Kaake, R. M., & Hüttenhain, R. (2021). Proteomic Approaches to Study SARS-CoV-2 Biology and COVID-19 Pathology. *Journal of Proteome Research, 20*(2), 1133–1152. DOI: 10.1021/acs.jproteome.0c00764 PMID: 33464917

Hadi, R., Poddar, A., Sonnaila, S., Bhavaraju, V. S. M., & Agrawal, S. (2024). Advancing CRISPR-Based Solutions for COVID-19 Diagnosis and Therapeutics. *Cells, 13*(21), 1794. DOI: 10.3390/cells13211794 PMID: 39513901

Hagiwara, S., Nishida, N., & Kudo, M. (2015). Antiviral therapy for chronic hepatitis B: Combination of nucleoside analogs and interferon. *World Journal of Hepatology, 7*(23), 2427–2431. DOI: 10.4254/wjh.v7.i23.2427 PMID: 26483864

He, Y., Shen, M., Wang, X., Yin, A., Liu, B., Zhu, J., & Zhang, Z. (2024). Suppression of Interferon Response and Antiviral Strategies of Bunyaviruses. *Tropical Medicine and Infectious Disease*, *9*(9), 205. DOI: 10.3390/tropicalmed9090205 PMID: 39330894

Hegazy, S. K., Tharwat, S., & Hassan, A. H. (2023). Comparing the efficacy of regen-cov, remdesivir, and favipiravir in reducing invasive mechanical ventilation need in hospitalized COVID-19 patients. *World Journal of Clinical Cases*, *11*(26), 6105–6121. DOI: 10.12998/wjcc.v11.i26.6105 PMID: 37731581

Heneghan, C. J., Onakpoya, I., Jones, M. A., Doshi, P., Del Mar, C. B., Hama, R., Thompson, M. J., Spencer, E. A., Mahtani, K. R., Nunan, D., Howick, J., & Jefferson, T. (2016). Neuraminidase inhibitors for influenza: A systematic review and meta-analysis of regulatory and mortality data. *Health Technology Assessment*, *20*(42), 1–242. DOI: 10.3310/hta20420 PMID: 27246259

Hernandez Carballo, I., Bakola, M., & Stuckler, D. (2022). The impact of air pollution on COVID-19 incidence, severity, and mortality: A systematic review of studies in Europe and North America. *Environmental Research*, *215*(Pt 1), 114155. DOI: 10.1016/j.envres.2022.114155 PMID: 36030916

Ho, D., Quake, S. R., McCabe, E. R. B., Chng, W. J., Chow, E. K., Ding, X., Gelb, B. D., Ginsburg, G. S., Hassenstab, J., Ho, C. M., Mobley, W. C., Nolan, G. P., Rosen, S. T., Tan, P., Yen, Y., & Zarrinpar, A. (2020). Enabling Technologies for Personalized and Precision Medicine. *Trends in Biotechnology*, *38*(5), 497–518. DOI: 10.1016/j.tibtech.2019.12.021 PMID: 31980301

Hoang, P. T., Luong, Q. X. T., Ayun, R. Q., Lee, Y., Oh, K. J., Kim, T., Lee, T. K., & Lee, S. (2024). A synergistic therapy against influenza virus A/H1N1/PR8 by a HA1 specific neutralizing single-domain VL and an RNA hydrolyzing scFv. *Frontiers in Microbiology*, *15*, 1355599. DOI: 10.3389/fmicb.2024.1355599 PMID: 38706966

Hoffmann Dahl, E., Mbala, P., Juchet, S., Touré, A., Montoyo, A., Serra, B., Kojan, R., D'Ortenzio, E., Blomberg, B., & Jaspard, M. (2024). Improving Ebola virus disease outbreak control through targeted postexposure prophylaxis. *The Lancet. Global Health*, *12*(10), e1730–e1736. DOI: 10.1016/S2214-109X(24)00255-9 PMID: 39270687

Howe, A. Y. M., Rodrigo, C., Cunningham, E. B., Douglas, M. W., Dietz, J., Grebely, J., Popping, S., Sfalcin, J. A., Parczewski, M., Sarrazin, C., de Salazar, A., Fuentes, A., Sayan, M., Quer, J., Kjellin, M., Kileng, H., Mor, O., Lennerstrand, J., Fourati, S., & Collaborators, S. H. A. R. E. D. (2022). Characteristics of hepatitis C virus resistance in an international cohort after a decade of direct-acting antivirals. *JHEP Reports : Innovation in Hepatology*, *4*(5), 100462. DOI: 10.1016/j.jhepr.2022.100462 PMID: 35434589

Hung, S. K., Wu, C. C., Singh, A., Li, J. H., Lee, C., Chou, E. H., Pekosz, A., Rothman, R., & Chen, K. F. (2023). Developing and validating clinical features-based machine learning algorithms to predict influenza infection in influenza-like illness patients. *Biomedical Journal*, *46*(5), 100561. DOI: 10.1016/j.bj.2022.09.002 PMID: 36150651

Hussein, M., Molina, M. A., Berkhout, B., & Herrera-Carrillo, E. (2023, January 13). A CRISPR–Cas Cure for HIV/AIDS. *International Journal of Molecular Sciences*, *24*(2), 1563. DOI: 10.3390/ijms24021563 PMID: 36675077

Hussen, B. M., Najmadden, Z. B., Abdullah, S. R., Rasul, M. F., Mustafa, S. A., Ghafouri-Fard, S., & Taheri, M. (2024). CRISPR/Cas9 gene editing: A novel strategy for fighting drug resistance in respiratory disorders. *Cell Communication and Signaling : CCS*, *22*(1), 329. DOI: 10.1186/s12964-024-01713-8 PMID: 38877530

Hyun, J., Lee, S. K., Kim, J. H., Cho, E.-J., Kim, H.-S., Kim, J.-S., Song, W., & Kim, H. S. (2023). Variant Analysis of the Thymidine Kinase and DNA Polymerase Genes of Herpes Simplex Virus in Korea: Frequency of Acyclovir Resistance Mutations. *Viruses*, *15*(8), 1709. DOI: 10.3390/v15081709 PMID: 37632051

Ikematsu, H., Baba, T., Saito, M. M., Kinoshita, M., Miyazawa, S., Hata, A., Nakano, S., Kitanishi, Y., & Hayden, F. G. (2024). Comparative Effectiveness of Baloxavir Marboxil and Oseltamivir Treatment in Reducing Household Transmission of Influenza: A Post Hoc Analysis of the BLOCKSTONE Trial. *Influenza and Other Respiratory Viruses*, *18*(5), e13302. DOI: 10.1111/irv.13302 PMID: 38706384

Jennings, L., Kellermann, T., Spinelli, M., Nkantsu, Z., Cogill, D., van Schalkwyk, M., Decloedt, E., van Zyl, G., Orrell, C., & Gandhi, M. (2022). Drug Resistance, Rather than Low Tenofovir Levels in Blood or Urine, Is Associated with Tenofovir, Emtricitabine, and Efavirenz Failure in Resource-Limited Settings. *AIDS Research and Human Retroviruses*, *38*(6), 455–462. DOI: 10.1089/aid.2021.0135 PMID: 34779228

Ji, J. J., Sun, Q. M., Nie, D. Y., Wang, Q., Zhang, H., Qin, F. F., Wang, Q. S., Lu, S. F., Pang, G. M., & Lu, Z. G. (2021). Probiotics protect against RSV infection by modulating the microbiota-alveolar-macrophage axis. *Acta Pharmacologica Sinica*, *42*(10), 1630–1641. DOI: 10.1038/s41401-020-00573-5 PMID: 33495515

Jiang, H., Zhu, X., Jiao, J., Yan, C., Liu, K., Chen, W., & Qin, P. (2025). CRISPR/dCas9-based hotspot self-assembling SERS biosensor integrated with a smartphone for simultaneous, ultrasensitive and robust detection of multiple pathogens. *Biosensors & Bioelectronics*, *270*, 116974. DOI: 10.1016/j.bios.2024.116974 PMID: 39586143

Jones, J. E., Le Sage, V., & Lakdawala, S. S. (2021). Viral and host heterogeneity and their effects on the viral life cycle. *Nature Reviews. Microbiology*, *19*(5), 272–282. DOI: 10.1038/s41579-020-00449-9 PMID: 33024309

Jose-Abrego, A., Roman, S., Rebello Pinho, J. R., Gomes-Gouvêa, M. S., & Panduro, A. (2023). High Frequency of Antiviral Resistance Mutations in HBV Genotypes A2 and H: Multidrug Resistance Strains in Mexico. *Journal of Clinical and Translational Hepatology*, *11*(5), 1023–1034. DOI: 10.14218/JCTH.2022.00135S PMID: 37577226

Katsiki, A. D., Karatzas, P. E., De Lastic, H. X., Georgakilas, A. G., Tsitsilonis, O., & Vorgias, C. E. (2024). DExplore: An Online Tool for Detecting Differentially Expressed Genes from mRNA Microarray. *Experimental Biology*, *13*(5), 351. DOI: 10.3390/biology13050351 PMID: 38785833

Kausar, S., Said Khan, F., Ishaq Mujeeb Ur Rehman, M., Akram, M., Riaz, M., Rasool, G., Hamid Khan, A., Saleem, I., Shamim, S., & Malik, A. (2021). A review: Mechanism of action of antiviral drugs. *International Journal of Immunopathology and Pharmacology*, *35*, 20587384211002621. DOI: 10.1177/20587384211002621 PMID: 33726557

Kefi, M., Cardoso-Jaime, V., Saab, S. A., & Dimopoulos, G. (2024). Curing mosquitoes with genetic approaches for malaria control. *Trends in Parasitology*, *40*(6), 487–499. DOI: 10.1016/j.pt.2024.04.010 PMID: 38760256

Khan, M. M., Ali, M. J., Hanif, H., Maqsood, M. H., Ahmad, I., Alvarez, J. E. G., Catana, M. A., & Lau, D. T. Y. (2022). The dilemma of cytomegalovirus and hepatitis B virus interaction. *Gastroenterology Report*, *10*, goac018. Advance online publication. DOI: 10.1093/gastro/goac018 PMID: 35663152

Khoo, T., Lam, D., & Olynyk, J. K. (2021). Impact of modern antiviral therapy of chronic hepatitis B and C on clinical outcomes of liver disease. *World Journal of Gastroenterology*, *27*(29), 4831–4845. DOI: 10.3748/wjg.v27.i29.4831 PMID: 34447229

Koujah, L., Shukla, D., & Naqvi, A. R. (2019). CRISPR–Cas based targeting of host and viral genes as an antiviral strategy. *Seminars in Cell & Developmental Biology*, *96*, 53–64. DOI: 10.1016/j.semcdb.2019.04.004 PMID: 30953741

Kumar, A., Combe, E., Mougené, L., Zoulim, F., & Testoni, B. (2024). Applications of CRISPR/Cas as a Toolbox for Hepatitis B Virus Detection and Therapeutics. *Viruses*, *16*(10), 1565. DOI: 10.3390/v16101565 PMID: 39459899

Kumar, N., Sharma, S., Kumar, R., Tripathi, B. N., Barua, S., Ly, H., & Rouse, B. T. (2020). Host-Directed Antiviral Therapy. *Clinical Microbiology Reviews*, *33*(3), e00168–e19. DOI: 10.1128/CMR.00168-19 PMID: 32404434

Lakhman, S. S., Ma, Q., & Morse, G. D. (2009). Pharmacogenomics of CYP3A: Considerations for HIV treatment. *Pharmacogenomics*, *10*(8), 1323–1339. DOI: 10.2217/pgs.09.53 PMID: 19663676

Landeras-Bueno, S., Wasserman, H., Oliveira, G., VanAernum, Z. L., Busch, F., Salie, Z. L., Wysocki, V. H., Andersen, K., & Saphire, E. O. (2021). Cellular mRNA triggers structural transformation of Ebola virus matrix protein VP40 to its essential regulatory form. *Cell Reports*, *35*(2), 108986. DOI: 10.1016/j.celrep.2021.108986 PMID: 33852858

Lane, M., Luke, C., Bresee, J., Dugan, V., Post, D., Schafer, J., Roberts, P., Wentworth, D., & Ison, M. (2024). Meeting report: Controlled human influenza virus infection model studies: Current status and future directions for innovation. *Influenza and Other Respiratory Viruses*, *18*(10), e13358. DOI: 10.1111/irv.13358 PMID: 39440405

Lange, S., Baehr, C., Cakman-Hinrichs, N. I., Cron, K., Fengels, H., Gregor, C., Matschenz, K., Petersen, J., Steinfurth, R., Stoehr, A., Unger, S., Gil Mir, M., Hernández, C., Schwenken, M., & Buggisch, P. (2024). ReLink strategy in diagnosed-but-untreated hepatitis C-positive patients in Germany: report from a single center. Die ReLink-Strategie bei diagnostizierten aber unbehandelten Hepatitis C-positiven Patient*innen in Deutschland: Bericht aus einem einzelnen Zentrum - Hepatitis C ReLink-Projekt Deutschland. *Zeitschrift für Gastroenterologie*, *62*(10), 1701–1707. DOI: 10.1055/a-2349-2767 PMID: 39013433

Langmia, I. M., Just, K. S., Yamoune, S., Brockmöller, J., Masimirembwa, C., & Stingl, J. C. (2021). CYP2B6 Functional Variability in Drug Metabolism and Exposure Across Populations-Implication for Drug Safety, Dosing, and Individualized Therapy. *Frontiers in Genetics*, *12*, 692234. DOI: 10.3389/fgene.2021.692234 PMID: 34322158

Lashgarian, H. E., Valibeik, A., Marzban, A., Karkhane, M., & Shahzamani, K. (2021). The Relationship Between HCV-NS5A Gene Mutations and Resistance to Combination Therapy in Patients with HCV- Genotype 1-B. *Reports of Biochemistry & Molecular Biology*, *10*(2), 233–242. DOI: 10.52547/rbmb.10.2.233 PMID: 34604413

Lawitz, E., Mangia, A., Wyles, D., Rodriguez-Torres, M., Hassanein, T., Gordon, S. C., Schultz, M., Davis, M. N., Kayali, Z., Reddy, K. R., Jacobson, I. M., Kowdley, K. V., Nyberg, L., Subramanian, G. M., Hyland, R. H., Arterburn, S., Jiang, D., McNally, J., Brainard, D., & Gane, E. J. (2013). Sofosbuvir for previously untreated chronic hepatitis C infection. *The New England Journal of Medicine*, *368*(20), 1878–1887. DOI: 10.1056/NEJMoa1214853 PMID: 23607594

Lazarevic, I., Banko, A., Miljanovic, D., & Cupic, M. (2024). Hepatitis B Surface Antigen Isoforms: Their Clinical Implications, Utilization in Diagnosis, Prevention and New Antiviral Strategies. *Pathogens (Basel, Switzerland)*, *13*(1), 46. DOI: 10.3390/pathogens13010046 PMID: 38251353

Lee, H. L., & Jang, J. W. (2024). Core protein inhibitors: Opportunities and challenges at the forefront of hepatitis B cure: Editorial on "Phase 1 trial of the safety, pharmacokinetics, and antiviral activity of EDP-514 in untreated viremic chronic hepatitis B patients". *Clinical and Molecular Hepatology*, *30*(4), 692–694. DOI: 10.3350/cmh.2024.0346 PMID: 38741237

Li, H., Zhang, N., Ma, X., Wang, Y., Yang, F., Wang, W., Huang, Y., Xie, Y., & Du, Y. (2025). Google Trends applications for COVID-19 pandemic: A bibliometric analysis. *Digital Health*, *11*, 20552076241310055. DOI: 10.1177/20552076241310055 PMID: 39758260

Li, S. (2024). Application of Artificial Intelligence in Early Diagnosis of Influenza A (H1N1) Virus Infection. *Molecular Pathogens*, *15*(1), 1–8. DOI: 10.5376/mp.2024.15.0001 PMID: 39860962

Li, Y., Muffat, J., Omer Javed, A., Keys, H. R., Lungjangwa, T., Bosch, I., Khan, M., Virgilio, M. C., Gehrke, L., Sabatini, D. M., & Jaenisch, R. (2019). Genome-wide CRISPR screen for Zika virus resistance in human neural cells. *Proceedings of the National Academy of Sciences of the United States of America*, *116*(19), 9527–9532. DOI: 10.1073/pnas.1900867116 PMID: 31019072

Lin, C. L., & Kao, J. H. (2015). Hepatitis B virus genotypes and variants. *Cold Spring Harbor Perspectives in Medicine*, *5*(5), a021436. DOI: 10.1101/cshperspect.a021436 PMID: 25934462

Lin, D. M., Koskella, B., & Lin, H. C. (2017). Phage therapy: An alternative to antibiotics in the age of multidrug resistance. *World Journal of Gastrointestinal Pharmacology and Therapeutics*, *8*(3), 162–173. DOI: 10.4292/wjgpt.v8.i3.162 PMID: 28828194

Ling, K. M., Stick, S. M., & Kicic, A. (2023). Pulmonary bacteriophage and cystic fibrosis airway mucus: Friends or foes? *Frontiers in Medicine*, *10*, 1088494. DOI: 10.3389/fmed.2023.1088494 PMID: 37265479

Liu, Z., Chen, X., Carter, W., Moruf, A., Komatsu, T. E., Pahwa, S., Chan-Tack, K., Snyder, K., Petrick, N., Cha, K., Lal-Nag, M., Hatim, Q., Thakkar, S., Lin, Y., Huang, R., Wang, D., Patterson, T. A., & Tong, W. (2022). AI-powered drug repurposing for developing COVID-19 treatments. *Reference Module in Biomedical Sciences*, B978-0-12-824010-6.00005-8. DOI: 10.1016/B978-0-12-824010-6.00005-8

Louie, J. K., Yang, S., Acosta, M., Yen, C., Samuel, M. C., Schechter, R., Guevara, H., & Uyeki, T. M. (2012). Treatment with neuraminidase inhibitors for critically ill patients with influenza A (H1N1)pdm09. *Clinical Infectious Diseases*, *55*(9), 1198–1204. DOI: 10.1093/cid/cis636 PMID: 22843781

Luo, Z., Liu, L. F., Jiang, Y. N., Tang, L. P., Li, W., Ouyang, S. H., Tu, L. F., Wu, Y. P., Gong, H. B., Yan, C. Y., Jiang, S., Lu, Y. H., Liu, T., Jiang, Z., Kurihara, H., Yu, Y., Yao, X. S., Li, Y. F., & He, R. R. (2020). Novel insights into stress-induced susceptibility to influenza: Corticosterone impacts interferon-β responses by Mfn2-mediated ubiquitin degradation of MAVS. *Signal Transduction and Targeted Therapy*, *5*(1), 202. DOI: 10.1038/s41392-020-00238-z PMID: 32943610

Macarrón Palacios, A., Korus, P., Wilkens, B. G. C., Heshmatpour, N., & Patnaik, S. R. (2024). Revolutionizing in vivo therapy with CRISPR/Cas genome editing: Breakthroughs, opportunities and challenges. *Frontiers in Genome Editing*, *6*, 1342193. DOI: 10.3389/fgeed.2024.1342193 PMID: 38362491

Mahajan, S., Choudhary, S., Kumar, P., & Tomar, S. (2021). Antiviral strategies targeting host factors and mechanisms obliging +ssRNA viral pathogens. *Bioorganic & Medicinal Chemistry*, *46*, 116356. DOI: 10.1016/j.bmc.2021.116356 PMID: 34416512

Maleki Varnosfaderani, S., & Forouzanfar, M. (2024). The Role of AI in Hospitals and Clinics: Transforming Healthcare in the 21st Century. *Bioengineering (Basel, Switzerland)*, *11*(4), 337. DOI: 10.3390/bioengineering11040337 PMID: 38671759

Malik, J. A., Ahmed, S., Yaseen, Z., Alanazi, M., Alharby, T. N., Alshammari, H. A., & Anwar, S. (2022). Association of SARS-CoV-2 and polypharmacy with gut–lung axis: From pathogenesis to treatment. *ACS Omega*, *7*(38), 33651–33665. DOI: 10.1021/acsomega.2c02524 PMID: 36164411

Maqsood, R., Smith, M. F., Holland, L. A., Sullins, R. A., Holland, S. C., Tan, M., Hernandez Barrera, G. M., Thomas, A. W., Islas, M., Kramer, J. L., Nordstrom, L., Mulrow, M., White, M., Murugan, V., & Lim, E. S. (2024). Influenza Virus Genomic Surveillance, Arizona, USA, 2023-2024. *Viruses*, *16*(5), 692. DOI: 10.3390/v16050692 PMID: 38793574

Marcinkowska-Swojak, M., Rakoczy, M., Podkowiński, J., Handschuh, J., Wojciechowski, P., & Handschuh, L. (2024). Od Sangera do sekwencjonowania genomów – przegląd technologii sekwencjonowania DNA [From Sanger to genome sequencing - an overview of DNA sequencing technologies]. *Postepy Biochemii*, *70*(2), 173–189. DOI: 10.18388/pb.2021_534 PMID: 39083466

Maserat, E. (2022). Integration of Artificial Intelligence and CRISPR/Cas9 System for Vaccine Design. *Cancer Informatics*, *21*, 11769351221140102. DOI: 10.1177/11769351221140102 PMID: 36452267

Maslennikov, R., Ivashkin, V., Efremova, I., Poluektova, E., & Shirokova, E. (2021). Probiotics in hepatology: An update. *World Journal of Hepatology*, *13*(9), 1154–1166. DOI: 10.4254/wjh.v13.i9.1154 PMID: 34630882

Matthew, A. N., Leidner, F., Lockbaum, G. J., Henes, M., Zephyr, J., Hou, S., Rao, D. N., Timm, J., Rusere, L. N., Ragland, D. A., Paulsen, J. L., Prachanronarong, K., Soumana, D. I., Nalivaika, E. A., Kurt Yilmaz, N., Ali, A., & Schiffer, C. A. (2021). Drug Design Strategies to Avoid Resistance in Direct-Acting Antivirals and Beyond. *Chemical Reviews*, *121*(6), 3238–3270. DOI: 10.1021/acs.chemrev.0c00648 PMID: 33410674

Mays, J. W., Powell, N. D., Hunzeker, J. T., Hanke, M. L., Bailey, M. T., & Sheridan, J. F. (2012). Stress and the anti-influenza immune response: Repeated social defeat augments clonal expansion of CD8(+)T cells during primary influenza A viral infection. *Journal of Neuroimmunology*, *243*(1-2), 34–42. DOI: 10.1016/j.jneuroim.2011.12.011 PMID: 22244573

Mazziotta, C., Tognon, M., Martini, F., Torreggiani, E., & Rotondo, J. C. (2023). Probiotics Mechanism of Action on Immune Cells and Beneficial Effects on Human Health. *Cells*, *12*(1), 184. DOI: 10.3390/cells12010184 PMID: 36611977

McCombie, W. R., McPherson, J. D., & Mardis, E. R. (2019). Next-Generation Sequencing Technologies. *Cold Spring Harbor Perspectives in Medicine, 9*(11), a036798. DOI: 10.1101/cshperspect.a036798 PMID: 30478097

Medina, C., García, A. H., Crespo, F. I., Toro, F. I., Mayora, S. J., & De Sanctis, J. B. (2023). A Synopsis of Hepatitis C Virus Treatments and Future Perspectives. *Current Issues in Molecular Biology, 45*(10), 8255–8276. DOI: 10.3390/cimb45100521 PMID: 37886964

Model, A. E., Lim, D., Nguyen, T. M., Sreekanth, V., & Choudhary, A. (2022, February). CRISPR-based therapeutics: Current challenges and future applications. *Trends in Pharmacological Sciences, 43*(2), 151–161. DOI: 10.1016/j.tips.2021.10.012 PMID: 34952739

Mohamed, H., Gurrola, T., Berman, R., Collins, M., Sariyer, I. K., Nonnemacher, M. R., & Wigdahl, B. (2022). Targeting CCR5 as a Component of an HIV-1 Therapeutic Strategy. *Frontiers in Immunology, 12*, 816515. DOI: 10.3389/fimmu.2021.816515 PMID: 35126374

Mohamed Abdelnajid, D., Elmowafy, A. Y., Rostaing, L., & Elrakaiby, M. T. (2023). Prediction of response to sofosbuvir-based therapy using serum interleukin-12 and single-nucleotide polymorphism of the interleukin 28B gene as predictive factors in HCV positive genotype-4 patients. *Medicine, 102*(28), e34125. DOI: 10.1097/MD.0000000000034125 PMID: 37443472

Mojahed, N., Mohammadkhani, M. A., & Mohamadkhani, A. (2022). Climate Crises and Developing Vector-Borne Diseases: A Narrative Review. *Iranian Journal of Public Health, 51*(12), 2664–2673. DOI: 10.18502/ijph.v51i12.11457 PMID: 36742229

Monoson, A., Schott, E., Ard, K., Kilburg-Basnyat, B., Tighe, R. M., Pannu, S., & Gowdy, K. M. (2023). Air pollution and respiratory infections: the past, present, and future. *Toxicological sciences: an official journal of the Society of Toxicology, 192*(1), 3–14. https://doi.org/DOI: 10.1093/toxsci/kfad003

Morey, J. N., Boggero, I. A., Scott, A. B., & Segerstrom, S. C. (2015). Current Directions in Stress and Human Immune Function. *Current Opinion in Psychology, 5*, 13–17. DOI: 10.1016/j.copsyc.2015.03.007 PMID: 26086030

Motomura, K., Kuwano, A., Tanaka, K., Koga, Y., Masumoto, A., & Yada, M. (2023, August 30). Potential Predictive Biomarkers of Systemic Drug Therapy for Hepatocellular Carcinoma: Anticipated Usefulness in Clinical Practice. *Cancers (Basel), 15*(17), 4345. DOI: 10.3390/cancers15174345 PMID: 37686621

Mueller Brown, K., Le Sage, V., French, A. J., Jones, J. E., Padovani, G. H., Avery, A. J., Schultz-Cherry, S., Rosch, J. W., Hiller, N. L., & Lakdawala, S. S. (2022). Secondary infection with *Streptococcus pneumoniae* decreases influenza virus replication and is linked to severe disease. *FEMS Microbes*, *3*, xtac007. Advance online publication. DOI: 10.1093/femsmc/xtac007 PMID: 35392116

Muteeb, G., Rehman, M. T., Shahwan, M., & Aatif, M. (2023, November 15). Origin of Antibiotics and Antibiotic Resistance, and Their Impacts on Drug Development: A Narrative Review. *Pharmaceuticals (Basel, Switzerland)*, *16*(11), 1615. DOI: 10.3390/ph16111615 PMID: 38004480

Naidoo, K., & Oliver, S. V. (2024). Gene drives: An alternative approach to malaria control? *Gene Therapy*. Advance online publication. DOI: 10.1038/s41434-024-00468-8 PMID: 39039203

Najafi, S., Tan, S. C., Aghamiri, S., Raee, P., Ebrahimi, Z., Jahromi, Z. K., Rahmati, Y., Sadri Nahand, J., Piroozmand, A., Jajarmi, V., & Mirzaei, H. (2022, April). Therapeutic potentials of CRISPR–Cas genome editing technology in human viral infections. *Biomedicine and Pharmacotherapy*, *148*, 112743. DOI: 10.1016/j.biopha.2022.112743 PMID: 35228065

Namasivayam, V., Vanangamudi, M., Kramer, V. G., Kurup, S., Zhan, P., Liu, X., Kongsted, J., & Byrareddy, S. N. (2019). The Journey of HIV-1 Non-Nucleoside Reverse Transcriptase Inhibitors (NNRTIs) from Lab to Clinic. *Journal of Medicinal Chemistry*, *62*(10), 4851–4883. DOI: 10.1021/acs.jmedchem.8b00843 PMID: 30516990

Nguyen, P. H. A., Fukushi, S., Yamada, S., Harada, S., Yoshikawa, T., Kinoshita, H., Kawahara, M., Ogawa, T., Ebihara, H., Moi, M. L., & Saijo, M. (2024). Analysis of antiviral drug properties of thymidine kinase of herpes B virus using recombinant herpes simplex virus 1. *Microbiology Spectrum*, *12*(1), e0309123. DOI: 10.1128/spectrum.03091-23 PMID: 38095468

Noorbakhsh, F., Abdolmohammadi, K., Fatahi, Y., Dalili, H., Rasoolinejad, M., Rezaei, F., Salehi-Vaziri, M., Shafiei-Jandaghi, N. Z., Gooshki, E. S., Zaim, M., & Nicknam, M. H. (2019). Zika Virus Infection, Basic and Clinical Aspects: A Review Article. *Iranian Journal of Public Health*, *48*(1), 20–31. DOI: 10.18502/ijph.v48i1.779 PMID: 30847308

Nyathi, S., Rezende, I. M., Walter, K. S., Thongsripong, P., Mutuku, F., Ndenga, B., Mbakaya, J. O., Aswani, P., Musunzaji, P. S., Chebii, P. K., Maina, P. W., Mutuku, P. S., Ng'ang'a, C. M., Malumbo, S. L., Jembe, Z., Vu, D. M., Mordecai, E. A., Bennett, S., Andrews, J. R., & LaBeaud, A. D. (2024). Molecular epidemiology and evolutionary characteristics of dengue virus 2 in East Africa. *Nature Communications, 15*(1), 7832. DOI: 10.1038/s41467-024-51018-0 PMID: 39244569

O' Kane, R., & Hathorn, E. (2023). Hepatitis C: Recent advances and practical management. *Frontline Gastroenterology, 14*(5), 415–421. DOI: 10.1136/flgastro-2022-102373 PMID: 37581179

Obayomi, O. V., Olaniran, A. F., & Owa, S. O. (2024). Unveiling the role of functional foods with emphasis on prebiotics and probiotics in human health: A review. *Journal of Functional Foods, 119*, 106337. DOI: 10.1016/j.jff.2024.106337

Ogaugwu, C. E., Agbo, S. O., & Adekoya, M. A. (2019). CRISPR in Sub-Saharan Africa: Applications and Education. *Trends in Biotechnology, 37*(3), 234–237. DOI: 10.1016/j.tibtech.2018.07.012 PMID: 30100229

Pal, A., & Kundu, R. (2020). Human Papillomavirus E6 and E7: The Cervical Cancer Hallmarks and Targets for Therapy. *Frontiers in Microbiology, 10*, 3116. DOI: 10.3389/fmicb.2019.03116 PMID: 32038557

Parambi, D. G. T., Alharbi, K. S., Kumar, R., Harilal, S., Batiha, G. E., Cruz-Martins, N., Magdy, O., Musa, A., Panda, D. S., & Mathew, B. (2022, January). Gene Therapy Approach with an Emphasis on Growth Factors: Theoretical and Clinical Outcomes in Neurodegenerative Diseases. *Molecular Neurobiology, 59*(1), 191–233. DOI: 10.1007/s12035-021-02555-y PMID: 34655056

Parkhitko, A. A., Jouandin, P., Mohr, S. E., & Perrimon, N. (2019). Methionine metabolism and methyltransferases in the regulation of aging and lifespan extension across species. *Aging Cell, 18*(6), e13034. DOI: 10.1111/acel.13034 PMID: 31460700

Peci, A., Winter, A. L., Li, Y., Gnaneshan, S., Liu, J., Mubareka, S., & Gubbay, J. B. (2019). Effects of Absolute Humidity, Relative Humidity, Temperature, and Wind Speed on Influenza Activity in Toronto, Ontario, Canada. *Applied and Environmental Microbiology, 85*(6), e02426–e18. DOI: 10.1128/AEM.02426-18 PMID: 30610079

Petersen, L. R., Brault, A. C., & Nasci, R. S. (2013). West Nile virus: Review of the literature. *Journal of the American Medical Association, 310*(3), 308–315. DOI: 10.1001/jama.2013.8042 PMID: 23860989

Przemska-Kosicka, A., Childs, C. E., Enani, S., Maidens, C., Dong, H., Dayel, I. B., Tuohy, K., Todd, S., Gosney, M. A., & Yaqoob, P. (2016). The effect of a synbiotic on the response to seasonal influenza vaccination is strongly influenced by the degree of immunosenescence. *Immunity & Ageing*, *13*(6), 6. Advance online publication. DOI: 10.1186/s12979-016-0061-4 PMID: 26985232

Qahwaji, R., Ashankyty, I., Sannan, N. S., Hazzazi, M. S., Basabrain, A. A., & Mobashir, M. (2024). Pharmacogenomics: A Genetic Approach to Drug Development and Therapy. *Pharmaceuticals (Basel, Switzerland)*, *17*(7), 940. DOI: 10.3390/ph17070940 PMID: 39065790

Quer, J., Colomer-Castell, S., Campos, C., Andrés, C., Piñana, M., Cortese, M. F., González-Sánchez, A., Garcia-Cehic, D., Ibáñez, M., Pumarola, T., Rodríguez-Frías, F., Antón, A., & Tabernero, D. (2022). Next-Generation Sequencing for Confronting Virus Pandemics. *Viruses*, *14*(3), 600. DOI: 10.3390/v14030600 PMID: 35337007

Rao, M. S., Van Vleet, T. R., Ciurlionis, R., Buck, W. R., Mittelstadt, S. W., Blomme, E. A. G., & Liguori, M. J. (2019). Comparison of RNA-Seq and Microarray Gene Expression Platforms for the Toxicogenomic Evaluation of Liver From Short-Term Rat Toxicity Studies. *Frontiers in Genetics*, *9*, 636. DOI: 10.3389/fgene.2018.00636 PMID: 30723492

Rayamajhi, N., Cheng, C. C., & Catchen, J. M. (2022). Evaluating Illumina-, Nanopore-, and PacBio-based genome assembly strategies with the bald notothen, Trematomus borchgrevinki. *G3 (Bethesda, Md.)*, *12*(11), jkac192. Advance online publication. DOI: 10.1093/g3journal/jkac192 PMID: 35904764

Resop, R. S., Fromentin, R., Newman, D., Rigsby, H., Dubrovsky, L., Bukrinsky, M., Chomont, N., & Bosque, A. (2020). Fingolimod inhibits multiple stages of the HIV-1 life cycle. *PLoS Pathogens*, *16*(8), e1008679. DOI: 10.1371/journal.ppat.1008679 PMID: 32790802

Reyed, R. M., & Prabhakar, P. K. (2025). An In-Depth Assessment of Dietary Methods and Nutritional Treatments to Improve Female Athletes. In Chatterjee, A., & Sarkar, T. (Eds.), *Examining Physiology, Nutrition, and Body Composition in Sports Science* (pp. 35–66). IGI Global., DOI: 10.4018/979-8-3693-6317-1.ch002

Russo, E., Fiorindi, C., Giudici, F., & Amedei, A. (2022). Immunomodulation by probiotics and prebiotics in hepatocellular carcinoma. *World Journal of Hepatology*, *14*(2), 372–385. DOI: 10.4254/wjh.v14.i2.372 PMID: 35317185

Saehlee, S., Seetaha, S., Klankaew, W., Srathong, P., Choowongkomon, K., & Choengpanya, K. (2024). Anti-Human Immunodeficiency Virus-1 Property of Thai Herbal Extract Kerra™. *Pharmaceuticals (Basel, Switzerland), 17*(7), 917. DOI: 10.3390/ph17070917 PMID: 39065767

Safarnezhad Tameshkel, F., Karbalaie Niya, M. H., Sohrabi, M., Panahi, M., Zamani, F., Imanzade, F., & Rakhshani, N. (2016). Polymorphism of IL-28B Gene (rs12979860) in HCV Genotype 1 Patients Treated by Pegylated Interferon and Ribavirin. *Iranian Journal of Pathology, 11*(3), 216–221. PMID: 27799970

Sallam, M., & Khalil, R. (2024). Contemporary Insights into Hepatitis C Virus: A Comprehensive Review. *Microorganisms, 12*(6), 1035. DOI: 10.3390/microorganisms12061035 PMID: 38930417

Salum, G. M., Dawood, R. M., Abd El-Meguid, M., Ibrahim, N. E., Abdel Aziz, A. O., & El Awady, M. K. (2019). Correlation between IL28B/TLR4 genetic variants and HCC development with/without DAAs treatment in chronic HCV patients. *Genes & Diseases, 7*(3), 392–400. DOI: 10.1016/j.gendis.2019.05.004 PMID: 32884993

Satam, H., Joshi, K., Mangrolia, U., Waghoo, S., Zaidi, G., Rawool, S., Thakare, R. P., Banday, S., Mishra, A. K., Das, G., & Malonia, S. K. (2023). Next-Generation Sequencing Technology: Current Trends and Advancements. *Biology (Basel), 12*(7), 997. DOI: 10.3390/biology12070997 PMID: 37508427

Satam, H., Joshi, K., Mangrolia, U., Waghoo, S., Zaidi, G., Rawool, S., Thakare, R. P., Banday, S., Mishra, A. K., Das, G., & Malonia, S. K. (2023). Next-Generation Sequencing Technology: Current Trends and Advancements. *Biology (Basel), 12*(7), 997. DOI: 10.3390/biology12070997 PMID: 37508427

Schiano Moriello, N., Pinchera, B., & Gentile, I. (2024). Personalized care approaches to hepatitis C therapy: Recent advances and future directions. *Expert Review of Anti-Infective Therapy, 22*(4), 139–151. DOI: 10.1080/14787210.2024.2328336 PMID: 38459735

Shah, N. J., Aloysius, M. M., Sharma, N. R., & Pallav, K. (2021). Advances in treatment and prevention of hepatitis B. *World Journal of Gastrointestinal Pharmacology and Therapeutics, 12*(4), 56–78. DOI: 10.4292/wjgpt.v12.i4.56 PMID: 34316384

Shyr, Z. A., Cheng, Y. S., Lo, D. C., & Zheng, W. (2021, October). Drug combination therapy for emerging viral diseases. *Drug Discovery Today, 26*(10), 2367–2376. DOI: 10.1016/j.drudis.2021.05.008 PMID: 34023496

Shyr, Z. A., Cheng, Y. S., Lo, D. C., & Zheng, W. (2021). Drug combination therapy for emerging viral diseases. *Drug Discovery Today*, *26*(10), 2367–2376. DOI: 10.1016/j.drudis.2021.05.008 PMID: 34023496

Siddiqui, D., Badar, U., Javaid, M., Farooqui, N., Shah, S. A., Iftikhar, A., Sultan, F., Mir, F., Furqan, S., Mahmood, S. F., & Abidi, S. H. (2023). Genetic and antiretroviral drug resistance mutations analysis of reverse transcriptase and protease gene from Pakistani people living with HIV-1. *PLoS One*, *18*(8), e0290425. DOI: 10.1371/journal.pone.0290425 PMID: 37616294

Sinclair, F., Begum, A. A., Dai, C. C., Toth, I., & Moyle, P. M. (2023). Recent advances in the delivery and applications of nonviral CRISPR/Cas9 gene editing. *Drug Delivery and Translational Research*, *13*(5), 1500–1519. DOI: 10.1007/s13346-023-01320-z PMID: 36988873

Singh, S., Kumar, R., Payra, S., & Singh, S. K. (2023). Artificial Intelligence and Machine Learning in Pharmacological Research: Bridging the Gap Between Data and Drug Discovery. *Cureus*, *15*(8), e44359. DOI: 10.7759/cureus.44359 PMID: 37779744

Sizar, O., Khare, S., & Goyal, A.. (2024 Jan). Vitamin D Deficiency. [Updated 2023 Jul 17] In *StatPearls* [Internet]. StatPearls Publishing., Available from https://www.ncbi.nlm.nih.gov/books/NBK532266/

Smith, S. J., Zhao, X. Z., Passos, D. O., Lyumkis, D., Burke, T. R.Jr, & Hughes, S. H. (2021). Integrase Strand Transfer Inhibitors Are Effective Anti-HIV Drugs. *Viruses*, *13*(2), 205. DOI: 10.3390/v13020205 PMID: 33572956

Stefanicka-Wojtas, D., & Kurpas, D. (2023, February 21). Personalized Medicine-Implementation to the Healthcare System in Europe (Focus Group Discussions). *Journal of Personalized Medicine*, *13*(3), 380. DOI: 10.3390/jpm13030380 PMID: 36983562

Strauss, M., Mičetić Turk, D., Lorber, M., Pogačar, M. Š., Koželj, A., Tušek Bunc, K., & Fijan, S. (2023). The Multi-Strain Probiotic OMNi-BiOTiC® Active Reduces the Duration of Acute Upper Respiratory Disease in Older People: A Double-Blind, Randomized, Controlled Clinical Trial. *Microorganisms*, *11*(7), 1760. DOI: 10.3390/microorganisms11071760 PMID: 37512932

Su, J., Yang, L., Sun, Z., & Zhan, X. (2024). Personalized Drug Therapy: Innovative Concept Guided With Proteoformics. *Molecular & Cellular Proteomics*, *23*(3), 100737. DOI: 10.1016/j.mcpro.2024.100737 PMID: 38354979

Suardi, C., Cazzaniga, E., Graci, S., Dongo, D., & Palestini, P. (2021). Link between Viral Infections, Immune System, Inflammation and Diet. *International Journal of Environmental Research and Public Health*, *18*(5), 2455. DOI: 10.3390/ijerph18052455 PMID: 33801527

Sudevan, N., Manrai, M., Tilak, T. V. S. V. G. K., Khurana, H., & Premdeep, H. (2024). Chronic hepatitis B and occult infection in chemotherapy patients - evaluation in oncology and hemato-oncology settings: The CHOICE study. *World Journal of Virology*, *13*(1), 89104. DOI: 10.5501/wjv.v13.i1.89104 PMID: 38616860

Talha, B., & Dhamoon, A. S. (2024 Jan). Ritonavir. [Updated 2023 Aug 8] In *StatPearls* [Internet]. StatPearls Publishing., Available from https://www.ncbi.nlm.nih.gov/books/NBK544312/

Tian, W. J., & Wang, X. J. (2023). Broad-Spectrum Antivirals Derived from Natural Products. *Viruses*, *15*(5), 1100. DOI: 10.3390/v15051100 PMID: 37243186

Tian, Y., Hu, D., Li, Y., & Yang, L. (2022). Development of therapeutic vaccines for the treatment of diseases. *Molecular Biomedicine*, *3*(1), 40. DOI: 10.1186/s43556-022-00098-9 PMID: 36477638

Tian, Y., Ran, H., Wen, X., Fu, G., Zhou, X., Liu, R., & Pan, T. (2023). Probiotics improve symptoms of patients with COVID-19 through gut-lung axis: A systematic review and meta-analysis. *Frontiers in Nutrition*, *10*, 1179432. DOI: 10.3389/fnut.2023.1179432 PMID: 37284648

Tsai, M. C., Hung, C. H., Lu, S. N., Wang, J. H., Chen, C. H., Kee, K. M., Chang, K. C., Chao, T. L., & Hu, T. H. (2021). The incidence of resistance-associated variants to NS5A in HCV subtypes 1a and 1b in Taiwan. *Biomedical Journal*, *44*(6, Suppl 1), S126–S131. DOI: 10.1016/j.bj.2020.08.004 PMID: 35123932

Udeze, A. O., Olaleye, D. O., & Odaibo, G. N. (2020). Polymorphisms and drug resistance analysis of HIV-1 isolates from patients on first line antiretroviral therapy (ART) in Southeastern Nigeria. *PLoS One*, *15*(4), e0231031. DOI: 10.1371/journal.pone.0231031 PMID: 32267869

Vakili, B., Shoaei, P., Shahzamani, K., Siadat, S. D., Shojaei, H., Esfandiari, Z., Nasri, E., Shabani, S., Zamani Moghadam, A., & Ataei, B. (2024). Gut-Lung Microbiota Characterization in Patients with Non-Small Cell Lung Carcinoma and COVID-19 Coinfection. *Archives of Iranian Medicine*, *27*(2), 62–71. DOI: 10.34172/aim.2024.11 PMID: 38619029

Varghese, N., Majeed, A., Nyalakonda, S., Boortalary, T., Halegoua-DeMarzio, D., & Hann, H. W. (2024). Review of Related Factors for Persistent Risk of Hepatitis B Virus-Associated Hepatocellular Carcinoma. *Cancers (Basel)*, *16*(4), 777. DOI: 10.3390/cancers16040777 PMID: 38398168

Villar, L. M., Del Campo, J. A., Ranchal, I., Lampe, E., & Romero-Gomez, M. (2013). Association between vitamin D and hepatitis C virus infection: A meta-analysis. *World Journal of Gastroenterology*, *19*(35), 5917–5924. DOI: 10.3748/wjg.v19.i35.5917 PMID: 24124339

von Delft, A., Hall, M. D., Kwong, A. D., Purcell, L. A., Saikatendu, K. S., Schmitz, U., Tallarico, J. A., & Lee, A. A. (2023). Accelerating antiviral drug discovery: Lessons from COVID-19. *Nature Reviews. Drug Discovery*, *22*(7), 585–603. DOI: 10.1038/s41573-023-00692-8 PMID: 37173515

Wang, H. (2024). Practical updates in clinical antiviral resistance testing. *Journal of Clinical Microbiology*, *62*(8), e0072823. DOI: 10.1128/jcm.00728-23 PMID: 39051778

Wang, H., Zhao, Y., Pei, Z., Zhao, J., Zhang, P., Zhang, X., Zhang, Z., & Lu, W. (2024). Effect of Mixed Probiotics on Alleviating H1N1 Influenza Infection and Regulating Gut Microbiota. *Foods*, *13*(19), 3079. DOI: 10.3390/foods13193079 PMID: 39410114

Wang, J., Zhu, Q., Xing, X., & Sun, D. (2024). A Mini-Review on the Common Antiviral Drug Targets of Coronavirus. *Microorganisms*, *12*(3), 600. DOI: 10.3390/microorganisms12030600 PMID: 38543651

Wang, R., Gu, Y., Chen, H., Tian, B., & Li, H. (2025). Uracil base PCR implemented for reliable DNA walking. *Analytical Biochemistry*, *696*, 115697. DOI: 10.1016/j.ab.2024.115697 PMID: 39442604

Wang, R., Xiong, K., Wang, Z., Wu, D., Hu, B., Ruan, J., Sun, C., Ma, D., Li, L., & Liao, S. (2023). Immunodiagnosis - the promise of personalized immunotherapy. *Frontiers in Immunology*, *14*, 1216901. DOI: 10.3389/fimmu.2023.1216901 PMID: 37520576

Wang, Y., Ling, L., Zhang, Z., & Marin-Lopez, A. (2022). Current Advances in Zika Vaccine Development. *Vaccines*, *10*(11), 1816. DOI: 10.3390/vaccines10111816 PMID: 36366325

Wang, Y., Wei, L., Jiang, D., Cong, X., Fei, R., Xiao, J., & Wang, Y. (2005). In vitro resistance to interferon of hepatitis B virus with precore mutation. *World Journal of Gastroenterology*, *11*(5), 649–655. DOI: 10.3748/wjg.v11.i5.649 PMID: 15655815

Wang, Y. H., Limaye, A., Liu, J. R., & Wu, T. N. (2022). Potential probiotics for regulation of the gut-lung axis to prevent or alleviate influenza in vulnerable populations. *Journal of Traditional and Complementary Medicine*, *13*(2), 161–169. DOI: 10.1016/j.jtcme.2022.08.004 PMID: 36970463

Wei, Y., Zhao, Z., & Ma, X. (2022). Description of CRISPR-Cas9 development and its prospects in human papillomavirus-driven cancer treatment. *Frontiers in Immunology*, *13*, 1037124. DOI: 10.3389/fimmu.2022.1037124 PMID: 36479105

Winkler, D. A. (2024). Computational repurposing of drugs for viral diseases and current and future pandemics. *Journal of Mathematical Chemistry*, *62*(11), 2844–2879. DOI: 10.1007/s10910-023-01568-3

Woodson, S. E., & Morabito, K. M. (2024). Continuing development of vaccines and monoclonal antibodies against Zika virus. npj. *Vaccines*, *9*(1), 91. DOI: 10.1038/s41541-024-00889-x PMID: 38789469

World Health Organization. (2025, January 7). Disease outbreak news: Trends of acute respiratory infection, including human metapneumovirus, in the Northern Hemisphere. Retrieved from https://www.who.int/emergencies/disease-outbreak-news/item/2025-DON550

Xiao, B., Wu, L., Sun, Q., Shu, C., & Hu, S. (2024). Dynamic analysis of SARS-CoV-2 evolution based on different countries. *Gene*, *916*, 148426. DOI: 10.1016/j.gene.2024.148426 PMID: 38575101

Xu, Z., Su, C., Xiao, Y., & Wang, F. (2022). Artificial intelligence for COVID-19: Battling the pandemic with computational intelligence. *Intelligent Medicine*, *2*(1), 13–29. DOI: 10.1016/j.imed.2021.09.001 PMID: 34697578

Xuan, Q., Wang, J., Nie, Y., Fang, C., & Liang, W. (2024). Research Progress and Application of Miniature CRISPR-Cas12 System in Gene Editing. *International Journal of Molecular Sciences*, *25*(23), 12686. DOI: 10.3390/ijms252312686 PMID: 39684395

Yang, Y., Zheng, Y., Zou, Q., Li, J., & Feng, H. (2024). Overcoming CRISPR-Cas9 off-target prediction hurdles: A novel approach with ESB rebalancing strategy and CRISPR-MCA model. *PLoS Computational Biology*, *20*(9), e1012340. DOI: 10.1371/journal.pcbi.1012340 PMID: 39226304

Yao, Z. Q., Schank, M. B., Zhao, J., El Gazzar, M., Wang, L., Zhang, Y., Hill, A. C., Banik, P., Pyburn, J. S., & Moorman, J. P. (2024). The potential of HBV cure: An overview of CRISPR-mediated HBV gene disruption. *Frontiers in Genome Editing*, *6*, 1467449. DOI: 10.3389/fgeed.2024.1467449 PMID: 39444780

Yu, Z. J., Mosher, E. P., & Bumpus, N. N. (2021). Pharmacogenomics of Antiretroviral Drug Metabolism and Transport. *Annual Review of Pharmacology and Toxicology*, *61*(1), 565–585. DOI: 10.1146/annurev-pharmtox-021320-111248 PMID: 32960701

Zahedipour, F., Zahedipour, F., Zamani, P., Jaafari, M. R., & Sahebkar, A. (2024). Harnessing CRISPR technology for viral therapeutics and vaccines: From preclinical studies to clinical applications. *Virus Research*, *341*, 199314. DOI: 10.1016/j.virusres.2024.199314 PMID: 38211734

Zeng, M. C., Jia, Q. J., & Tang, L. M. (2021). *rpoB* gene mutations in rifampin-resistant *Mycobacterium tuberculosis* isolates from rural areas of Zhejiang, China. *The Journal of International Medical Research*, *49*(3), 300060521997596. DOI: 10.1177/0300060521997596 PMID: 33715498

Zhang, H., Wang, J., Zhang, K., Shi, J., Gao, Y., Zheng, J., He, J., Zhang, J., Song, Y., Zhang, R., Shi, X., Jin, L., & Li, H. (2024). Association between heavy metals exposure and persistent infections: The mediating role of immune function. *Frontiers in Public Health*, *12*, 1367644. DOI: 10.3389/fpubh.2024.1367644 PMID: 39104887

Zhang, K., Wan, P., Wang, L., Wang, Z., Tan, F., Li, J., Ma, X., Cen, J., Yuan, X., Liu, Y., Sun, Z., Cheng, X., Liu, Y., Liu, X., Hu, J., Zhong, G., Li, D., Xia, Q., & Hui, L. (2024). Efficient expansion and CRISPR-Cas9-mediated gene correction of patient-derived hepatocytes for treatment of inherited liver diseases. *Cell Stem Cell*, *31*(8), 1187–1202.e8. DOI: 10.1016/j.stem.2024.04.022 PMID: 38772378

Zhang, M., Wang, S., Wilffert, B., Tong, R., van Soolingen, D., van den Hof, S., & Alffenaar, J. W. (2018). The association between the NAT2 genetic polymorphisms and risk of DILI during anti-TB treatment: A systematic review and meta-analysis. *British Journal of Clinical Pharmacology*, *84*(12), 2747–2760. DOI: 10.1111/bcp.13722 PMID: 30047605

Zhang, M.-L., Li, H.-B., & Jin, Y. (2024). Application and perspective of CRISPR/Cas9 genome editing technology in human diseases modeling and gene therapy. *Frontiers in Genetics*, *15*, 1364742. DOI: 10.3389/fgene.2024.1364742 PMID: 38666293

Zhang, Q., Dong, G., Meng, W., Chen, Z., Cao, Y., & Zhang, M. (2022). Perceived Stress and Psychological Impact Among Healthcare Workers at a Tertiaty Hospital in China During the COVID-19 Outbreak: The Moderating Role of Resilience and Social Support. *Frontiers in Psychiatry*, *12*, 570971. DOI: 10.3389/fpsyt.2021.570971 PMID: 35281206

Zhang, Y., Wang, M., Huang, M., & Zhao, J. (2024). Innovative strategies and challenges mosquito-borne disease control amidst climate change. *Frontiers in Microbiology*, *15*, 1488106. DOI: 10.3389/fmicb.2024.1488106 PMID: 39564491

Zhen, S., Qiang, R., Lu, J., Tuo, X., Yang, X., & Li, X. (2023). CRISPR/Cas9-HPV-liposome enhances antitumor immunity and treatment of HPV infection-associated cervical cancer. *Journal of Medical Virology*, *95*(1), e28144. DOI: 10.1002/jmv.28144 PMID: 36121194

Zhou, Y., Peng, S., Wang, H., Cai, X., & Wang, Q. (2024). Review of Personalized Medicine and Pharmacogenomics of Anti-Cancer Compounds and Natural Products. *Genes*, *15*(4), 468. DOI: 10.3390/genes15040468 PMID: 38674402

Zhou, Y., Peng, S., Wang, H., Cai, X., & Wang, Q. (2024). Review of Personalized Medicine and Pharmacogenomics of Anti-Cancer Compounds and Natural Products. *Genes*, *15*(4), 468. DOI: 10.3390/genes15040468 PMID: 38674402

Žigrayová, D., Mikušová, V., & Mikuš, P. (2023). Advances in Antiviral Delivery Systems and Chitosan-Based Polymeric and Nanoparticulate Antivirals and Antiviral Carriers. *Viruses*, *15*(3), 647. DOI: 10.3390/v15030647 PMID: 36992356

Chapter 4
Antiviral Drug Discovery in Emerging Economics: Challenges and Opportunities

Abhiruj Navabhatra
https://orcid.org/0000-0003-4129-1302
College of Pharmacy, Rangsit University, Thailand

Chonnipha Yangchaiya
College of Pharmacy, Rangsit University, Thailand

ABSTRACT

Emerging infectious diseases underscore the urgent need for new antiviral drugs because of their significant impact on global health. Although important advancements such as HAART for HIV and DAAs for hepatitis C have been made, challenges such as drug resistance, limited vaccine effectiveness, and complex regulatory landscapes persist. Traditional drug discovery methods are resource intensive, prompting a shift toward innovative techniques such as PROTACs, RIBOTACs, CRISPR-based strategies, and nanotechnologies, which address issues of resistance and drug delivery. Financial and regulatory challenges remain critical, particularly in developing countries where neglected diseases are prevalent. Collaborative efforts, including public–private partnerships, have successfully addressed funding gaps, as demonstrated during the COVID-19 pandemic. This chapter explores innovative antiviral approaches, including RNA-based therapies and biomaterials, while emphasizing sustainable development, regulatory reforms, and capacity building to address challenges in resource-limited settings.

DOI: 10.4018/979-8-3693-9276-8.ch004

Copyright © 2025, IGI Global Scientific Publishing. Copying or distributing in print or electronic forms without written permission of IGI Global Scientific Publishing is prohibited.

INTRODUCTION

New infectious diseases continue to pose significant threats to global health, accounting for 20% of deaths worldwide, with viruses responsible for 70% of these fatalities (Collaborators, 2022). Historical outbreaks, such as smallpox, the 1918 Spanish flu, and HIV, have had catastrophic impacts, and the COVID-19 pandemic has emerged as one of the deadliest in recent history (Szliszka et al., 2009). Despite advances in medical science, current treatment and prevention strategies, including antiviral therapies and vaccines, are limited by the rise of drug-resistant strains and limited vaccine efficacy against emerging viral variants. These challenges underscore the urgent need for innovative approaches in antiviral drug discovery (Kiso et al., 2024). Conventional approaches to drug discovery, such as random screening, require considerable time and resources. Consequently, the development of novel strategies has become imperative (Gangwal et al., 2024). Approaches such as proteolysis-targeting chimeras (PROTACs), ribonuclease-targeting chimeras (RIBCACs), targeted covalent inhibitors, and advanced drug delivery systems offer promising alternatives (Zakharova & Chudinov, 2024).

Recent research has increasingly focused on viral and host targets, yet the complete eradication of viral infections remains a challenge, with hepatitis C being a notable exception owing to the success of direct-acting antivirals. Antiviral drug development generally involves two key strategies: targeting viral proteins, which has led to the approval of over 80 antiviral drugs, albeit with issues of drug resistance, and targeting host factors, which offer broad-spectrum potential but carry a greater risk of adverse effects (He et al., 2024). The successful development of antivirals requires innovative collaborative models, such as public–private partnerships and consortia. Initiatives such as DNDi and the Bill and Melinda Gates Foundation's Tuberculosis Drug Accelerator highlight the effectiveness of such collaborations, and they are expected to significantly increase the efficiency of antiviral drug discovery (Adamson et al., 2021; Namba-Nzanguim et al., 2022). This chapter aims to provide a comprehensive analysis of contemporary antiviral drug development, with a particular focus on strategies to address a range of viral infections, including Ebola virus disease (EBOD), influenza, hepatitis, and COVID-19. The chapter identifies key targets for antiviral drug discovery and critically examines the challenges, potential solutions, and methodologies aimed at accelerating the development of effective antiviral therapies.

CHALLENGES IN ANTIVIRAL DRUG DISCOVERY

Drug discovery and development are lengthy and expensive and are made more challenging by the high failure rate of compounds in preclinical research. Only three out of ten drugs that reach clinical trials recover their initial investment. Therefore, an efficient drug discovery pipeline is crucial for profitability and continued investment in R&D. Efforts to control drug costs—driven by governments, industry, and civil society—further strain profit margins, contributing to the lack of focus on infectious disease research in pharmaceutical R&D (Adamson et al., 2021). Antiviral research has traditionally prioritized viruses of economic importance in developed nations, but the COVID-19 pandemic has revealed how emerging viruses can severely impact global economies and disrupt daily life (Mofijur et al., 2021).

Financial Constraints: Funding and investment challenges

Despite advancements in antiviral drugs, diseases such as cutaneous leishmaniasis (CL), which primarily affects developing countries, remain neglected. There is no safe or effective treatment for CL, and developing such therapies is costly and time-consuming, with estimates ranging from \$1.4–\$2.87 billion (Surur et al., 2020). Pharmaceutical companies, driven by profits, tend to prioritize diseases in high-income countries, creating a "fatal imbalance" in drug development. Drug research and development are complex, expensive, and slow, even with ample resources. Over 90% of drug candidates fail between phase I trials and regulatory approval, resulting in significant financial losses. During pandemics, factors such as access to high-quality chemical compounds, calculated risks, revised therapeutic profiles, and adequate funding can accelerate the process, alongside strong organizational commitment (Sun et al., 2022). For COVID-19, early animal models, such as ferrets and hamsters, were used but posed challenges. Later, mouse models proved more accessible and cost-effective, and nonhuman primates helped predict vaccine efficacy. A major hurdle in antiviral discovery is the limited availability of high-quality chemical compounds. The rapid development of antivirals such as remdesivir, molnupiravir, and nirmatrelvir shows the value of leveraging existing compounds to speed up the process (Figure 1) (von Delft et al., 2023). Future pandemic preparedness should focus on developing chemical probes that target viral proteins, as the viral proteome is smaller and requires less investment than the cost of a pandemic. Pharmaceutical development is shaped by a cycle of innovation, return on investment, and transition to generics. Patents, especially "composition of matter" patents, protect investments but can limit access to therapies, particularly in low- and middle-income countries (LMICs). Trade secrets can stifle innovation, especially where multiple therapies are needed for viral threats or resistance. The current pharmaceutical market model has flaws, including high prices during patent monopolies, which limit access in LMICs and for rare diseases. With respect to antimicrobial resistance (AMR), new antibiotics are not available commercially due to their low price and sales volume, despite the urgent need. The lack of investment in AMR and pandemic-related diseases before COVID-19 highlights the limitations of the current pharmaceutical model. Structural changes are necessary to ensure more resilient and equitable drug development for future global health challenges (Namba-Nzanguim et al., 2022).

Figure 1. The molecular structures of antiviral drugs that were rapidly developed and distributed in response to the COVID-19 pandemic

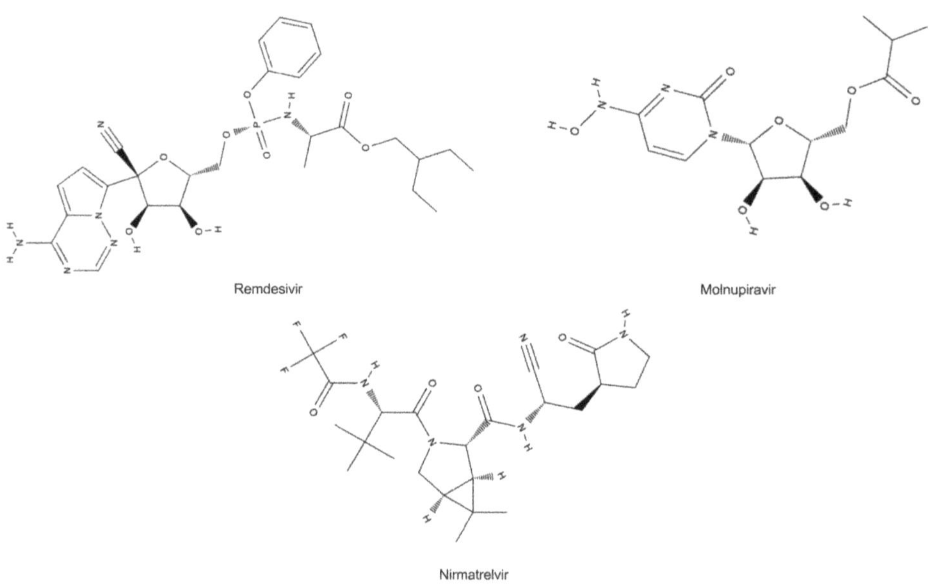

Advances and Challenges in Antiviral Drug Development and Biomaterial Applications

The presence of viral infections dates back to early human civilization, with scientific advancements, particularly in virus visualization, greatly contributing to the progress of antiviral therapy. The chemical synthesis of antiviral drugs has significantly strengthened the fight against viral diseases. However, the development and application of new antiviral agents remain complex and are often limited by issues such as drug resistance and cross-resistance (Du et al., 2024). As viruses are obligate intracellular parasites that are dependent on the host cell machinery for replication, achieving effective targeting without damaging host cells is difficult. Current antiviral therapies are often limited by suboptimal efficacy, resistance, and adverse effects, necessitating innovative approaches. Decades of biomaterials research have produced numerous functional therapeutics, including antiviral biomaterials, which are drawing increasing attention owing to their distinctive properties (Matthew et al., 2021). The ongoing emergence and mutation of viruses such as SARS-CoV, MERS-CoV, and SARS-CoV-2 suggest that biomaterials could serve as valuable complementary tools to traditional antiviral therapies. Engineered biomaterials, such as polymer chains, nanofibers, hydrogels, carbon dots, and metal nanoparticles, can capture viruses or inhibit viral entry by interacting with viral surface proteins (Table 1). Even if glycopeptide- and polysaccharide-modified biomaterials show good results in laboratory tests, their effectiveness in actual living systems is still not clear. The reversible nature of virus–biomaterial interactions underscores the need to develop irreversible binding materials that can be effectively cleared by the immune system for enhanced efficacy. Metal nanoparticles offer broad-spectrum antiviral properties but are limited by several challenges, including long-term toxicity and unknown clearance mechanisms. The translation of these technologies to clinical use demands safe, eco-friendly production and comprehensive knowledge of their safety, biodistribution, and pharmacokinetics. Biomaterials often require antiviral drug incorporation to be functional, offering design flexibility for reduced toxicity and controlled release. Further research into targeted delivery and *in vivo* behavior is critical, as infection-induced vascular permeability may support targeted delivery approaches. Combining biomaterials with diverse antiviral mechanisms could enhance treatment efficacy. Despite these advancements, significant challenges, such as safety evaluations and potential long-term effects, hinder their clinical application. Nontoxic agents such as modified cyclodextrins show promise. However, most biomaterials have not advanced beyond *in vitro* stages, with conventional antivirals continuing to dominate clinical treatment. Addressing COVID-19 requires understanding the evolving nature of SARS-CoV-2 and developing drugs that target viral entry and replication. Promising candidates such as EKL1C and GC-376 exhibit potential but require further validation. Comprehensive preclinical studies employing diverse animal models and assessing variant efficacy are essential for advancing antiviral drug development. The screening of new drugs should focus on evaluating target effectiveness and safety and incorporating traditional therapeutic approaches. The global variability of virus strains necessitates the use of multiple animal models to better simulate human infection and optimize treatment outcomes (Sha et al., 2023; Varahachalam et al., 2021).

Table 1. Examples of biomaterials that have been developed with strategies to exhibit antiviral properties (Kumari and Chatterjee, 2021)

Materials	Viruses tested	Results
Al 6063 Alloy Surfaces (Nanostructured, 23 nm ± 2 nm)	Respiratory syncytial virus (RSV), rhinovirus (RV)	More effective against RV than RSV; observed a 3–4 log10 reduction in virus viability.
Al 6063 Alloy Surfaces (Nanostructured, 23 nm ± 2 nm)	Severe acute respiratory syndrome coronavirus 2 (SARS-CoV-2)	Demonstrated antiviral activity with a 5-log reduction in viral viability.
PTFE Nanoparticles Sintered to PP Microfibers	Adenovirus type 4 and 7a	Reduced attachment of adenoviruses by 99.2–99.6%.
Polyethylenimines (PEIs) Derivatives (Hydrophobic Polycationic Coatings)	Influenza A virus (H1N1)	Killed influenza virus with 100% efficiency; achieved a 4-log reduction in viral load.
PEIs (Hydrophobic Polycationic Coatings)	Poliovirus, rotavirus	Disinfected aqueous solutions of enveloped and nonenveloped viruses effectively.
QACs (Hyperbranched Polymers with Quaternary Amines)	Poliovirus Sabin1, Influenza A (H1N1)	Demonstrated virucidal activity against enveloped influenza viruses.
QBESt, QBAm (Cationic Polymeric Coatings, Zwitterionic Structure)	Influenza A (H1N1) virus	Reduced viral burden by over 1,000-fold within 30 minutes.
Copper (CuO Nanoparticles Impregnated in N95 Masks)	Human influenza A (H1N1), avian influenza	Exhibited potent antiviral and biocidal properties.
Silver (PHBV Films Coated with AgNP Fibers)	Feline calicivirus (FCV), murine norovirus (MNV)	No infectious FCV detected; MNV titers reduced by 0.86 log10.

Infrastructure Weaknesses: Operational and Logistical Hurdles

Operational and logistical challenges present significant impediments to the efficiency of antiviral drug discovery. The identification of effective viral targets is inherently complex, given the intricate nature of viral structures and life cycles (Shi et al., 2021). Addressing these complexities necessitates extensive research and development efforts, which are both resource intensive and costly. The drug development process encompasses multiple stages, including discovery, preclinical testing, clinical trials, and regulatory approval. Each stage poses distinct challenges that can result in delays and increased expenditures (Reddy & Singh, 2023). For example, the clinical development of antiviral agents requires rigorous proof of concept for antiviral activity, often demonstrated through the short-term administration of investigational drugs to individuals with chronic infections (von Delft et al., 2023). Additionally, the emergence of new viral germs, such as SARS-CoV-2, highlights the urgent need for quick creation and use of antiviral drugs. However, the existing drug discovery pipeline remains slow and inefficient, emphasizing the critical need for innovative strategies to accelerate the development of effective antiviral agents. In conclusion, the operational and logistical barriers to antiviral drug discovery are multifaceted, encompassing scientific, technical, and procedural challenges that collectively hinder the timely development of antiviral medications (Khorram-Manesh et al., 2024). Addressing these challenges will require a concerted effort from researchers, policymakers, and industry stakeholders to establish a robust framework that ensures global preparedness for future viral threats.

Regulatory and Bureaucratic Obstacles: Navigating Complex Regulatory Environments

Navigating the complex regulatory landscape in antiviral drug discovery presents considerable challenges that impede the timely development and deployment of effective therapies. The stringent requirements imposed by regulatory agencies, although essential for ensuring drug safety and efficacy, often result in extended timelines and increased costs (G. R. Painter et al., 2021). For example, the clinical development of antiviral agents demands rigorous proof of concept for antiviral activity, which is typically achieved through the short-term administration of investigational drugs to chronically infected individuals. The COVID-19 pandemic has highlighted the urgent need for rapid therapeutic interventions (Mei & Tan, 2021; Tompa et al., 2021). However, the imperative to develop and deploy antiviral therapies often clashes with the meticulous and time-intensive regulatory processes currently in place. This incongruity underscores the necessity for adaptive regulatory frameworks capable of expediting the approval of antiviral drugs while maintaining robust safety standards (Chitalia & Munawar, 2020). Innovative strategies, such as drug repurposing, have emerged as practical approaches to accelerate antiviral discovery (von Delft et al., 2023). By identifying new therapeutic applications for existing drugs, this approach circumvents certain early-stage development hurdles, facilitating more expedient transitions to clinical evaluation. Moreover, the integration of artificial intelligence (AI) and machine learning (ML) within the drug discovery pipeline holds significant potential for optimizing regulatory processes. These technologies increase the efficiency of identifying viable antiviral candidates and predicting their safety profiles, thereby enabling more informed and data-driven regulatory decisions (Robinson, 2021). In conclusion, while regulatory and bureaucratic barriers are inherent to the antiviral drug discovery process, the adoption of flexible regulatory strategies and the incorporation of advanced technological tools can effectively address these challenges. Such initiatives are essential for bolstering global preparedness and ensuring timely and effective responses to emerging viral threats. Figure 2 provides a visual representation of the challenges associated with antiviral drug discovery, highlighting key obstacles in the development process.

Figure 2. Illustration of the challenges associated with antiviral drug discovery

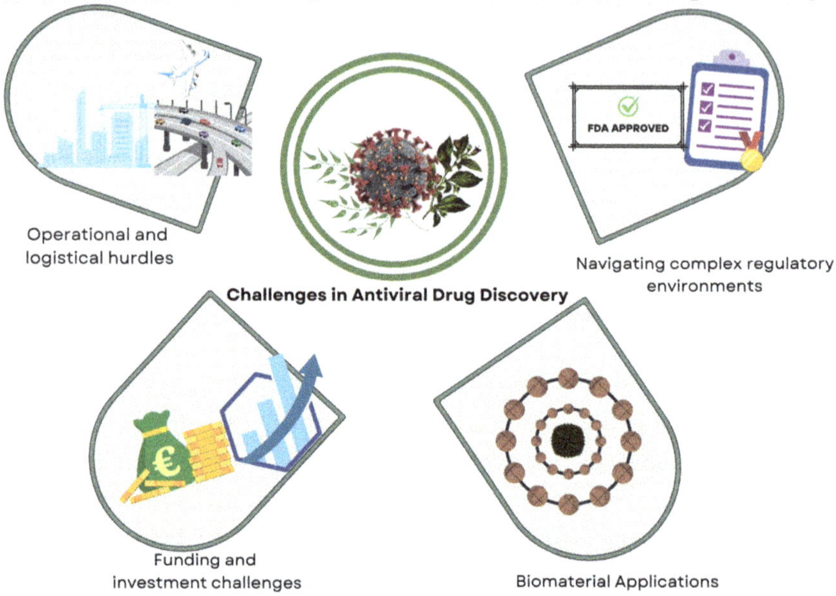

CURRENT LANDSCAPE OF ANTIVIARAL RESEARCH IN EMERGING ECONOMICS

Overview of Ongoing Research and Development Efforts

Historical examples demonstrate the effectiveness of interdisciplinary collaboration in the development of drugs and vaccines, highlighting the need for similar global efforts to prepare for future pandemics. Collaborative research involving the academic, government, industry, and philanthropic sectors is essential for identifying gaps, enhancing coordination, and aligning scientific progress with commercial goals. Models such as CARB-X illustrate successful approaches for integrated pandemic preparedness. Effective response requires established regulatory partnerships and contributions from industry in manufacturing, quality control, and safety (Meganck & Baric, 2021). While intellectual property can hinder development, adaptable management frameworks can balance rights with equitable global access. The severe impact of the COVID-19 pandemic underscores the importance of a coordinated preparedness strategy, as past outbreaks of SARS-CoV and MERS-CoV emphasize the urgency of proactive measures. Comprehensive global efforts and broad-spectrum antivirals are vital for future security. Small interfering RNAs (siRNAs) show great promise as a novel class of antiviral agents because of their ability to selectively target and degrade viral RNA within host cells, preventing replication. As part of the RNA interference (RNAi) pathway, siRNAs harness a natural cellular process, offering high specificity and targeting previously "undruggable" genes through posttranscriptional gene silencing (PTGS) and transcriptional gene silencing (TGS). Despite evidence of their antiviral efficacy, siRNAs have progressed slowly in clinical trials. While the U.S. FDA has approved six siRNA therapies for metabolic disorders, no antiviral siRNA has completed clinical testing or received regulatory approval. Challenges include delivery, cell targeting, stability, off-target effects, and market feasibility. Alnylam's ALN-RSV01, the first antiviral siRNA in clinical trials, did not meet its primary endpoint, yet research and development continue to fuel optimism for its future use. Ongoing research focuses on overcoming barriers to translating preclinical success into clinical application. Although preclinical data show strong efficacy, the clinical results have been limited and mixed across 17 trials. While small interfering RNA (siRNA) therapies are successful at treating metabolic diseases, antiviral use poses unique challenges. Recent advances in delivery and stability suggest that progress is being made in addressing these issues. mRNA vaccines such as Pfizer-BioNTech's BNT162b2 and Moderna's mRNA-1273 have proven effective, although variant-specific responses remain critical. Long-term studies are needed to assess safety and immunity durability. Research is advancing mRNA technologies through novel delivery methods, optimized sequences, and scalable production (Chokwassanasakulkit et al., 2024; Kumari et al., 2022). The biphasic nature of COVID-19 necessitates targeted therapies for the viral and inflammatory phases. Variants have led to shortages of treatments, requiring strategic distribution and adaptation of therapeutic use. The receptor-binding domain (RBD) of SARS-CoV-2 remains a focal point for vaccine and antibody development, although mutations continue to challenge immunity. Ongoing analyses are key for guiding next-generation vaccine strategies. Repurposing existing drugs has proven practical, with treatments such as molnupiravir and paxlovid gaining approval. Agreements ensuring royalty-free sales during the pandemic have supported equitable access. Effective COVID-19 treatments should prevent severe disease and remain effective against variants. The emergence of highly transmissible coronaviruses, including SARS, MERS, and COVID-19, highlights the need for sustained investment in surveillance, public health, vaccination, clinical care, and research. As recovery progresses, addressing long-term effects and advancing broad-spectrum treatments are essential for maintaining global health resilience (G. R. Painter et al., 2021; W. P. Painter et al., 2021). Table 2 lists examples of the current strategies being employed in antiviral drug discovery for COVID-19 (Mei & Tan, 2021).

Table 2. Examples of the current strategies for antiviral drug discovery for COVID-19

Strategy	Description	Examples/Notes
Empirical Drug Repositioning	Repurposing existing drugs with proven safety profiles for COVID-19 treatment.	A direct-acting antiviral (e.g., Remdesivir) and a host-targeting agent (e.g., Ivermectin, chloroquine) fall into this category.
Direct-Acting Antivirals (DAAs)	Target viral proteins to inhibit replication and spread.	Examples: Remdesivir (targets RNA polymerase), Protease inhibitors like lopinavir-ritonavir.
Host-Targeting Antivirals	Modulate virus–host interactions to block viral entry or replication.	Examples: ACE2 inhibitors (e.g., glycyrrhizin), TMPRSS2 inhibitors (e.g., camostat mesylate).
High-Throughput Screening	Screening large libraries of compounds for activity against SARS-CoV-2.	Examples: apilimod, MDL-28170, imatinib identified through screening.
Artificial Intelligence	AI-assisted discovery to identify potential drug candidates and mechanisms.	Examples: Drugs like Baricitinib (Janus kinase inhibitor), network medicine approaches suggesting combinations.
Therapeutic Antibodies	Neutralizing antibodies targeting the spike protein to block viral entry.	Examples: REGN-COV2 cocktail, LY-CoV555, JMB2002.
Immuno-Regulators	Target host immune response to reduce inflammation and cytokine storm effects.	Examples: Dexamethasone, Tocilizumab (IL-6 inhibitor).
Traditional Chinese Medicine	Use of plant-based natural compounds with antiviral properties.	Examples: Lianhuaqingwen capsule, compounds like Baicalin and Scutellarin.

Case Studies: Successes and Setbacks

Antiviral research has achieved significant advancements, exemplified by transformative breakthroughs such as highly active antiretroviral therapy (HAART) for HIV, which has markedly reduced AIDS-related mortality. Nevertheless, substantial challenges persist, as evidenced by the limited efficacy of certain treatments. For example, high-dose oral favipiravir did not significantly improve early COVID-19 clearance (Luvira et al., 2024). Despite the notable progress of antiretroviral therapy (ART) in suppressing HIV replication and preserving $CD4^+$ T-cell function, the quest for a definitive cure remains unfulfilled. The sterilizing cure of the Berlin patient has catalyzed sustained research efforts, yet setbacks, such as those documented in the Boston and Essen patients, underscore the intricate challenges inherent in HIV eradication. While these outcomes may be viewed as setbacks, they have nonetheless yielded critical insights that continue to refine therapeutic strategies. Innovative therapeutic modalities, including chemotherapy and hematopoietic stem cell transplantation utilizing CCR5 Δ32 mutation-matched donors, as exemplified by London and Düsseldorf patients, underscore the potential for novel interventions. Importantly, however, these approaches are not universally applicable. Recent advancements, such as long-acting slow-effective release ART (LASER ART) and gene-editing technologies such as CRISPR/Cas9, have aimed to address persistent challenges associated with HIV latency and viral reservoirs. Despite their promise, significant hurdles remain, including the virus's ability to persist in latent cellular compartments and anatomical reservoirs, as well as the necessity of replacing infected cells with HIV-resistant variants. Furthermore, efforts to restore normal immune function are impeded by residual inflammation and coagulation abnormalities following treatment. One of the most notable recent developments is the intact proviral DNA assay (IPDA), which facilitates the estimation of intact proviral DNA from a limited number of $CD4^+$ T cells. While this tool represents a significant leap forward, questions regarding its precision and practical application remain unresolved. Emerging cure strategies, such as bone marrow transplantation, the "kick-and-kill" method, CRISPR/Cas9-mediated genome editing, and the "block-and-lock" approach, offer considerable promise. Combination therapies, in particular, are highly encouraging avenues for achieving comprehensive viral eradication. The pursuit of an HIV cure, while formidable, is buoyed by cases such as the Berlin patient and the study of elite controllers (ECs)—individuals who naturally suppress viral replication owing to unique host factors and specific HLA alleles. ECs provide a compelling model for developing therapies aimed at achieving either a disease-free or treatment-free state (Kalidasan & Theva Das, 2020). In the context of emerging economies, antiviral research exemplifies a dynamic interplay of challenges and innovations, particularly in response to COVID-19 and other viral threats. Efforts have concentrated on the repurposing of FDA-approved pharmaceuticals and the exploration of novel therapeutic approaches, including gene editing and immunomodulators, to address immediate public health needs while simultaneously increasing the capacity for future preparedness (Barghash et al., 2024). Many advances have also been made in the development of antiviral vaccines and drugs against respiratory viruses such as SARS-CoV-2 and influenza. Lessons gleaned from past outbreaks have informed the design of proactive strategies to mitigate the impact of future pandemics (Muthukutty et al., 2024). The advancement of broad-spectrum antiviral therapies and the strategic repurposing of existing drugs highlight the robust framework in place for addressing emerging viral diseases. RNA-based therapeutics and combination treatment regimens have become pivotal in combating both current and future viral threats. Enhanced epidemiological surveillance systems have further strengthened early detection and rapid response capabilities, emphasizing the critical importance of integrating diagnostics, therapeutic interventions, and public health measures. Collectively, these efforts underscore the adaptability and continued evolution of antiviral research within emerging economies, highlighting both achievements and areas requiring further innovation. This comprehensive approach not only addresses present challenges but also fortifies global health security, ensuring preparedness for future threats.

OPPORTUNITIES FOR GROWTH AND INNOVATION

Leveraging Local Biodiversity and Traditional Knowledge

Leveraging local biodiversity and traditional knowledge presents significant potential for advancing antiviral drug research and discovery, especially in regions endowed with unique flora and fauna. Biodiversity in tropical and subtropical regions, for example, harbors a rich array of bioactive compounds with promising antiviral properties (Adegboye et al., 2021). Traditional knowledge systems, which have utilized these natural resources for medicinal purposes over generations, offer invaluable insights that can guide researchers in the identification of effective compounds for antiviral development (Mehta & Pradhan, 2024). Empirical studies have demonstrated that natural compounds derived from plants, fungi, and marine organisms exhibit antiviral activity against pathogens such as HIV, influenza, and coronaviruses, highlighting the largely untapped potential of local ecosystems (Riccio et al., 2020). Engaging with indigenous communities is essential to ensure adherence to ethical practices, equitable benefit sharing, and the preservation of traditional knowledge. Such partnerships foster a respectful exchange of knowledge and support sustainable harvesting practices, thereby safeguarding biodiversity while advancing scientific research (Albuquerque et al., 2022). Furthermore, the integration of modern techniques—such as bioinformatics and high-throughput screening—with traditional knowledge can expedite the discovery process, increasing both its efficiency and specificity. By focusing on the unique resources within local ecosystems and respecting indigenous expertise, researchers can significantly expand the pipeline for antiviral agents. This approach not only enhances antiviral research but also supports biodiversity conservation and cultural preservation, providing a holistic framework for sustainable drug discovery (Namba-Nzanguim et al., 2022; Tiwari & Singh, 2023).

Public–private Partnerships: A Gateway for Innovation

Public–private partnerships (PPPs) represent a pivotal mechanism for fostering innovation in antiviral research and development. By combining resources, expertise, and infrastructure, PPPs enable the public and private sectors to address the high costs and inherent risks associated with antiviral drug discovery collaboratively, which often discourages investment in this essential field (Moon et al., 2022). Such collaborations integrate the foundational scientific knowledge of public research institutions with the advanced technological capabilities, clinical trial expertise, and efficient manufacturing processes offered by private companies (Davis et al., 2021). Evidence suggests that PPPs can accelerate drug development by facilitating early-stage research, streamlining regulatory pathways, and enhancing the scalability of production. Additionally, PPPs promote knowledge sharing and foster an open innovation environment, allowing stakeholders to exchange data, insights, and intellectual property in ways that expedite scientific breakthroughs while minimizing duplication of effort (Barbosu, 2024). Notably, high-profile partnerships established during the development of COVID-19 vaccines have underscored the transformative potential of PPPs, demonstrating how shared objectives can lead to substantial progress in addressing global health challenges (Cook & Lauer, 2021). Moreover, by creating pathways to equitable access, PPPs can help reduce healthcare disparities, particularly in low- and middle-income countries, ensuring that antiviral treatments reach those in greatest need (Ghasemi et al., 2022). Through carefully structured partnerships, stakeholders across the public and private sectors can foster a sustainable innovation ecosystem, thus paving the way for the development of next-generation antiviral therapies to address both existing and emerging viral threats (Samira et al., 2024).

International Collaboration and Support

Since 2005, the world has faced numerous significant public health emergencies, with the COVID-19 pandemic having the most profound health and economic impacts. These crises reveal that actions by individual nations can have extensive global repercussions, highlighting the necessity for coordinated international responses that outperform unilateral efforts. The EU's response to COVID-19 exemplifies the tension between national interests and the need for sustained cooperation to achieve global benefits. While the EU has shown commendable governance and leadership through initiatives such as a proposed pandemic treaty, reconciling the diverse priorities of member states with collective interests has proven challenging, particularly in health policy, often leading to political indecision and conflicting actions. Future threats, such as new variants of SARS-CoV-2, will continue to test global resilience. The post-COVID-19 era requires addressing setbacks to reduce poverty and inequality, as health systems and resources have been severely strained, necessitating significant recovery time, especially if austerity measures are imposed. Robust multilateral collaboration is essential for mitigating such shocks and rethinking governance. COVID-19 and previous pandemics emphasize the value of multilateralism in research, knowledge sharing, vaccine distribution, and travel regulation. Effective multilateral action, however, depends on principles such as trust-based relationships and mutual responsibility, which are often lacking, as shown by countries taking transactional approaches, such as export restrictions. The absence of robust frameworks has forced reliance on economically damaging measures such as travel bans. The pandemic revealed the need for adaptable coordination mechanisms and proactive preparation for future crises. The success of frameworks such as the International Health Regulations demonstrates that enhanced governance in multilateral institutions is crucial, facilitating contributions and support aligned with countries' capacities and needs. The EU is advancing regional collaboration through the European Health Union and initiatives such as HERA, which are designed for emergency preparedness and coordination with bodies such as the EMA, ECDC, and WHO. This leadership, mirrored in regional actions globally, underscores a growing focus on collective health security and preparedness (Anderson et al., 2021; Jit et al., 2021).

Innovative Funding Models

Drug discovery and development are lengthy, costly, and face high attrition rates, with only a few drugs recouping capital investment after clinical trials. Thus, an efficient drug discovery pipeline is essential to ensure profitability and sustain R&D investment. Market pressures to contain drug prices, combined with fragmented funding, often deprioritize infectious disease research, with a focus on economically significant viruses. COVID-19, however, has underscored the profound impact of emerging viruses on global economies and society. The limited number of approved antivirals, concentrated on a narrow range of viruses, reveals the need for comprehensive funding across the drug development continuum. Short-term funding and limited support for neglected viruses hinder drug advancement. To enhance discovery, the field should explore new starting points, such as natural products. COVID-19 has fostered public–private partnerships, exemplified by the COVID-19 therapeutic accelerator and agreements such as Gilead–Cipla's affordable generic remdesivir for developing countries. COVID-19 has also exposed systemic vulnerabilities in Africa, highlighting the need for local R&D and clinical trials. The pandemic has shifted life sciences funding, with increased government and philanthropic investments in areas traditionally funded by the private sector. This shift's endurance remains uncertain. Traditional industry-led R&D, often focused on niche markets, has revealed limitations, and the pandemic has accelerated public support for critical health innovations, such as those in the NIH Small Business Innovation Research Program (Robinson, 2021; Robinson, 2023). The study by Bhaven et al. (2021) analyzes how the COVID-19 pandemic reshaped biomedical innovation policies and mechanisms, detailing the pandemic's impact on traditional "push" (research funding) and "pull" (private sector incentives) strategies, which led to transformative changes in public–private collaboration for technology development. Key findings include shifts in funding priorities, with pre-COVID-19 biomedical innovation relying mainly on government funding for early-stage research, whereas the private sector managed late-stage development driven by patent-based profits. The pandemic, however, prompted governments and nonprofits to directly fund late-stage research, manufacturing, and procurement, moving away from patent incentives—a shift exemplified by the U.S. "Operation Warp Speed." Additionally, global collaboration and funding models, including NIH-led ACTIV and BARDA partnerships, helped coordinate research and reduce redundancy, whereas international bodies such as CEPI and COVAX focused on equitable vaccine distribution, despite noted coordination gaps with major players such as China. The pandemic also revealed a shift in the role of patents; traditionally, patents incentivized private sector investment, but their importance waned under the urgent need for rapid response. Governments employed advance purchase agreements to secure vaccine doses, bypassing patents, with COVAX pooling resources to ensure affordable vaccine access for low- and middle-income countries. Further challenges arose in manufacturing and access, where limitations in knowledge transfer and manufacturing capacity highlighted the need for complex technology-sharing frameworks, as limited infrastructure in some regions raised concerns over global vaccine accessibility. For future biomedical innovation, the authors propose that elements of the COVID-19 response, such as advance purchase agreements and increased public investment in late-stage research, could model future health crises, offering a coordinated approach to addressing high-priority health challenges. Ultimately, the article suggests that the COVID-19 innovation system might serve as a blueprint for balancing public health priorities with efficient innovation, advocating for a lasting reevaluation of global biomedical research and access strategies (Sampat & Shadlen, 2021).

CASE STUDIES OF INNOVATIVE APPROACHES

Open-source Platforms in India and China

During the COVID-19 pandemic, while the world focused on vaccines, competition unfolded between India and China in the pharmaceutical sector. India retained its title as "the pharmacy of the world," supplying 60,000 generic brands and accounting for 40% of U.S. generics, 25% of U.K. medicines, and 80% of antiviral drugs for AIDS. However, despite its extensive pharmaceutical infrastructure, India lags behind China in innovative drug development and new chemical entity (NCE) research, ranking 40th in the Global Innovation Index 2023 compared with China's 12th position (Kumar, 2024). India's progress in drug innovation began after joining the World Trade Organization in 1995 and introducing product patents in 2005. Major Indian pharmaceutical companies embark on innovative R&D, but high risks, costs, and limited industry-academia collaboration hinder sustained progress, leading many companies to scale back or exit their R&D efforts. India recently secured approvals for enmetazobactam, a β-lactamase inhibitor, from both the EMA and U.S. FDA, marking a notable achievement (Singh, 2023). However, China remains ahead, contributing to 10% of new drug launches worldwide in 2018 and securing the approval of six U.S. FDAs. China's dominance in scientific research was further highlighted in 2022 when it became the largest contributor to high-quality natural science research, while India ranked 10th (CICC Research, 2024). Despite investing less than 1% of its GDP in R&D, compared with China's 2.4% and the U.S.'s 3.4%, India continues to show potential in advancing its pharmaceutical sector (Kumar, 2024). India's potential in drug discovery remains significant, with the resources and talent to reassert itself in innovative drug development. Renewed focus and investment could help India achieve its goal of becoming a global leader in pharmaceutical innovation while maintaining its dominance in generic medicine production. India's drug discovery efforts gained recognition in 1997 when Dr. Reddy's Laboratories (DRL) introduced its first preclinical small molecule, a PPARγ agonist for diabetes, which was later licensed to Novo Nordisk as balaglitazone (DRF 2593). This marked the first time an Indian pharmaceutical company licensed a new chemical entity (NCE) to a Western firm, inspiring other Indian companies to start NCE research programs. DRL's second molecule, ragaglitazar, was also licensed to Novo Nordisk, but both drugs were eventually discontinued because of toxicological concerns. DRL, however, became the first Indian company to advance an in-house NCE to phase III clinical trials with balaglitazone, although the program was later halted over safety risks similar to those of GSK's rosiglitazone. Despite these challenges, DRL's early successes and the 2005 introduction of India's product patent law encouraged major Indian pharmaceutical companies—including Ranbaxy, Lupin, Sun Pharma, and Zydus—to launch drug discovery initiatives. Zydus Life Sciences notably succeeded in developing saroglitazar (ZYH1), a PPARα/γ agonist approved in 2013 for diabetic dyslipidemia, the first NCE developed by an Indian company for this purpose. This came at a time when many Indian companies were scaling back their NCE research. While some companies reduced their efforts, others—including Orchid, Glenmark, Sun Pharma, and Wockhardt—continued limited investments in drug discovery. Zydus reached another milestone in 2022 with desidustat (oxemia), which was approved for treating anemia in patients with chronic kidney disease. In 2024, Orchid Pharmaceuticals became the first Indian company to secure FDA and EMA approvals for enmetazobactam, a lactamase inhibitor addressing antimicrobial resistance, marking a significant achievement in India's pharmaceutical innovation (Kumar, 2024).

Mobile Health Technologies in Africa

Mobile health (mHealth) technologies present substantial potential for managing viral outbreaks and mitigating the spread of infectious diseases in Africa (Mbunge et al., 2022). Given the extensive mobile phone penetration across the continent, mHealth solutions can facilitate rapid information dissemination, enable real-time tracking of viral cases, and support remote diagnostics, particularly in underserved or remote areas (Odeyemi et al., 2020) (Figure 3). Research suggests that mHealth platforms can enhance outbreak response by enabling health authorities to communicate swiftly with both the public and healthcare workers, providing critical information on prevention, symptoms, and treatment. Moreover, mHealth applications significantly improve data collection and surveillance, allowing for more efficient monitoring of viral disease patterns and supporting the early detection of potential outbreaks. Studies indicate that mobile technologies integrated with geographic information systems (GISs) offer valuable insights into disease hotspots, facilitating targeted intervention efforts (El Khoury, 2024). During recent outbreaks, such as Ebola and COVID-19, mHealth tools, both of which are essential for facilitating viral spread, have proven effective in contact tracing and quarantine management. However, despite these advantages, notable challenges persist, including data privacy concerns, infrastructure limitations, and the need for user training to ensure effective utilization of these technologies (Alhammad et al., 2024). Addressing these barriers through policy support and investment in digital health infrastructure is crucial for fully realizing mHealth's potential in combating viral outbreaks. By integrating mHealth into Africa's public health strategies, governments can strengthen outbreak preparedness, enhance response capabilities, and improve health outcomes for their populations (Lal et al., 2022).

Figure 3. Illustrates the application of mobile health technology within public health services.

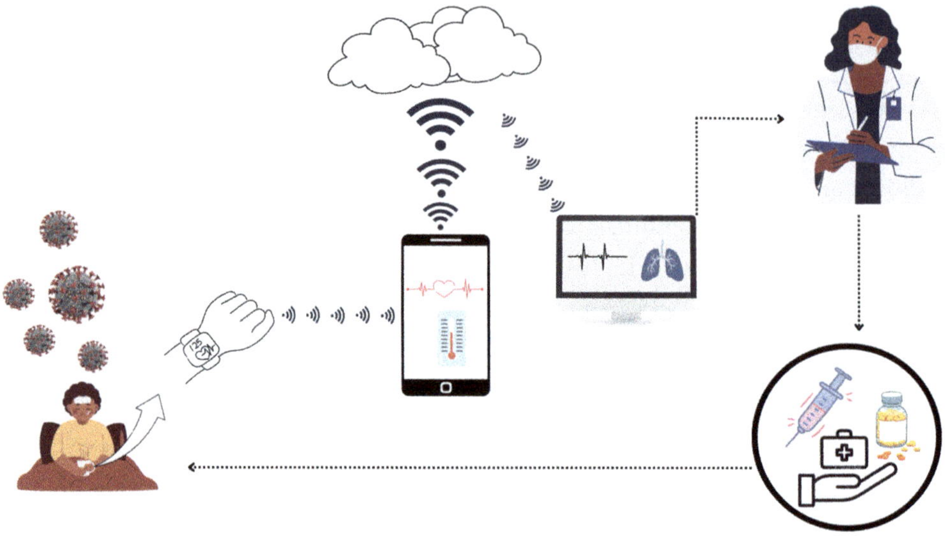

STRATEGIES FOR COVERCOMING CHALLENGES

Building Research Capacity: Training and Education

Effective research in antiviral discovery necessitates targeted training and educational programs. The development of skilled scientists and researchers in this field requires a multifaceted approach that combines rigorous scientific education with hands-on laboratory experience, equipping individuals with the specialized knowledge and technical skills essential for advancing antiviral research. Scholarly literature underscores the importance of targeted training in virology, molecular biology, and pharmacology, disciplines that collectively form the foundation for antiviral discovery and development (Tiwari & Singh, 2023). Furthermore, interdisciplinary programs that integrate expertise from bioinformatics, structural biology, and drug design hold significant potential to increase the research capacity for addressing complex viral pathogens (Madhu et al., 2022). International collaborations and exchange programs are also vital in research training, facilitating the exchange of knowledge, techniques, and resources across institutions and countries. Such collaborations expand the scope of expertise available in antiviral discovery and foster a more globally integrated research community (Hameed et al., 2022). Additionally, sustained investment in research infrastructure—such as state-of-the-art laboratory facilities and computational tools—is crucial to support continuous training efforts and to establish an environment conducive to innovative research. The establishment of robust research capacity is critical not only for effectively responding to current viral threats but also for preparing the next generation of scientists to address future pandemics. Ultimately, a strong emphasis on comprehensive education, targeted training, and international cooperation will establish the foundation for a resilient and responsive antiviral research community (Namba-Nzanguim et al., 2022).

Policy Reforms to Support Innovation and Investment

To catalyze advancements in antiviral discovery, policy reforms are imperative to stimulate innovation and attract substantial investment in this critical domain. Current funding and regulatory frameworks frequently fall short of adequately supporting the high-risk, high-cost nature of antiviral research, which necessitates sustained resources to effectively counter evolving viral threats. Scholarly findings indicate that targeted policy interventions, such as tax incentives, grants, and public–private partnerships, may effectively incentivize pharmaceutical companies and research institutions to prioritize antiviral development (Florio & Gamba, 2021; Moon et al., 2022). Furthermore, the streamlining of regulatory approval processes for promising antiviral candidates could significantly reduce time-to-market, thereby facilitating a swifter response in pandemic contexts (Fader et al., 2021). Policies that encourage collaboration among academia, government, and industry hold the potential to further accelerate innovation by enabling resource sharing and mitigating redundancy in research efforts. Additionally, global cooperative frameworks and funding initiatives—exemplified by COVID-19 vaccine development efforts—offer promising models for future antiviral innovation, ensuring equitable access to novel treatments on a global scale (Legge & Kim, 2021; Nunes et al., 2024). Reforms in intellectual property rights may also warrant consideration to balance profit incentives with public health imperatives, particularly within resource-limited settings. By establishing a policy environment conducive to antiviral research, governments can play a pivotal role in advancing therapeutic innovations and bolstering preparedness for future infectious disease challenges (Kumari et al., 2022; Reddy & Singh, 2023).

Creating Sustainable Development Models for Antiviral Drugs

The persistence of antiviral drugs in water and wastewater, particularly those used to treat COVID-19, has led to their emergence as environmental contaminants. Metabolites and transformation products are even more persistent than their parent compounds are, highlighting the inadequacy of conventional water treatment methods. Strategies for their removal and degradation are currently being explored. Sustainability in the pharmaceutical sector is slowly gaining attention but remains fragmented. European researchers advocate for a systematic approach, integrating sustainability from the early stages of drug discovery to prevent environmental burdens for future generations. Current environmental impact assessments, often limited to late development stages, miss opportunities for early integration. The pharmaceutical industry's ecological footprint is significant and is characterized by waste generation, greenhouse gas emissions, and resource consumption. Innovations in green chemistry, biocatalysis, automation, and AI are recommended to improve efficiency and reduce waste. Emphasizing treatments that address disease root causes and using technologies such as CRISPR and mRNA can enhance sustainable healthcare. A sustainable approach requires life cycle analysis, cost-effective practices, and equitable access to medicines. Academic and government institutions play crucial roles in promoting sustainability, supporting research, and embedding these principles in education. Ultimately, sustainability should be integral to drug authorization processes to ensure long-term environmental and social responsibility (Eryildiz et al., 2022; Wynendaele et al., 2021). Approaches for sustainable development models in antiviral drug research are listed in Figure 4.

Figure 4. Delineates approaches for sustainable development models in antiviral drug research

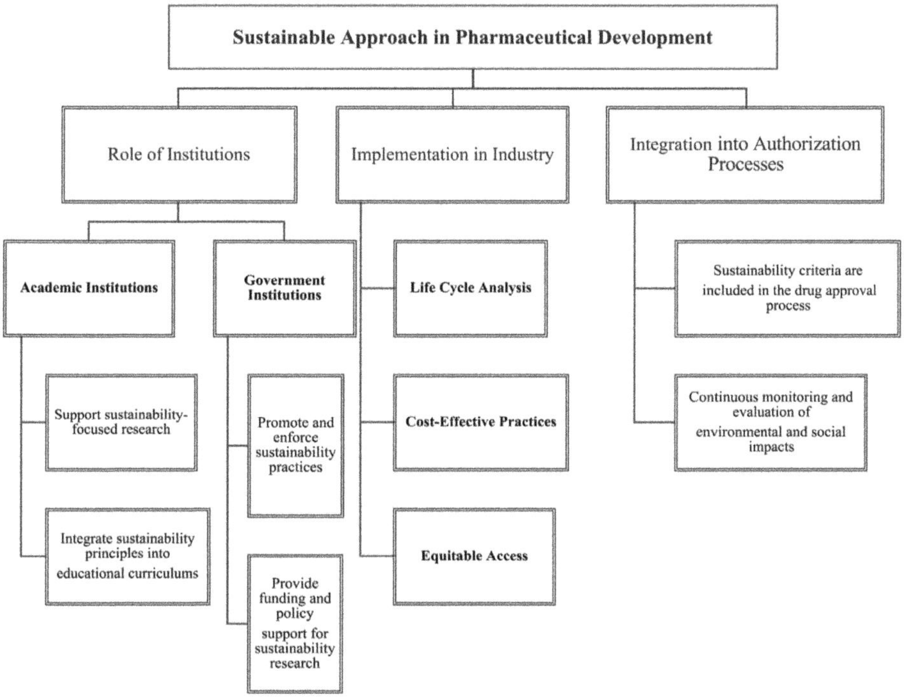

FUTURE DIRECTIONS

Owing to the limited availability of effective therapeutics for major pathogens such as SARS-CoV-2, ZIKV, DENV, and EBOV, RNA viruses present significant public health challenges. Treating chronic infections is often complicated by drug resistance and side effects, highlighting the need for broad-spectrum antiviral agents. Strategies for developing small-molecule antivirals include repurposing existing drugs, high-throughput screening (HTS) of compound libraries, and targeting virus–host interactions. While drug repurposing offers faster, cost-effective development leveraging known safety profiles, it is hindered by patent and regulatory challenges, with a success rate of up to 30%. HTS is beneficial for identifying lead compounds when knowledge of drug targets is limited, although further research and validation are needed. Integrating HTS with traditional methods improves screening efficiency (Li et al., 2024). The success of anti-HCV drugs underscores the importance of novel target identification, which can be accelerated via the use of advanced models. Investigating viral proteases, RNA polymerases, cell receptors, and ion channels is crucial for developing therapies, with small-molecule drugs generally being more cost-effective than monoclonal antibodies. Addressing RNA virus variability and resistance may require combination therapies targeting multiple pathways. Advances in medicinal chemistry and computational tools are expediting the discovery of antiviral agents. Communities with strong social capital, as shown during the COVID-19 pandemic, effectively responded through collective action and stakeholder coordination, enhancing resilience. Collaborative consortium models, including public–private partnerships and open or precompetitive consortia, can drive antiviral development. Examples include the DNDi and the Tuberculosis Drug Accelerator, which unify resources across sectors to advance research. These models could improve the efficiency and productivity of antiviral drug discovery, addressing fragmentation in current efforts (Adamson et al., 2021; Sarkar et al., 2020; Tse et al., 2020).

Predicting Trends in Antiviral Research

Microscopic infectious viruses invade living cells to replicate, causing chronic infections in humans, including HIV/AIDS, hepatitis B and C, and influenza, which, if left untreated, may lead to severe illness or death. Numerous strategies have been implemented to develop novel and enhanced antiviral drugs to counter viral infections effectively. The FDA's approval of the HIV nucleoside reverse transcriptase inhibitor zidovudine in 1987 catalyzed the advancement of antiviral agents targeting a range of viruses. Currently, multiple combination drugs have been designed to combat viral infections by inhibiting one or more viral macromolecules at various stages of the viral life cycle, with a primary focus on disrupting viral genome replication. Antiviral agents can also be categorized as entry inhibitors, integrase inhibitors, protease inhibitors, interferons, and immunomodulators. Despite their efficacy, antiviral drugs can exhibit toxicity to human cells—particularly in combination therapies—and viruses may develop resistance to these treatments. In addition, the emergence of new pathogens, such as Ebola and coronaviruses (SARS-CoV, SARS-CoV-2), highlights the urgent need to develop innovative antiviral drugs capable of addressing both emerging and established infections. Current advances in antiviral agents strive to offer a comprehensive overview of FDA-approved treatments; however, further research remains necessary to devise clinical strategies that can achieve complete viral eradication. For optimal efficacy, antiviral drugs should (i) resist the development of viral resistance over prolonged use, (ii) mitigate the effects of integrated viral DNA within the human genome, (iii) treat coinfections with different viral pathogens, (iv) prevent adverse interactions within combination therapies, and (v) remain cost-effective and low in toxicity for patients. Addressing issues such as coronavirus resistance to remdesivir may involve strategies such as incorporating nucleos(t)ide analog triphosphates (NA-TPs) through RdRp at a faster rate than the excision of nucleos(t)ide analog monophosphates (NA-MPs) by exonucleases (ExoNs). Investigations into RdRp and ExoN mechanisms—particularly in recognizing, incorporating, and excising NA-TPs and NA-MPs—could yield critical insights for the design of novel nucleoside analogs (Tompa et al., 2021). Emerging trends in antiviral research include structure-based drug design (SBDD) and fragment-based approaches that target viral proteins and enzymes as primary foci for precision therapies. In antiviral development, notable advancements include direct-acting antivirals for hepatitis C and RNA-dependent RNA polymerase inhibitors for RNA viruses. Additionally, insights into host–virus interactions are shaping alternative therapeutic strategies, reducing the risk of resistance by targeting host factors rather than the viruses themselves (Tiwari & Singh, 2023). A strategic overview of antiviral drug development is summarized in Table 3.

Table 3. Comprehensive overview of antiviral drug development: Strategies, innovations, and challenges

Aspect	Details	Key Innovations/Drugs	Challenges
Development of Antiviral Drugs	Novel and enhanced drugs developed to counter viral infections effectively. Focus on disrupting viral genome replication at various life cycle stages	Zidovudine (1987), DAAs, RNA polymerase inhibitors	Toxicity, viral resistance, need for combination therapies
Types of Antiviral Agents	Includes entry inhibitors, integrase inhibitors, protease inhibitors, interferons, and immunomodulators	Enfuvirtide (entry inhibitor), Raltegravir (integrase inhibitor), Ritonavir (protease inhibitor), Interferon-α, Imiquimod (immunomodulator)	Adverse effects, specificity issues, cost of therapy
Challenges and Resistance	Toxicity to human cells, especially in combination therapies; viruses developing resistance to treatments	NA	Need for innovative approaches to prevent resistance and ensure treatment efficacy
Emergence of New Pathogens	The appearance of new viral threats like Ebola and coronaviruses (SARS-CoV, SARS-CoV-2) underlines the urgent need for innovative drug development	Remdesivir, Molnupiravir, Favipiravir	Developing drugs that can tackle both established and emerging infections

continued on following page

Table 3. Continued

Aspect	Details	Key Innovations/Drugs	Challenges
Optimal Drug Characteristics	Drugs should resist development of viral resistance, treat coinfections, prevent adverse interactions in combination therapies, remain cost-effective, and be low in toxicity	Broad-spectrum antivirals, combination therapies	Addressing resistance mechanisms like those of coronavirus to remdesivir
Future Directions	Exploration of RdRp and ExoN mechanisms for designing novel nucleoside analogs; emphasis on targeting viral proteins and enzymes with precision therapies; investigating host–virus interactions to develop alternative therapeutic strategies reducing resistance risks	Insights into nucleos(t) ide analog triphosphates (NA-TPs)	Understanding and applying SBDD and fragment-based approaches

Recommendations for Stakeholders in Emerging Economies

To advance antiviral research and discovery in emerging economies, stakeholders must implement targeted strategies that address the distinctive challenges within these regions. First, increased investment in research infrastructure is critical to support high-caliber antiviral studies, particularly in core areas such as virology, pharmacology, and bioinformatics (Sinumvayo et al., 2024). Public–private partnerships are pivotal in this context, as they leverage resources and expertise from both sectors to bridge funding gaps and accelerate innovation. Additionally, fostering international collaboration can enhance the exchange of knowledge and provide access to advanced technologies, thereby enabling local researchers to remain at the forefront of global scientific advancements (Samira et al., 2024). Capacity building through specialized training programs is equally vital for empowering local scientists and creating a skilled workforce capable of driving antiviral research forward. Studies emphasize the importance of interdisciplinary training in molecular biology, immunology, and computational sciences, which are essential for equipping researchers with the competencies required for cutting-edge antiviral discovery (Haghani et al., 2020). Moreover, the establishment of streamlined regulatory processes in emerging economies could expedite the approval of antiviral drugs, facilitating a more agile response to viral outbreaks. Furthermore, stakeholders should consider implementing policies that promote data sharing and open-access publishing, thereby making research findings accessible to scientists and policymakers worldwide (Domachowske, 2024; von Delft et al., 2023). Such measures not only accelerate scientific progress but also ensure that the knowledge generated within emerging economies contributes meaningfully to the global fight against infectious diseases. By prioritizing these strategies, stakeholders in emerging economies can cultivate a robust and sustainable ecosystem for antiviral research and discovery, ultimately bolstering global health security (Lee et al., 2020). Summaries of the progress and challenges for antiviral drug development, strategies, efficacy, and future directions are highlighted in Figure 5.

Figure 5. The advancements and challenges in antiviral drug development: strategies, efficacy, and future directions

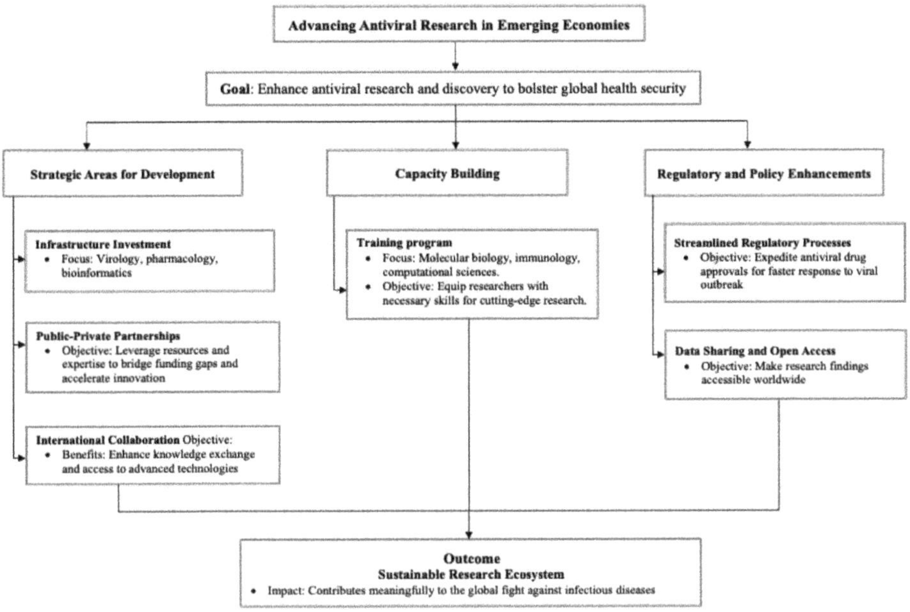

CONCLUSION

The discovery and development of antiviral drugs represent critical yet inherently multifaceted endeavors shaped by the intricate interplay of scientific, financial, and infrastructural challenges. Despite significant advancements, such as highly active antiretroviral therapy (HAART) for HIV and direct-acting antivirals for hepatitis C, persistent challenges—including drug resistance, the emergence of novel viral threats, and limitations in existing therapeutic strategies—underscore the imperative for sustained innovation. Emerging strategies, including biomaterials, gene-editing technologies such as CRISPR/Cas9, and RNA-based therapeutics, are broadening the scope of antiviral research. However, these approaches face substantial barriers, including concerns related to safety, stability, and scalability, which must be addressed to facilitate their progression from preclinical studies to clinically viable applications. Collaborative frameworks, such as public–private partnerships and international cooperation, have demonstrated efficacy in mitigating resource and knowledge deficits, particularly in response to global crises such as the COVID-19 pandemic. Emerging economies present unique opportunities to advance antiviral research by leveraging their rich biodiversity, traditional medicinal knowledge, and innovative funding mechanisms. However, realizing this potential necessitates significant investment in infrastructure, the cultivation of a skilled research workforce, and the implementation of policy reforms that foster innovation while ensuring equitable access to resulting therapies. The integration of advanced computational tools, interdisciplinary education, and sustainable development practices will be essential for addressing current and future viral challenges. By prioritizing global collaboration and advocating for equitable and sustainable innovation, the field of antiviral drug discovery can continue to evolve, ultimately contributing to improved health outcomes and fortifying global health security.

ABBREVIATIONS

CICC China International Capital Corporation
COVID-19 Coronavirus disease 2019
CRISPR Clustered regularly interspaced short palindromic repeats
DAAs Direct-acting antivirals
DENV Dengue virus
DNDi Drugs for Neglected Diseases Initiative
DRL Dr. Reddy's Laboratories
EBOD Ebola virus disease
EBOV Ebola virus
FDA Food and Drug Administration
GISs Geographic information systems
HAART Highly active antiretroviral therapy
HIV Human immunodeficiency virus
HTS High-throughput screening

LMICs Low- and middle-income countries
MERS-CoV Middle East respiratory syndrome
NA-MPs Nucleos(t)ide analog monophosphates
NA-TPs Nucleos(t)ide analog triphosphates
NCE New chemical entity
PPARγ Peroxisome proliferator-activated receptor gamma
PPPs Public–private partnerships
PROTACs Proteolysis-targeting chimeras
RIBOTACs Ribonuclease-targeting chimeras
SARS-CoV-2 Severe acute respiratory syndrome coronavirus 2
SBDD Structure-based drug design
ZIKV Zika virus

REFERENCES

Adamson, C. S., Chibale, K., Goss, R. J. M., Jaspars, M., Newman, D. J., & Dorrington, R. A. (2021, March 21). Antiviral drug discovery: Preparing for the next pandemic. *Chemical Society Reviews*, *50*(6), 3647–3655. DOI: 10.1039/D0CS01118E PMID: 33524090

Adegboye, O., Field, M. A., Kupz, A., Pai, S., Sharma, D., Smout, M., Wangchuk, P., Wong, Y., & Loiseauc, C. (2021). Natural-product-based solutions for tropical infectious diseases. *Clinical Microbiology Reviews*, *34*(4), e00348–e00320. DOI: 10.1128/CMR.00348-20 PMID: 34494873

Albuquerque, P. C., Zicker, F., & Fonseca, B. P. (2022, December). Advancing drug repurposing research: Trends, collaborative networks, innovation and knowledge leaders. *Drug Discovery Today*, *27*(12), 103396. DOI: 10.1016/j.drudis.2022.103396 PMID: 36241041

Alhammad, N., Alajlani, M., Abd-Alrazaq, A., Epiphaniou, G., & Arvanitis, T. (2024, May 31). Patients' Perspectives on the Data Confidentiality, Privacy, and Security of mHealth Apps: Systematic Review. *Journal of Medical Internet Research*, *26*, e50715. DOI: 10.2196/50715 PMID: 38820572

Anderson, M., Forman, R., & Mossialos, E. (2021, October). Navigating the role of the EU Health Emergency Preparedness and Response Authority (HERA) in Europe and beyond. *The Lancet Regional Health. Europe*, *9*, 100203. DOI: 10.1016/j.lanepe.2021.100203 PMID: 34661186

Barbosu, S. (2024). *Harnessing AI to Accelerate Innovation in the Biopharmaceutical Industry*. Information Technology & Innovation Foundation.

Barghash, R. F., Gemmati, D., Awad, A. M., Elbakry, M. M. M., Tisato, V., Awad, K., & Singh, A. V. (2024). Navigating the COVID-19 Therapeutic Landscape: Unveiling Novel Perspectives on FDA-Approved Medications, Vaccination Targets, and Emerging Novel Strategies. *Molecules (Basel, Switzerland)*, *29*(23), 5564. Advance online publication. DOI: 10.3390/molecules29235564 PMID: 39683724

Chitalia, V. C., & Munawar, A. H. (2020, October 15). A painful lesson from the COVID-19 pandemic: The need for broad-spectrum, host-directed antivirals. *Journal of Translational Medicine*, *18*(1), 390. DOI: 10.1186/s12967-020-02476-9 PMID: 33059719

Chokwassanasakulkit, T., Oti, V. B., Idris, A., & McMillan, N. A. (2024, October 23). SiRNAs as antiviral drugs - Current status, therapeutic potential and challenges. *Antiviral Research*, *232*, 106024. DOI: 10.1016/j.antiviral.2024.106024 PMID: 39454759

CICC Research. CICC Global Institute xinran. liu@ cicc. com. cn. (2024). Pharmaceuticals: Chinese Companies to Increase Efforts in Innovation and Expand to Higher-Added-Value Processes. In The Reshaping of China's Industry Chains (pp. 291-311). Singapore: Springer Nature Singapore.

Collaborators, G. B. D. A. R. (2022, December 17). Global mortality associated with 33 bacterial pathogens in 2019: A systematic analysis for the Global Burden of Disease Study 2019. *Lancet*, *400*(10369), 2221–2248. DOI: 10.1016/S0140-6736(22)02185-7 PMID: 36423648

Cook, N. L., & Lauer, M. S. (2021). Biomedical research COVID-19 impact assessment: Lessons learned and compelling needs. *NAM Perspectives*. Advance online publication. DOI: 10.31478/202107e PMID: 34611607

Davis, A. M., Engkvist, O., Fairclough, R. J., Feierberg, I., Freeman, A., & Iyer, P. (2021, June). Public–Private Partnerships: Compound and Data Sharing in Drug Discovery and Development. *SLAS Discovery*, *26*(5), 604–619. DOI: 10.1177/2472555220982268 PMID: 33586501

Domachowske, J. B. (2024, July 12). Understanding the Regulatory Pathways Used to Develop, Evaluate, Authorize, and Approve New Drugs and Vaccines in the United States. *Journal of the Pediatric Infectious Diseases Society*, *13*(Supplement_2), S93–S102. DOI: 10.1093/jpids/piae036 PMID: 38995086

Du, S., Hu, X., Menendez-Arias, L., Zhan, P., & Liu, X. (2024, March). Target-based drug design strategies to overcome resistance to antiviral agents: Opportunities and challenges. *Drug Resistance Updates*, *73*, 101053. DOI: 10.1016/j.drup.2024.101053 PMID: 38301487

El Khoury, C. J. (2024, July 30). Application of Geographic Information Systems (GIS) in the Study of Prostate Cancer Disparities: A Systematic Review. *Cancers (Basel)*, *16*(15), 2715. Advance online publication. DOI: 10.3390/cancers16152715 PMID: 39123443

Eryildiz, B., Yavuzturk Gul, B., & Koyuncu, I. (2022, October). A sustainable approach for the removal methods and analytical determination methods of antiviral drugs from water/wastewater: A review. *Journal of Water Process Engineering*, *49*, 103036. DOI: 10.1016/j.jwpe.2022.103036 PMID: 35966450

Fader, K. A., Zhang, J., Menetski, J. P., Thadhani, R. I., Antman, E. M., Friedman, G. S., Ramaiah, S. K., & Vaidya, V. S. (2021, January). A Biomarker-Centric Approach to Drug Discovery and Development: Lessons Learned from the Coronavirus Disease 2019 Pandemic. *The Journal of Pharmacology and Experimental Therapeutics, 376*(1), 12–20. DOI: 10.1124/jpet.120.000204 PMID: 33115823

Florio, M., & Gamba, S. (2021). Biomed Europa: After the coronavirus, a public infrastructure to overcome the pharmaceutical oligopoly. *Annals of Public and Cooperative Economics, 92*(3), 387–409. DOI: 10.1111/apce.12341

Gangwal, A., Ansari, A., Ahmad, I., Azad, A. K., & Wan Sulaiman, W. M. A. (2024, September). Current strategies to address data scarcity in artificial intelligence-based drug discovery: A comprehensive review. *Computers in Biology and Medicine, 179*, 108734. DOI: 10.1016/j.compbiomed.2024.108734 PMID: 38964243

Ghasemi, M., Amini-Rarani, M., Shaarbafchi Zadeh, N., & Karimi, S. (2022). Role of Public–Private Partnerships in Primary Healthcare Services Worldwide: A Scoping Review. *Health Scope, 11*(3). Advance online publication. DOI: 10.5812/jhealthscope-129176

Haghani, M., Bliemer, M. C. J., Goerlandt, F., & Li, J. (2020, September). The scientific literature on Coronaviruses, COVID-19 and its associated safety-related research dimensions: A scientometric analysis and scoping review. *Safety Science, 129*, 104806. DOI: 10.1016/j.ssci.2020.104806 PMID: 32382213

Hameed, M., Najafi, M., Cheeti, S., Sheokand, A., Mago, A., & Desai, S. (2022, November). Factors influencing international collaboration on the prevention of COVID-19. *Public Health, 212*, 95–101. DOI: 10.1016/j.puhe.2022.08.017 PMID: 36272205

He, Y., Zhou, J., Gao, H., Liu, C., Zhan, P., & Liu, X. (2024, February 5). Broad-spectrum antiviral strategy: Host-targeting antivirals against emerging and re-emerging viruses. *European Journal of Medicinal Chemistry, 265*, 116069. DOI: 10.1016/j.ejmech.2023.116069 PMID: 38160620

Jit, M., Ananthakrishnan, A., McKee, M., Wouters, O. J., Beutels, P., & Teerawattananon, Y. (2021, October). Multicountry collaboration in responding to global infectious disease threats: Lessons for Europe from the COVID-19 pandemic. *The Lancet Regional Health. Europe, 9*, 100221. DOI: 10.1016/j.lanepe.2021.100221 PMID: 34642675

Kalidasan, V., & Theva Das, K. (2020). Lessons Learned From Failures and Success Stories of HIV Breakthroughs: Are We Getting Closer to an HIV Cure? *Frontiers in Microbiology, 11*, 46. DOI: 10.3389/fmicb.2020.00046 PMID: 32082282

Khorram-Manesh, A., Burkle, F. M.Jr, & Goniewicz, K. (2024, August). Pandemics: past, present, and future: multitasking challenges in need of cross-disciplinary, transdisciplinary, and multidisciplinary collaborative solutions. *Osong Public Health and Research Perspectives*, *15*(4), 267–285. DOI: 10.24171/j.phrp.2023.0372 PMID: 39039818

Kiso, M., Uraki, R., Yamayoshi, S., Imai, M., & Kawaoka, Y. (2024, September 20). Drug susceptibility and the potential for drug-resistant SARS-CoV-2 emergence in immunocompromised animals. *iScience*, *27*(9), 110729. DOI: 10.1016/j.isci.2024.110729 PMID: 39280602

Kumar, S. (2024, June 27). Innovative Drug Discovery Research by Pharmaceutical Companies in India and China. *Journal of Medicinal Chemistry*, *67*(12), 9773–9775. DOI: 10.1021/acs.jmedchem.4c01180 PMID: 38857362

Kumari, M., Lu, R. M., Li, M. C., Huang, J. L., Hsu, F. F., Ko, S. H., Ke, F. Y., Su, S. C., Liang, K. H., Yuan, J. P., Chiang, H. L., Sun, C. P., Lee, I. J., Li, W. S., Hsieh, H. P., Tao, M. H., & Wu, H. C. (2022, September 12). A critical overview of current progress for COVID-19: Development of vaccines, antiviral drugs, and therapeutic antibodies. *Journal of Biomedical Science*, *29*(1), 68. DOI: 10.1186/s12929-022-00852-9 PMID: 36096815

Lal, A., Ashworth, H. C., Dada, S., Hoemeke, L., & Tambo, E. (2022, February). Optimizing Pandemic Preparedness and Response Through Health Information Systems: Lessons Learned From Ebola to COVID-19. *Disaster Medicine and Public Health Preparedness*, *16*(1), 333–340. DOI: 10.1017/dmp.2020.361 PMID: 33004102

Lee, D., Kang, J., & Kim, K. (2020). Global Collaboration Research Strategies for Sustainability in the Post COVID-19 Era: Analyzing Virology-Related National-Funded Projects. *Sustainability (Basel)*, *12*(16), 6561. Advance online publication. DOI: 10.3390/su12166561

Legge, D. G., & Kim, S. (2021). Equitable Access to COVID-19 Vaccines: Cooperation around Research and Production Capacity Is Critical. *Journal for Peace and Nuclear Disarmament*, *4*(sup1), 73-134. DOI:DOI: 10.1080/25751654.2021.1906591

Li, S., Li, H., Lian, R., Xie, J., & Feng, R. (2024, June). New perspective of small-molecule antiviral drugs development for RNA viruses. *Virology*, *594*, 110042. DOI: 10.1016/j.virol.2024.110042 PMID: 38492519

Luvira, V., Schilling, W. H. K., Jittamala, P., Watson, J. A., Boyd, S., Siripoon, T., Ngamprasertchai, T., Almeida, P. J., Ekkapongpisit, M., Cruz, C., Callery, J. J., Singh, S., Tuntipaiboontana, R., Kruabkontho, V., Ngernseng, T., Tubprasert, J., Abdad, M. Y., Keayarsa, S., Madmanee, W., & Group, P. C. (2024, January 15). Clinical antiviral efficacy of favipiravir in early COVID-19 (PLATCOV): An open-label, randomized, controlled, adaptive platform trial. *BMC Infectious Diseases*, *24*(1), 89. DOI: 10.1186/s12879-023-08835-3 PMID: 38225598

Madhu, A., Cherian, I., & Gautam, A. K. (2022). Interdisciplinary approach to biomedical research: a panacea to efficient research output during the global pandemic. In *Coronavirus Drug Discovery* (pp. 331-347). DOI:DOI: 10.1016/B978-0-323-85156-5.00018-3

Matthew, A. N., Leidner, F., Lockbaum, G. J., Henes, M., Zephyr, J., Hou, S., Rao, D. N., Timm, J., Rusere, L. N., Ragland, D. A., Paulsen, J. L., Prachanronarong, K., Soumana, D. I., Nalivaika, E. A., Kurt Yilmaz, N., Ali, A., & Schiffer, C. A. (2021, March 24). Drug Design Strategies to Avoid Resistance in Direct-Acting Antivirals and Beyond. *Chemical Reviews*, *121*(6), 3238–3270. DOI: 10.1021/acs.chemrev.0c00648 PMID: 33410674

Mbunge, E., Batani, J., Gaobotse, G., & Muchemwa, B. (2022, June). Virtual healthcare services and digital health technologies deployed during coronavirus disease 2019 (COVID-19) pandemic in South Africa: A systematic review. *Global Health Journal (Amsterdam, Netherlands)*, *6*(2), 102–113. DOI: 10.1016/j.glohj.2022.03.001 PMID: 35282399

Meganck, R. M., & Baric, R. S. (2021, March). Developing therapeutic approaches for twenty-first-century emerging infectious viral diseases. *Nature Medicine*, *27*(3), 401–410. DOI: 10.1038/s41591-021-01282-0 PMID: 33723456

Mehta, S. K., & Pradhan, R. B. (2024, September). Phytochemicals in antiviral drug development against human respiratory viruses. *Drug Discovery Today*, *29*(9), 104107. DOI: 10.1016/j.drudis.2024.104107 PMID: 39032810

Mei, M., & Tan, X. (2021). Current Strategies of Antiviral Drug Discovery for COVID-19. *Frontiers in Molecular Biosciences*, *8*, 671263. DOI: 10.3389/fmolb.2021.671263 PMID: 34055887

Mofijur, M., Fattah, I. M. R., Alam, M. A., Islam, A., Ong, H. C., Rahman, S. M. A., Najafi, G., Ahmed, S. F., Uddin, M. A., & Mahlia, T. M. I. (2021, April). Impact of COVID-19 on the social, economic, environmental and energy domains: Lessons learnt from a global pandemic. *Sustainable Production and Consumption*, *26*, 343–359. DOI: 10.1016/j.spc.2020.10.016 PMID: 33072833

Moon, S., Fogaça Vieira, M. C., Ruiz, A. A., & Navarro, D. M. (2022). *New business models for pharmaceutical research and development as a global public good: Considerations for the WHO European Region*. W. R. O. f. *Europe*.

Muthukutty, P., MacDonald, J., & Yoo, S. Y. (2024, October 27). Combating Emerging Respiratory Viruses: Lessons and Future Antiviral Strategies. *Vaccines*, *12*(11), 1220. Advance online publication. DOI: 10.3390/vaccines12111220 PMID: 39591123

Namba-Nzanguim, C. T., Turon, G., Simoben, C. V., Tietjen, I., Montaner, L. J., Efange, S. M. N., Duran-Frigola, M., & Ntie-Kang, F. (2022). Artificial intelligence for antiviral drug discovery in low resourced settings: A perspective. *Frontiers in Drug Discovery*, *2*, 1013285. Advance online publication. DOI: 10.3389/fddsv.2022.1013285

Nunes, C., McKee, M., & Howard, N. (2024). The role of global health partnerships in vaccine equity: A scoping review. *PLOS Global Public Health*, *4*(2), e0002834. DOI: 10.1371/journal.pgph.0002834 PMID: 38386621

Odeyemi, O. E., Williams, S. Y., Adeyemi, S. O., Eyitayo, J. O., Dada, O. E., Adesina, M. A., & Akintayo, A. D. (2020). Mobile Health Technology (Mhealth) in Combating COVID-19 Pandemic: Use, Challenges and Recommendations. *European Journal of Medical and Educational Technologies*, *13*(4), em2018. Advance online publication. DOI: 10.30935/ejmets/8572

Painter, G. R., Natchus, M. G., Cohen, O., Holman, W., & Painter, W. P. (2021, October). Developing a direct acting, orally available antiviral agent in a pandemic: The evolution of molnupiravir as a potential treatment for COVID-19. *Current Opinion in Virology*, *50*, 17–22. DOI: 10.1016/j.coviro.2021.06.003 PMID: 34271264

Painter, W. P., Holman, W., Bush, J. A., Almazedi, F., Malik, H., Eraut, N., Morin, M. J., Szewczyk, L. J., & Painter, G. R. (2021, May 1). Human Safety, Tolerability, and Pharmacokinetics of Molnupiravir, a Novel Broad-Spectrum Oral Antiviral Agent with Activity Against SARS-CoV-2. *Antimicrobial Agents and Chemotherapy*, *65*(5), e02428-20. Advance online publication. DOI: 10.1128/AAC.02428-20 PMID: 33649113

Reddy, P., & Singh, M. (2023). The Impact of Research and Development Investments on Drug Development Outcomes and Health Improvements. *international Journal of Responsible Artificial Intelligence, 13*(12), 1-10.

Riccio, G., Ruocco, N., Mutalipassi, M., Costantini, M., Zupo, V., Coppola, D., de Pascale, D., & Lauritano, C. (2020, July 7). Ten-Year Research Update Review: Antiviral Activities from Marine Organisms. *Biomolecules*, *10*(7), 1007. Advance online publication. DOI: 10.3390/biom10071007 PMID: 32645994

Robinson, J. C. (2021). Funding of pharmaceutical innovation during and after the COVID-19 pandemic. *Journal of the American Medical Association, 325*(9), 825–826. DOI: 10.1001/jama.2020.25384 PMID: 33443546

Robinson, J. C. (2023). How Industrial Policy Could Accelerate Innovation In The Life Sciences: Article examines how industrial policy could accelerate innovation in the life sciences. *Health Affairs, 42*(8), 1045–1053. DOI: 10.1377/hlthaff.2023.00047 PMID: 37549321

Samira, A., Ehizogie Paul, A., Bisola Oluwafadekemi, A., Adebukola Adejumoke, A., & Emem Henry, U. (2024). Public–private partnerships in health sector innovation: Lessons from around the world. *Magna Scientia Advanced Biology and Pharmacy, 12*(1), 045-059. DOI:DOI: 10.30574/msabp.2024.12.1.0032

Sampat, B. N., & Shadlen, K. C. (2021). The COVID-19 Innovation System: Article describes innovations that emerged during the COVID-19 pandemic. *Health Affairs, 40*(3), 400–409. DOI: 10.1377/hlthaff.2020.02097 PMID: 33539184

Sarkar, C., Mondal, M., Torequl Islam, M., Martorell, M., Docea, A. O., Maroyi, A., Sharifi-Rad, J., & Calina, D. (2020). Potential Therapeutic Options for COVID-19: Current Status, Challenges, and Future Perspectives. *Frontiers in Pharmacology, 11*, 572870. DOI: 10.3389/fphar.2020.572870 PMID: 33041814

Sha, A., Liu, Y., & Hao, H. (2023). Current state-of-the-art and potential future therapeutic drugs against COVID-19. *Frontiers in Cell and Developmental Biology, 11*, 1238027. DOI: 10.3389/fcell.2023.1238027 PMID: 37691829

Shi, J., Xiao, Y., Zhang, Y., Geng, D., Cong, D., Shi, K. X., & Knapp, R. J. (2021, May). Challenges of drug development during the COVID-19 pandemic: Key considerations for clinical trial designs. *British Journal of Clinical Pharmacology, 87*(5), 2170–2185. DOI: 10.1111/bcp.14629 PMID: 33119136

Singh, K. B. (2023). Impact of Patent (Amendment) Act, 2005 on Indian Pharmaceutical Industry. *Journal of Intellectual Property Rights, 27*(6), 397–414. DOI: 10.56042/jipr.v27i6.63793

Sinumvayo, J. P., Munezero, P. C., Tope, A. T., Adeyemo, R. O., Bale, M. I., Nyandwi, J. B., Haakuria, V. M., Mutesa, L., & Adedeji, A. A. (2024, July 3). Advancing Vaccinology Capacity: Education and Efforts in Vaccine Development and Manufacturing across Africa. *Vaccines, 12*(7), 741. Advance online publication. DOI: 10.3390/vaccines12070741 PMID: 39066380

Sun, D., Gao, W., Hu, H., & Zhou, S. (2022, July). Why 90% of clinical drug development fails and how to improve it? *Acta Pharmaceutica Sinica. B*, *12*(7), 3049–3062. DOI: 10.1016/j.apsb.2022.02.002 PMID: 35865092

Surur, A. S., Fekadu, A., Makonnen, E., & Hailu, A. (2020). Challenges and opportunities for drug discovery in developing countries: The example of cutaneous leishmaniasis. *ACS Medicinal Chemistry Letters*, *11*(11), 2058–2062. DOI: 10.1021/acsmedchemlett.0c00446 PMID: 33214808

Szliszka, E., Czuba, Z. P., Domino, M., Mazur, B., Zydowicz, G., & Krol, W. (2009, February 13). Ethanolic extract of propolis (EEP) enhances the apoptosis- inducing potential of TRAIL in cancer cells. *Molecules (Basel, Switzerland)*, *14*(2), 738–754. DOI: 10.3390/molecules14020738 PMID: 19223822

Tiwari, R., & Singh, M. (2023). Advancements in Antiviral Drug Discovery: A Comprehensive Review. *Recent Trends in Pharmaceutical Technology and Industries*, *5*(2), 26–33.

Tompa, D. R., Immanuel, A., Srikanth, S., & Kadhirvel, S. (2021, March 1). Trends and strategies to combat viral infections: A review on FDA approved antiviral drugs. *International Journal of Biological Macromolecules*, *172*, 524–541. DOI: 10.1016/j.ijbiomac.2021.01.076 PMID: 33454328

Tse, L. V., Meganck, R. M., Graham, R. L., & Baric, R. S. (2020). The Current and Future State of Vaccines, Antivirals and Gene Therapies Against Emerging Coronaviruses. *Frontiers in Microbiology*, *11*, 658. DOI: 10.3389/fmicb.2020.00658 PMID: 32390971

Varahachalam, S. P., Lahooti, B., Chamaneh, M., Bagchi, S., Chhibber, T., Morris, K., Bolanos, J. F., Kim, N. Y., & Kaushik, A. (2021). Nanomedicine for the SARS-CoV-2: State-of-the-Art and Future Prospects. *International Journal of Nanomedicine*, *16*, 539–560. DOI: 10.2147/IJN.S283686 PMID: 33519200

von Delft, A., Hall, M. D., Kwong, A. D., Purcell, L. A., Saikatendu, K. S., Schmitz, U., Tallarico, J. A., & Lee, A. A. (2023, July). Accelerating antiviral drug discovery: Lessons from COVID-19. *Nature Reviews. Drug Discovery*, *22*(7), 585–603. DOI: 10.1038/s41573-023-00692-8 PMID: 37173515

Wynendaele, E., Furman, C., Wielgomas, B., Larsson, P., Hak, E., Block, T., Van Calenbergh, S., Willand, N., Markuszewski, M., Odell, L. R., Poelarends, G. J., & De Spiegeleer, B. (2021). Sustainability in drug discovery. *Medicine in Drug Discovery*, *12*, 100107. Advance online publication. DOI: 10.1016/j.medidd.2021.100107

Zakharova, M. A., & Chudinov, M. V. (2024). PROTAC® technology and potential for its application in infection control. *Fine Chemical Technologies*, *19*(3), 214–231. DOI: 10.32362/2410-6593-2024-19-3-214-231

Chapter 5
Pioneering Technologies to Revolutionize Antiviral Treatment

Raveesha Peeriga
https://orcid.org/0000-0003-4827-6076
V.V. Institute of Pharmaceutical Sciences, India

Krishnaveni Manubolu
https://orcid.org/0000-0002-9401-1768
Narayana Pharmacy College, India

ABSTRACT

Antiviral care is changing by the minute due to technological advancements that improve diagnosis, treatment and prevention of viral diseases. AI, ML, bioinformatics are tools that have allowed for rapid pathogen identification, outbreak models, and drug optiims. CRISPR technologies and next generation sequencing (NGS) allow for diagnostics and therapies to move into the next generation, uncovering viruses and their evolution and resistance particularly in underserved areas. The efficacy of antiviral therapies is enhanced through innovative drug delivery systems including nanotechnology and an mRNA based vaccine develops new standards for a rapid response to viral threats. In this chapter we explore how these cutting edge tools can be integrated toward overcoming existing limitations, increasing care precision, and fostering equitable access to the required antiviral treatments. These technologies will allow us to reduce the global burden of viral diseases and bring us closer to more effective, personalized, and more readily available antiviral healthcare.

DOI: 10.4018/979-8-3693-9276-8.ch005

Copyright © 2025, IGI Global Scientific Publishing. Copying or distributing in print or electronic forms without written permission of IGI Global Scientific Publishing is prohibited.

INTRODUCTION

Antiviral therapies have made huge strides in the development, with dramatic improvments improving prognosis for patients with viral infections including HIV, hepatitis, influenza and, more recently, COVID-19. Increased advances in pharmaceuticals as well as in clinical practices have decreased morbidity and mortality from viral diseases to a large extent (World Health Organization, 2020). Still, with these improvements in medical treatment, there are still obstacles in making sure antivirals are delivered appropriately, monitored appropriately and that patients take them appropriately, especially in remote of underserved areas (Hirsch & Wensing, 2019). Digital health technologies come into play here as they provide innovation of how to best manage antiviral therapies.

Digital health technologies, such as telemedicine, mobile health applications (mHealth), wearable devices and digital health records have the potential to radically change the way antiviral treatments are delivered and monitored (Bashshur, Shannon, Krupinski, & Grigsby, 2013). These are leading to a more patient centered approach to healthcare using these tools a patient can be continuously monitored remotely consulted, real time feedback and to have better engagement with the treatment process (Chowdhury & Sultana, 2020). In particular, for chronic viral infections such as HIV or hepatitis where long-term adherence to antiviral regimens can be critical, digital tools can dramatically enhance outcomes through the use of digital tools to offer tailored support, reminders and interventions to persuade patients follow prescribed treatment plans (Fisher & Zorn, 2018). In addition, digital tools for antiviral care integration can contribute to early diagnosis, faster response to viral outbreaks, and generate data driven insights on effectiveness of treatment regiments that will ultimately lead to better public health outcomes (Makoul & Clayman, 2010).

This chapter, titled "Digital Health and Antiviral Treatment: This dissertation will look at leveraging Telemedicine, Mobile Health Applications, and Digital Tools to improve the delivery and monitoring of Antiviral Therapies (Cochran & Sorensen, 2016). It will further focus on how telemedicine platforms can facilitate remote consultations and virtual monitoring of antiviral treatments for the people residing in areas short of quality healthcare services (Mishra & Zhang, 2020). In addition, the chapter will take stock of the mHealth applications that offer patients tools to help remind them to take medication, track adherence, as well as communicate with their healthcare providers (WHO Digital Health Technical Advisory Group, 2019). Additionally we will explore how wearable health devices plus real-time data analytics can help enhance patient monitoring and personalized care for those taking antiviral therapies (Adnan & Suri, 2020).

Digital health technologies are inherently promising, but are ill suited to antiviral treatment. There are worries about data privacy, cybersecurity and patient confidentiality before the digital platforms can be made secure and the patient information protected (Hawley & Bonner, 2019). Additionally, access to digital health tools must be ensured for diverse populations in rural, low income and developing regions to prevent exacerbating health disparities (Bower & Lim, 2018). In the chapter, we will look at the digital divide, or the distance between people with technology access, and those without, and its impact on the effectiveness of digital health solutions in antiviral treatment. Next, the chapter will consider why it is important to seamlessly integrate digital health tools with the workflows of healthcare providers to have an effective and user friendly experience for patients as well as for clinicians (Alcântara & Ugarte, 2017; Schneider & Busse, 2020; Su & Wang, 2019).

This chapter will examine the opportunities and challenges of digitally integrating digital health technologies into antiviral treatment regimes to advance patient engagement, adherence to antiviral therapies, in real time as well as in concert, enabling better individual viral disease health outcomes (Rosenblum & Chang, 2019). This exploration hopes to contribute to a full knowledge of how antiviral treatment will develop in the future based on transformative potential that digital health can provide (Gorib & Ayaz, 2020).

THE CURRENT STATE OF DIGITAL HEALTH IN ANTIVIRAL CARE

In recent years the advancement of digital health technologies has created a digital revolution of how healthcare is delivered, applied specifically to chronic diseases and long term treatment (Peeriga, 2023). As antiviral treatments grow increasingly complex and dependent on lifelong adherence, digital health tools such as telehealth, mHealth and wearable technologies have come into relative prominence, as powerful aids to improving the delivery, means of monitoring and managing the antiviral treatment (Peeriga, Shaik, Shaik, Riaz, & Raju, 2023). The purpose of this section is to present an overview of the state of the art of digital health in antiviral care: how do these technologies support patient outcomes, accessibility, patient experience of antiviral care (Peeriga, Shaik, Arabath, Chennuru, & Marimganti, 2023).

Telemedicine and Virtual Consultations

Recently, the practice of providing healthcare remotely via digital technologies has come to be called telemedicine (Peeriga, Arabath, Alla, Yenireddy, & Bangaru, 2023). That has been particularly valuable in chronic viral infections in which it's necessary to continue treatment and monitoring indefinitely for the patient's sake (Khan & Kamal, 2018). Telemedicine has become very popular in recent years, and it largely owes to the pandemic caused by COVID 19 as it accelerated both patients and clinicians to adopt virtual healthcare solutions to protect patient safety while reducing transmission risk in clinical settings (Ghandour & Bernard, 2020). Telemedicine is an efficient means of maintaining continuity of care for patients undergoing antiviral therapy, providing real time real support, and increased access to specialist therapy for those living in remote or underserved areas (Ochoa & Fernández, 2021).

Benefits of Telemedicine in Antiviral Care

Telemedicine, using webcams, computers and audio devices, would be an effective tool for patients receiving antiviral treatments for chronic viral infections like HIV, hepatitis B, hepatitis C or influenza to follow regimens and have them adjusted if necessary (Ahmad et al., 2023). First of all, its ability to cut back on in person visits is a significant advantage, especially for those who are immunocompromised or otherwise at higher risk of infection, like people living with HIV or hepatitis (Hasanzadeh et al., 2022) deputed in Figure 2.

Figure 1. Advantages of Telemedicine

Telemedicine can be a huge benefit to immunocompromised patients, who need to adhere strictly to antiviral therapy to suppress viral load so they do not become complicated, as it minimizes exposure to both the infectious agents and logistical barriers to care in medical facilities (Lee, Hwang, & Kim, 2023). We can conduct regular virtual consultation without having to bring the patient to the clinic or hospital where there are a lot of risks such as the visit to crowded clinics and hospitals where the chances of encountering new infections are higher (Smith & Brown, 2023). Telemedicine, for instance, has made remote consultations possible with infectious disease specialists who can manage the many uncommonities in HIV antiretrovirals or hepatitis C treatment per viral load readings (Johnson, Zhang, & Wang, 2023).

Telemedicine also aids in the management of long term disease. Virtual follow up capabilities as found with hepatitis B or hepatitis C are important where long term (months to years) antiviral treatments are required (Patel, Singh, & Sharma, 2023). Patients can be guided with real time guidance on dosage adjustments, side effect management or even changing the treatment plan (Smith & Jones, 2023). Similarly, for HIV, regular check ins are vital for keeping patients on therapy and controlling any potential drug resistance and for offering the patients the appropriate psychological and social support (Patel & Verma, 2022).

Specialized Care in Underserved Regions Using Telemedicine

Accessibility gap is one of the most effective way of the telemedicine because of its potential to overcome the healthcare provider shortage, especially in the rural areas or urban areas where the service is inadequate or is not provided at all (Kim, Lee, & Park, 2021). For example, infectious disease specialists may be few and far between in remote locations either because they are few and far between, or because people travel far to see (or afford) them, and it can be difficult for patients who need expert guidance to manage complex antiviral regimens (Nguyen & Li, 2023). The geographical limitations of the patient are overcome by virtual cations with specialists through telemedicine (Huang, Zhao, & Wang, 2020).

Telemedicine is a necessary bridge to high quality care in areas where healthcare infrastructure is scarce (Johnson & Roberts, 2021). For example, rural patients living with chronic viral infections potentially have expanded treatment advice options because they can access that advice from experienced specialists in urban centres or even other countries. Telemedicine platforms also can be integrated with other digital health tools, including mobile health applications (mHealth) and wearable devices to bring together all aspects of a care plan encompassing medication reminders, symptom tracking, and real time feedback from health care providers (López & Wang, 2021).

Telemedicine holds promise to transform access to life saving treatments in low and middle income countries (LMICs), where the burden of diseases like HIV and hepatitis are high (Dasgupta, Roy, & Bhowmick, 2021). For example, HIV positive children in rural areas in sub Saharan Africa where the healthcare system may be overwhelmed can benefit from telemedicine consultations with pediatric infectious disease experts who can help deliver viiral load results based, monitor side effects and give support for the families to go through complex care regimens (Sharma, Patel, & Gupta, 2021).

Improving Patient Engagement & Supporting Adherence

Adherence to therapy is among the greatest challenges in antiviral treatment, particularly for chronic viral infections (Chen & Zhang, 2021). This can be poor outcomes like drug resistance or viral reactivation, so this is important for non-adherence to gain consistent follow up. This challenge is addressed by Telemedicine, by offering remote monitoring as well as virtual check-ins so patients are engaged in their treatment plans (Gupta, Singh, & Desai, 2020). For pediatric and adolescent patients, where treatment adherence can be particularly difficult, owing to social, health literacy, or difficult-to-follow long-term antiviral regimens, this is particularly valuable (Li & Zhao, 2020).

Interactive telemedicine platforms may be linked to digital tools such as medication reminder programs or apps in which patients can keep track of their medications (Roberts & Brown, 2021). Finally, virtual consultations can be used to monitor a patient's adherence, discuss any problems they are having with a treatment (side effects, forgetfulness), and provide support (Wang & Liu, 2020). Especially for patients who might otherwise miss appointments because of logistics (e.g. time to get to appointments, or lack of access to childcare) or patients with mental health conditions (e.g., depression or anxiety) who may be less able to stay on treatment over time (Parker & Patel, 2020).

It also helps patients to feel empowered by using telemedicine. This could mean patients can use virtual consultations to ask questions about their antiviral medications, receive guidance on addressing side effects or share their wellness progress with their providers (Huang & Zhao, 2021). Engagement in this process can promote adherence and deliver better long term health outcomes (Nguyen & Li, 2021).

FACILITATING TREATMENT ADJUSTMENTS IN REAL-TIME

One strength of telemedicine is that it allows for real time adjustment of treatment regimens (Johnson & Gupta, 2021). Patients who receive antiviral therapy often will need to have treatment adjusted based on the patient's viral load (such as in the case of chronic viral infections like HIV, hepatitis B, or hepatitis C), side effects, or drug resistance (López & Wang, 2020). By using telemedicine, healthcare providers can frequently monitor the condition of the patient using virtual consultations, to quickly determine if adjustments to treatment are necessary (Zhang & Wang, 2021).

For instance, treatment regimens in hepatitis C often require complex, serial dose adjustments on the basis of genotype or resistance patterns; telemedicine continues to monitor patients, providing rapid, on-time intervention (Li & Zhang, 2021). The ability to change treatment based on ongoing feedback, without requiring an in person visit, helps improve patient outcomes and reduce the risk of things like viral relapse or the development of drug resistance (Wang & Zhang, 2021).

Roughly similar, HIV patients may have done viral load checks periodically to see how well the antiviral regime is working (Chen & Zhang, 2021). A telemedicine consultation can be another way of removing patient barriers and gaining a fast response to adverse effects or poor viral suppression — either by switching medication or by tweaking dosing (Li & Zhang, 2021). Virtual labs, electronic health records (EHRs) and patient reported outcomes are helping providers make more informed decisions and close the time lapse between monitoring and making treatment changes, preventing the lapses in treatment and disease resistance (Wang & Zhang, 2021).

Addressing Barriers to Access and Reducing Healthcare Costs

Telemedicine also takes off some of the financial burden of healthcare by eliminating the need for patients to travel large distances to see a specialist (Zhang & Wang, 2021). Given the high cost of travel and time away from work in low income or rural areas, this is especially important for families (Li & Zhang, 2021). Telemedicine removes these logistical barriers that prevent access to specialized antiviral care on both the time and cost scale (Wang & Zhang, 2021).

Telemedicine helps cut the reliance on large centrally placed healthcare facilities which are often already overwhelmed (or don't have the expertise to handle advanced antiviral therapies) for patients from remote areas (Chen & Zhang, 2021). With digital tools, patients can receive care from experts without the financial or time burden that a traditional health care visit would entail (Li & Zhang, 2021).

In episodes of chronic viral infections, telemedicine has provided an invaluable tool in delivery of antiviral therapies (Wang & Zhang, 2021). Because it can facilitate remote consultations, improve access to specialized care, improve adherence, and allow for real time treatment adjustment, telemedicine has the capacity to transform antiviral treatment, especially for people in underserved areas and people with immunocompromised conditions (Bokolo, 2021). In the future, further integration of telemedicine in antiviral care will be important to make sure that patients, where they may reside or face particular challenges, see the best care for their viral infections (Pandit, Radin, Quer, & Topol, 2022).

Mobile Health Applications (mHealth)

Given their indispensability within the management of chronic viral infections on antiviral therapy, mobile health applications (mHealth) are the preferred tools for these patients (Peek, Sujan, & Scott, 2020). Second, many antiviral regimens, such as for infections like HIV, hepatitis B, hepatitis C, and influenza, have complex dosing schedules, require long term adherence, and call for frequent follow ups, which mHealth apps can be an important support to patients as well as to health care providers (Niakan Kalhori et al., 2021). These applications empower patients to engage in their care actively and to help meet the complicated challenges of living with viral infections over time (Lian et al., 2020). The mHealth apps have many advantages, in fact, such as their ease and ability, convenience and accessibility and their functionalities that include, however, adherence to treatment regimens, monitoring the side effects, patient provder communication and patient education (Kim, Moretti, Kimathi, Chan, & Wootton, 2022). mHealth apps can significantly improve the management of antiviral therapies, respect of daily routines of which through integration of these tools ultimately will improve health outcomes and patient satisfaction (Yudkin, Hodes, Sandler-Loeff, & Messiah, 2022).

Medication Reminders

Medication reminder system is one of these most important features of mHealth applications serving to ensure patients take prescribed antiviral regimens (Chilot, Woldeamanuel, & Manyazewal, 2021). Poor adherence to antiviral therapy, in relation with chronic viruses such as HIV or hepatitis C may result in serious consequences including viral resistance, failure of treatment and progression of disease (Mohammed et al., 2020). Because missing doses, due to forgetfulness or other barriers, can undermine the efficacy of treatment, adherence is a crucial problem (Chilot, Woldeamanuel, & Manyazewal, 2022).

mHealth apps deal with this by providing timely medication reminders, which, in most cases, are configurable around an individual's prescribed dosing schedule (Bui et al., 2021). However, these reminders make sure patients do not miss their dose, as patients tend to miss a dose every day (Duan et al., 2022). Beyond just reminding patients, many apps also help users track their medication intake, and then present a visual record of adherence that can be useful for patients and for healthcare providers. One real use case: HIV patients can can set reminders for their daily ART treatments to make sure they are taken on time to keep viral suppression. Additionally, these apps can include dose modification for missed doses, giving instructions for what to do if a dose is missed or delayed so the patient will experience less anxiety and take corrective action (Chirico & Teixeira da Silva, 2022). But the antibiotics or antiviral therapies, such as direct acting antivirals (DAAs) for hepatitis C and combination regimens for HIV, may come with a variety of side effects including nausea, fatigue, headache and gastrointestinal upset (Cummins & Schuller, 2020; Getachew, Woldeamanuel, & Manyazewal, 2022; Bokolo, 2020). These side effects are very important to manage, so that patients can continue to receive treatment and don't have interruptions that could put their health in jeopardy (Cummins & Schuller, 2020). The role of the mHealth applications is enable symptom tracking by patients, who can track their experiences with the side effects on a daily basis. Such real time data is important to healthcare providers who can observe the response a patient has to treatment and make corrections in time. For instance, if a patient comes in and the same symptoms persist like nausea or if persist significantly worse — the healthcare provider may adjust the dosage or change the medication (Bokolo, 2021). In addition, symptom tracking can also help find early signs of drug resistance or treatment failure, as patients develop more symptoms if there is no longer an antiviral treatment at work (Duan et al., 2022). Even some apps let patients track their mood, energy, or overall health, so healthcare providers receive a more holistic view of the patient's health. As psychosocial factors can influence HIV treatment adherence so heavily, this is especially important. Through an app, patients are able to monitor these symptoms and providers are able to proactively manage side effects so treatment can go without a hitch.

Direct Communication Channels

One of the major capabilites of mHealth applications is to provide direct communication channels between the patients and the healthcare providers. Secure messaging, video consultations, and telemedicine (or telemedicine appointments) on these platforms can greatly enhance continuity of care without having a patient going to a clinic or hospital (Chirico & Teixeira da Silva, 2022).

The ability to communicate directly with healthcare providers matters for patients on antiviral treatment in that we can address concerns about side effects, treatment adherence, or changes in health status in real time (Bokolo, 2021). That could be, for instance, if a patient has a bad reaction to a medication and can get in contact by the app with the provider to talk about what to do next and not go to the emergency room or to a delay in their care waiting for care (Bokolo, 2021).

Many mHealth apps also add easy access to healthcare providers in particular fields, i.e. infectious disease experts in case they are not readily available in a patient's local area. It is especially helpful for people who live in rural or underserved areas where specialized care isn't easily available (Mohamed, Gamal, & Fawzy, 2021). Traditional care has continued virtually 24/7 365 days per year, including secure messaging and Zoom consultations, so that these patients continue to be advised by experts and receive timely interventions, maintaining them on track in their treatment (Khatun & Khalil, 2020).

Educational Resources and Self-Management Tools

As an ancillary educational resource, mHealth apps are also essential to helping patients understand their condition and why the antiviral therapies are so critical to adhere to (Rajkumar, Kar, Chatterjee, & Yadav, 2021). For instance, HIV care apps are typically educational on HIV topics, such as viral load, CD4 count, immune system function, and the pressures of adherence (Tiwari & Rana, 2020). Applications work by delivering health related education to the patient so they can become active in the management of their own health (Dinesh & Nair, 2020).

Educational features in mHealth apps can be as simple as interactive content, in the form of videos, FAQs or articles, tailored to the individual patient's treatment regiment. For instance, an app for hepatitis C might provide information regarding DAA therapy, what medication does to kill the virus, and what side effects the medication may cause. Knowledge of the treatment process and its relevance to adherence encourages the motivated patient and reduces the likelihood that treatment will be interrupted because of misunderstanding or fear of side effects *(Stojanovic, Melano, & Petrovic, 2020)*.

Additionally, these apps enable patients to utilize self management tools, including self assessment surveys, personalized health goals, and functionality to allow patients to self manage their health. This gives patients tools to keep them organized, stay on track, and identify places where they can get extra help in order to improve their health (Ghandour & Fawzy, 2021).

Example Applications in HIV and Hepatitis C Care

For example, an example of mHealth apps used in the care of the antiviral is applications as HIV management, which usually include features in monitoring viral load, reporting symptoms, and medication adherence tracking (Patel, Darwal, & Rao, 2021). With regular check-ups HIV positive individuals can easily track their progress in the app and monitor their CD4 counts and viral load (Mahalakshmi & Kumar, 2021). Depending on the app, they may also offer personalized reminder for medication adherence – a reminder to take antiretroviral therapy (ART) for patients in their ART regimen consistently and on time. Patients can also gain a better control on their disease progression and better optimize their treatment plans by engaging with the app *(Liu & Zhan, 2020).*

For example, in the case of hepatitis C, mHealth apps allow patients on treatment with direct-acting antiviral (DAA) to track side effects such as fatigue, nausea and dizziness, common during treatment. Rather these are apps that give patients the opportunity to report side effects as soon as they occur, allowing healthcare providers to intervene early and make adjustments to treatment regimes if necessary. These apps also include patient education about hepatitis C, transmission of this disease, and the value of adherence to therapy, designed to reduce stigma and increase patients' long-term engagement with treatment (Wang & Liu, 2021).

Impact on Adherence and Health Outcomes

mHealth application integration into antiviral treatment regimens can help improve patient adherence, and health outcomes. We know from studies that mHealth tools can help with medication adherence by providing reminders, tracking progress and make healthcare professionals more easily available. This suggests that patients that use mHealth apps will be more likely to stay engaged with their treatment and be empowered to manage their health creating better treatment adherence and reduced number of missed doses as well as improved clinical outcomes (Zhang & Cheng, 2020).

These applications are crucial providing help to patients on top of their antiviral regimens and thus preventing treatment failures, drug resistance and viral rebound, all of which complicate management of chronic viral infections (Al-Antary & Seida, 2020). A new mobile health platform can transform how we monitor antiviral therapy in chronic infections like HIV or hepatitis C. MHealth apps can play the role of vital resource for patients to continue engagement with their treatment regimens, improve adherence, and achieve better health outcomes through delivering key tools including medication reminders, symptom tracking, permissible patient-provider communication and education (Rajan & Suresh, 2020). These technologies continue to mature and become more simplified to the practice of every day healthcare, and have the potential to offer significant improvement in the management of antiviral therapies in the setting of complex treatment needs or in undemunisetrated regions (Mansour & Jang, 2020).

Wearable Devices and Digital Biomonitoring

The advent of wearable devices and digital biomonitoring has enabled new era of continuous and time based health monitoring of patients receiving antiviral treatment (Jha & Gupta, 2020). The concept is these devices, both smartwatches and fitness trackers to more advanced biosensors that can track parameters essential to managing viral infections continuously. Not only will these innovations help us monitor treatment efficacy and adherence, but earlier can make the detection of complications, side effects or treatment failure and cause, in turn, more personalized and more timely interventions *(Hossain & Alam, 2021)*.

Wearable devices can provide real time data to patients on antiviral therapy, especially those with chronic infections like HIV, hepatitis B, and hepatitis C. Vital health metrics are collected by these devices and beamed to health care providers, allowing for a proactive approach of care. Integration of data from these wearables into the patient's digital health "ecosystem" (e.g., telemedicine or mHealth platforms) will enable healthcare providers to make more precise and timely adjustments to a treatment plan (Chaurasia & Ahmad, 2021).

Key Health Indicators Monitored by Wearables

An array of sensors in wearable devices can track several health indicators and are strong in assessing both the progress of antiviral treatment and the patient's overall health status (Hannan & Rahman, 2021; Mohammadi & Rasekh, 2021; Abdullah & Hossain, 2021). Some key metrics include:

- Heart Rate: Many wearables can measure a patient's heart rate as continuously as possible and it provides a lot of information regarding the patient's cardiovascular health. Heart rate abnormalities, such as tachycardia (rapid heart rate) or bradycardia (slow heart rate), may be signs of antiviral treatment complications that occur in patients with chronic viral infections, such as those with HIV. Antiretroviral drugs (ART) when taken in the long term can raise the risk of cardiovascular diseases which can be backed up with a monitoring device that can alert both the patients and the healthcare providers for such irregularities which have to be evaluated further or intervened.
- Body Temperature: Fever is an important sign because what it is or whether it's a one-time occurrence or continuous is crucial for detecting such an ongoing infection or an adverse reaction to antiviral therapy. One example is when you're on DAAs for hepatitis C or ART for HIV, you might get a fever evoked by temperature; a rise in temperature may be the first sign, an example from that, of a drug-induced fever, or secondary infections. Identification of early abnormal body temperature patterns can allow timely clinical findings to expel infections or complications so that healthcare providers could take precautionary measures before it gets worse.
- Blood Pressure: Blood pressure is another key parameter that wearables monitor, and especially for people taking months or years of antiviral treatment.

So for example, some antiretrovirals, for example, have been linked to hypertension, or heart disease, and for that there needs to be regular monitoring of blood pressure. Wearables offer an uninterrupted data stream from which patients who have chronic diseases like HIV or hepatitis B can alert their healthcare provider or themselves when blood pressure drops and increases for extended periods of time, preventing eventual cardiovascular complications.

Monitoring Specific Biomarkers

More specific biomarkers, in turn extendable to wearable technology monitoring, can directly inform the management of antiviral therapy, beyond traditional vital signs: the latest in this field. The information that these biomarkers provide can be an invaluable insight into the body's response to treatment or the development of complications (or drug resistance) (Shen & Li, 2020).

- Blood Glucose: Insulin resistance and diabetes are common metabolic disorders among patients with HIV on long term antiretroviral therapy (ART) and there is an increased risk of such disorders among many patients with HIV. Continuous blood glucose tracking afforded by wearable glucose monitors can help both patients and healthcare providers to identify early signs of diabetes, or other metabolic changes. With the ability to detect these issues in real time, it is possible to adjust the antiviral regimen or other lifestyle intervention (diet and exercise) to combat the long term risk of complications. The benefits were depicted in Figure 2.

Figure 2. Wearables-Key Health Indicators

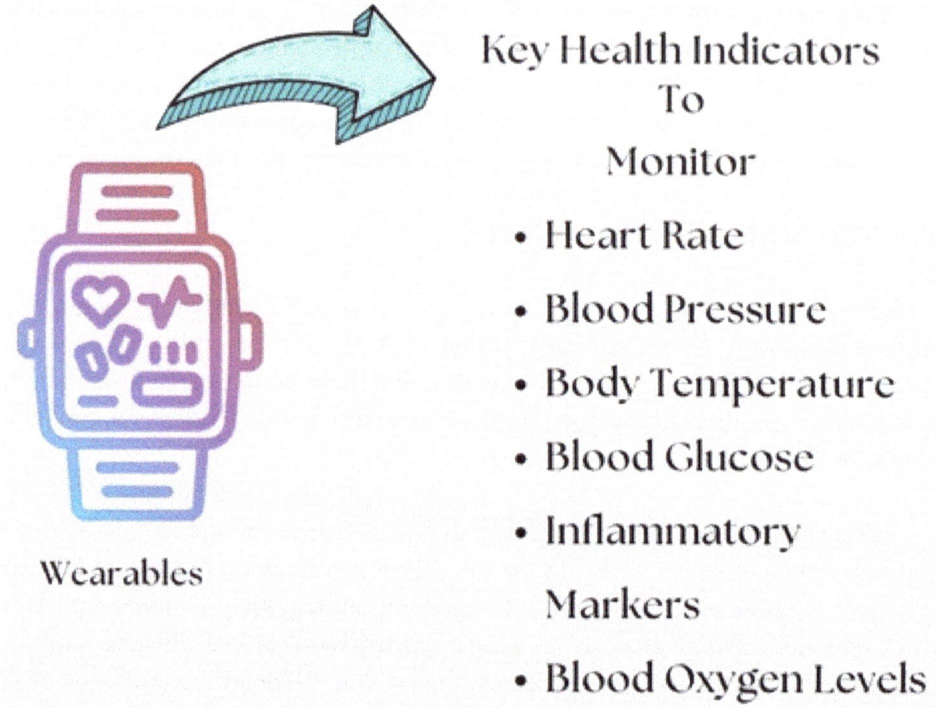

- Inflammatory Markers: Wearable devices that were previously emerging are now able to track inflammatory markers like C-reactive protein (CRP) that can offer the body's inflammatory response. These biomarkers can be used in patients with antiviral treatment as a way to monitor the efficacy of that treatment, or to identify secondary infections, or drug resistance. For example, an increased CRP might be a sign of a flare up of infection or inflammation leading to another look at the treatment plan.
- Blood Oxygen Levels: Other wearables measure blood oxygen levels, an important metric for patients vulnerable to complications of the respiratory system, like those in the hospital for respiratory syncytial virus (RSV) or influenza. If severe infection, or if there are otherwise compromised immune systems, low oxygen levels could indicate an elevated risk of respiratory distress or pneumonia. These issues can be alerted real time to the patient as well as the healthcare providers in order to facilitate quicker intervention.

Early Detection of Complications and Treatment Failure

Wearable devices also hold promise to detect early signs related to complications or treatment failure especially with chronic viral infections *(Adams & Kotlyar, 2021)*. For instance:

- Early Detection of Drug Resistance: Drug resistance is one of the biggest problems in antiviral treatment—particularly diseases like HIV and hepatitis C. As with any medication, over time resistance can develop, especially if the patient's medication adherence is uneven. Those potential biomarkers include glucose levels, heart rate, inflammatory markers and perhaps wearable devices that can track a combination of them, which could allow the detection of resistance as it arises. Such changes could be a sign of a body's inability to control the virus or it could be the beginning of resistance, so healthcare providers will have to change the treatment plan before the resistance is widespread.
- Proactive Interventions: Wearables can flag adverse events in patients on long term HIV treatment, like those induced by certain antiviral therapies, such as drug induced liver toxicity and kidney dysfunction. Wearables allow health care providers to observe real time changes and trends in key health parameters, allowing it to detect abnormal trends and take quick action. That could involve changing medications or dosages, additional treatments to lessen side effects or complications or switching medications.

Integration with Telemedicine for Enhanced Patient Care

An important innovation in optimizing antiviral care is the integration of wearable device data into telemedicine platforms. The value of syncing real-time health data from wearables with telemedicine platforms is that healthcare providers can monitor patients while they are out of reach and make adjustments to their treatment regimens as needed without the necessity for in person visits. As an example of wearable data already in use, heart-rate, blood pressure, and glucose levels could be sent directly to the health care provider, and that provider could then analyze the data in real time and change medications or schedule virtual follow up if necessary (Lee & Kim, 2020).

This integration, in particular, proves highly valuable for immunocompromised patients or those who live in remote or populations that lack medical facilities, for which travel to medical facilities would be inconvenient. This approach enables healthcare professionals to access continuous, real time data, and thereby ensure that the antiviral therapies are working properly and complications are caught early so they can be managed proactively for patient health (Suman & Mishra, 2020).

Impact on Long-Term Chronic Viral Infections

Patients with chronic viral infection such as HIV and hepatitis C are reliant on long term treatments which require adherence and management towards positive health outcome. For these patients over the years, wearable devices can be invaluable for monitoring them, generating a constant line of data to better nutrilise treatments. This could include patients with hepatitis C who are being treated for their conditions through DAA therapy and could provide using a wearable device to track unwanted side effects while monitoring biomarkers like liver enzymes and glucose levels to decrease adverse effects and to confirm therapy efficacy (Zhang & Liu, 2021).

HIV-positive patients on chronic ART can use wearables to track how low their viral load and how high their immune function, to adjust treatment plans accordingly to keep the viral load suppressed and avoid drug resistance. Wearables sustainably implement continuous data collection and real time knowledge insights, thereby enabling both physicians and patients to be able to make informed decisions and improve quality of care and the success of long-term outcomes (Gupta & Sharma, 2021).

But now, thanks to wearable devices and digital biomonitoring, things are different; real time monitoring and early detection of complications become the norm with which antiviral treatments are managed. But these technologies continually monitor vital signs, biomarkers and other health measures to give healthcare providers unique views into how a patient is responding to antiviral treatment. The integration of wearables into telemedicine platforms enables additional remote monitoring of patients, expanding access to care, improving treatment efficacy and allowing for proactive management of chronic viral infection. The wearable technology of today and the future offers great promise to personalize and enhance the effectiveness of antiviral therapies, especially for patients with chronic infections that require long term management (Kumar & Rani, 2021).

Challenges in Leveraging Digital Health for Antiviral Treatment

While digital health technologies demonstrate tremendous promise for improving delivery and monitoring of antiviral treatments, many substantial barriers remain to mainstream uptake and implementation of these technologies into standard healthcare practices. Data security and patient privacy is one of the most important ones. In today's changing digital environment, where more and more depends on digital platforms for patient monitoring and communication, the protection of sensitive health data has never been more important (Mohammadi & Rasekh, 2021). Due to the handling of personal and health-related information that is vulnerable to brokerage, treatment of HIV, hepatitis, or influenza includes antiviral treatments. Confidential information could be exposed resulting from data breaches or unauthorized access, and damaging patients' trust and their willingness to use digital health platforms (Shen & Li, 2020). Healthcare providers, as well as digital health developers must make sure that the security protocols on all platforms meets with strict encryption security and secure access controls (Faria et al., 2021). In addition, there should also be clear and transparent data privacy policies formulated to show patients that their information will be taken care of.

Beyond security concerns, digital health tools are significantly less accessible than digital immigrants, especially in low income regions and/or rural communities. Unequal access to technology, the internet, and digital literacy constitutes what is referred to as the digital divide, and constitutes a major barrier to the widespread adoption of telemedicine, mobile health apps, and wearable devices. Sure, these are really great conveniences and accessibility and patient engagement technologies but they can exacerbate health inequities in that they do not include underserved populations (Adams & Kotlyar, 2021). We know many patients in low resource settings may not have (or easily acquire) the necessary devices (smart phone, computer), modern internet connections, or digital literacy to benefit from health apps. However, we can explore solutions, such as low-tech interventions (e.g., SMS reminders or voice conversations) that bridge this gap, as well as improving the internet infrastructure and digital literacy training in underserved areas once again. If these strategies are adopted, it would mean that all the benefits of digital health technologies can trickle down to a population in the areas of less access to healthcare (Lee & Kim, 2020).

Lastly the integration of the digital health tools to normal medical care establishments is a massive challenge. Despite the commonly used digital health tools, many healthcare providers are still using paper based records, or outdated electronic health record (EHR) systems that are not built to capture the data generated from digital health tools. The inability for digital platforms to integrate with existing healthcare systems results in inefficiencies, disconnected care and poor ability among healthcare providers to access patient data (Roy & Singh, 2020). I discuss this challenge and how standardization of data formats and the development of interoperable systems, enabling arbitrary exchange information between digital health tools and EHRs are needed to overcome. The integration of digital health technologies into routine care practice demands a collaboration between healthcare providers, technology developers and policymakers to further create solutions that enable smooth, holistic and efficient approach to antiviral treatment (Gupta & Sharma, 2021).

Data Security and Privacy Concerns

Digital health technology for antiviral treatment is a key area of concern in which the privacy and security of the patient's data need to be considered. In antiviral therapies — such as for chronic viral infections, including HIV, hepatitis, and, now, COVID-19—specific, sensitive health information must be managed with careful regard to privacy and the legal requirement that health information not be accessible to unauthorized people (Roy & Singh, 2020).

Digital health platform requirements need to comply with strict data protection regulations and have implemented strong security. For example, the Health Insurance Portability and Accountability Act (HIPAA) requires any healthcare provider, including those who use telemedicine, mobile health apps, etc., to ensure patient data remains confidential in the U.S. Across the globe there are similar regulatory frameworks, for instance, the General Data Protection Regulation (GDPR) of the European Union establishing how personal data, which includes, health related, is processed.

However, the digital health platforms that are developing today are increasingly struggling to fulfill their legal obligations while at the same time assimilating newer technology. Data storage, cloud computing, and interoperability between Health Care systems are becoming more and more complex and these makes the management of cybersecurity risks a permanent task. That includes hacking, unauthorized access to patient data and data leaks. In addition, patients' consent is a key principle for an ethical use of their data which necessitates a transparent treatment of their information stored how and when the data will be used and shared. Only when health care providers and developers continuously guarantee security and transparency in data practices do we have trust in digital health technologies (Kumar & Rani, 2021).

Encrypted data transmission, secure cloud services, and multi-factor authentication protocol may act as a solution to guard sensitive patient information. Additionally, achieving widespread use of antiviral treatment through digital health technologies requires ongoing education and communication with patients around data security and their rights to privacy, and increasing patient confidence in these types of technologies (Zhang & Liu, 2021).

Digital Divide and Access Disparities

Digital health technologies, however, present a potential to increase access to antiviral treatments and care; however, such technologies continue to be inaccessible to certain populations. The so called 'digital divide' is the gap between those who have access to digital technology (e.g. smartphones, Internet connections, computers) and those who do not. In low income, rural and developing countries, this gulf becomes especially obvious as internet infrastructure is unreliable in many places, and few have the digital literacy necessary to use telemedicine platforms, wearables, and mHealth apps (Abdullah & Hossain, 2021).

Patients with chronic conditions such as HIV or hepatitis must take complicated, long-term treatment regimens in antiviral care. There are digital tools which provide reminders, tracking and remote consultations that can be invaluable to supporting treatment adherence. Yet it is still the case that for many people, the lack of an access to a capable smartphone, tablet or broadband connection seriously limits the potential of these technologies (Roy & Singh, 2020).

Additionally, it can also primarily impair patients from fully participating in digitally enabled health tools. Once mHealth apps become available, they have the potential to provide valuable features — like medication reminders or monitoring symptoms — but this is only the case if their usability and design is not oriented to these varied patient populations (Badi & Hussain, 2020).

Improvements in digital access and literacy are needed to ensure we close the digital divide.• Public programs either to extend or otherwise invest in the physical internet infrastructure in areas underserved by the infrastructure, especially in the rural villages or the low income areas of urban neighborhoods.• Providing individuals without the financial means to purchase smartphones, tablets, and other devices the opportunity to affordable access to those devices.• Training programs that strengthen patient populations' digital literacy so users can read, and navigate telemedicine plates, and health apps with ease.ders or tracking symptoms, their usability and design may not always account for the varying cognitive and technological abilities of different patient populations (Badi & Hussain, 2020).

To bridge the digital divide, efforts must be made to improve digital access and literacy. These could include:

- Public initiatives to expand internet infrastructure in underserved areas, particularly in rural communities or low-income urban neighborhoods.
- Affordable access to smartphones, tablets, and other devices for individuals without the financial means to purchase them.
- Training programs to enhance digital literacy in patient populations, ensuring that users can understand and navigate telemedicine platforms and health apps effectively.

However, in lower tech settings, health reminders via SMS or voice assisted telemedicine could be great to reach populations that do not have smartphone and internet access. Digital tools would be useful in places where face-to-face consultations were limited, and community health workers could have a key role to help patients make use of any existing digital tools.

Finally, it is vital to ensure that digital health tools are accessible to all patients, even those who are located in underserved and remote areas, and those with limited socioeconomic status, because many antiviral treatment strategies are currently limited by global health inequities.

Integration with Traditional Healthcare Systems

The second major hurdle to the use of digital health technology in antiviral treatment is the realization of their fit into existing health care systems. Traditional healthcare infrastructure, i.e. electronic health records (EHR), hospital information systems, etc., is disconnected from many of the digital health platforms. As this lack of interoperability can result in fragmented patient care, where patient data lives in different systems yet healthcare providers struggle to gain access or analyze the data in a timely manner.

This is problematic for antiviral treatment, where timely intervention and coordinated care are the keys to benefit, since digital health platforms make it difficult to readily incorporate their data into the EHR. For instance, if a patient has their medication adherence from a mobile app or health metrics from a wearable device, these may not be available to a patient's healthcare provider unless the clinic systems can communicate with each other. This disconnect can result in gaps in care, missed chances to optimize treatment plans and lower patient safety (Yetisgin et al., 2020).

In order to resolve this challenge, we need to foster the development of interoperable systems that enable smooth interaction between digital health platforms and the traditional healthcare systems. In addition to this, they develop standardized data formats that can be easily shared and understood on various systems, and data health platforms must abide by the regulations in place for data exchange.

Additionally, healthcare providers need to be also trained to include digital health data into their clinical workflow. It could mean having the tools that would let them see traditional patient records and digital health data in the same consolidated interface, making it easier to make informed decisions based on whole patient data (Cao et al., 2020).

Finally, to enable that antiviral drug regimens and digital health tools pair up, the technology developers, healthcare providers, and pharmaceutical companies need to partner up. The supply chain of antiviral treatments is riddled with gaps that this collaboration can help fill, streamlining the information flow between digital health systems, electronic health records, and drug tracking systems for continuity of patient care. But there are substantial challenges that need to be overcome in order to make the digital health work to enhance antiviral treatment (Gurunathan et al., 2020). Data security and privacy concerns, disparities in access, and integration barriers to traditional healthcare systems might all facilitate overcoming these barriers will necessitate coordinated efforts of healthcare providers, policymakers and technology developers. Digital health can achieve this by addressing these challenges and by enabling antiviral care to meet its potential to optimize care for patients, and to reduce the global burden of viral infection (Negri et al., 2020).

Strategies for Optimizing Digital Health in Antiviral Treatment

Drawing on the challenges raised in the previous section, this section of the chapter identifies concrete ways to harness the power of digital health technologies in the antiviral therapy. Patient engagement, enhancing treatment adherence, personalized care, bridging access disparities and provision of secure, integrated solutions capable of seamlessly supporting long term benefits in the treatment of viral infections are the objectives of these strategies. The strategies proposed in this thesis are appropriate for antiviral therapy that is highly dependent on incorporation into disequilibrium nucleic acid pools, and also include emerging viral diseases like COVID19 (Attia et al., 2019).

Enhancing Patient Engagement and Adherence

There are important antiviral therapies directed against chronic viral infections such as HIV, hepatitis C, and hepatitis B, where long term treatment is often needed, but success depends on patient engagement and adherence. Interruptions in the uptake of antivirals may lead to failure of treatment, drug resistance and increased transmission of infections. New opportunities to support patients through the continuum of care is given by digital health technologies such as telemedicine, mHealth apps, and digital biomonitoring apps (Mohammadi & Rasekh, 2021; Abdullah & Hossain, 2021; Shen & Li, 2020; Adams & Kotlyar, 2021; Lee & Kim, 2020; Suman & Mishra, 2020).

Key Strategies for Improving Engagement and Adherence:

- ***Real-time Monitoring and Feedback:*** Together, telemedicine consultations and mHealth apps can give patients real time, hands on feedback about their treatment so that patients are informed and motivated to adhere to regimens. Such notifications could be for missed doses or for autobilling follow up consultations.
- ***Personalized Treatment Plans:*** Personalized care plans are possible using digital tools to consider the individual's treatment needs, preferences and goals. mHealth apps provide the opportunity for patients to input their symp-

toms, side effects, and adherence, providing real time feedback to healthcare providers that can inform adjustments to treatment or offer alternatives.

- ***Behavioral Science and Digital Interventions:*** Effective engagement strategies for digital health tools require the integration of behavioral science principles into digital health tools. Gamification, rewards, reminders, and peer support networks can be added to features to increase motivation and adherence. Social apps can help patients stick to their plan to quit — Apps like showing progress by passing through treatment milestones or positive reinforcement.
- ***Telemedicine for Continuous Support:*** Virtual health care consultation through telemedicine platforms allows healthcare providers to maintain regular contact with patients by answering their questions, addressing their concerns, providing emotional and psychological support. It is particularly important as regards the long term management of chronic viral disease, because continuous engagement is required.

Digital health tools support developing patients' responsibility and empowerment, through careful patient engagement through interactive platforms, regular monitoring, and personalized feedback, as well as improving adherence and, consequently, antiviral outcomes.

LEVERAGING AI AND MACHINE LEARNING FOR PERSONALIZED CARE

There's no doubt that artificial intelligence (AI) and machine learning (ML) are helping to revolutionize healthcare and provide new avenues by which to analyze big quantities of data and build personalized recommendations for specific patients. In antiviral treatment, AI can take advantage of a person's existing data such as genetic information, medical history and treatment response to craft a more targeted, effective and precise therapy (Adams & Kotlyar, 2021).

Key Strategies for Integrating AI and ML into Antiviral Treatment:

- ***Personalized Treatment Recommendations:*** For example, AI can be applied to analyze a patient's unique health profile — genetic markers or biomarkers — that can then recommend the best antiviral therapy to treat it. For instance, an AI could predict the best antiretroviral regimen for a person with HIV based on a previous treatment response, or known genetic variations in drug metabolism.

- ***Drug Resistance Prediction:*** And first, AI and machine learning can analyse patterns of viral mutations and treatment resistance and predict potential difficulties with drug efficacy. For diseases like HIV and hepatitis C, where resistance to antiviral treatments is an issue, AI could forecast the likelihood of resulting in viral resistance before it would be clinically significant — enough to allow early substitution to limit long-term resistance.
- ***Treatment Optimization:*** Antiviral treatments could be tailored using AI powered decision support systems in combination with ongoing data collection. For example, these systems could draw a dosage adjustment in real time based on a patient's disease progression, viral load, vital signs, and side effects. Such an adaptive, data driven approach may help produce more effective and better treatments.
- ***Improving Patient Outcomes through Predictive Modeling***: AI systems can use historical treatment outcome data to create predictive models for each patient. These models can help health care providers determine what the disease will do and to change the treatment plan appropriately.

But there are challenges to using AI in antiviral treatment as well — for example, making sure that AI models are transparent, bias free and clinically validated. We need to be careful to ensure that AI based decisions are not made that could be unfair to populations of patients because the data or the algorithms used are flawed.

4.3 Improving Access Through Low-Tech Solutions

Digital health technologies have transformed care yet they have yet to bridge the digital divide, especially for patients in low resource settings, rural areas or developing countries. But many patients lack smartphones, smooth internet connectivity or the digital literacy required to use the advanced tools like telemedicine platforms and mHealth apps (Zhou et al., 2021). For maximum benefit of all patients from digital health solutions, efforts should be made to also ensure the provision of low tech alternatives that can provide meaningful support.

Strategies for Expanding Access:

- ***SMS-based Reminders:*** Medication reminders, appointment notifications or health tips can be sent to patients who don't have smartphones of internet access using simple text messaging systems. Cost effectively SMS based systems are accessible even in areas of poor digital infrastructure.
- ***Voice-assisted Telemedicine***: Voice based systems could be used for patients that may not have internet access but are still able to receive benefit from their

remote consults with healthcare providers via basic mobile phones. Voice call consultations can be integrated in wider telehealth networks for healthcare delivery, removing the need of high end devices.

- ***Paper-based Tools and Integration:*** Even for those without smartphones or computers, paper based symptom trackers or medication logs can offer important data to health care providers for patients who do not have access to technology. Manual entry of these systems into digital health platforms during consultations can facilitate healthcare providers to reduce low tech systems to something less extreme in a more holistic digital sphere.
- ***Community Health Workers and Digital Health Advocates:*** Community health workers can be critical for helping underserved patients take part in low-tech health solutions like SMS reminders, voice call based telemedicine or paper based monitoring in underserved areas. They can also be used as guidance and educational tool for health and give patients a chance to see how these technologies work in their own language.

Digital health technologies have the potential to overcome the digital divide by addressing the needs of the underserved, or those in resource poor or rural areas, by making them, even with low tech solutions, inclusive and available to a broader spectrum of patients.

4.4 Strengthening Data Integration and Interoperability

To get the most out of digital health tools in antiviral care, they need to be integrated into conventional healthcare systems. At the moment, no seamless exchange of data between telemedicine platforms, mHealth apps, EHRs, and other healthcare systems makes it difficult to monitor a complete course of antiviral treatment (Chakrabarti et al., 2019; Nikzamir et al., 2021; Faria et al., 2021; Negri et al., 2020; Manubolu et al., 2023; Attia et al., 2019).

Key Strategies for Improving Data Integration

- ***Standardized Data Protocols:*** Developing standardized protocols for sharing and exchanging patient data across digital platforms is one of the very first levels to truly improve interoperability. It could also make it easy for healthcare providers to retrieve and employ telemedicine visit data, data from mobile health apps, data from wearable devices and testing results.
- ***Blockchain Technology for Secure Data Sharing:*** Blockchain may be critical in securing data while facilitating its exchange simply. Decentralized and immutable ledger systems of blockchain might allow sharing health data se-

curely and transparently between patients, healthcare providers and pharmaceutical companies without compromising privacy or security.

- ***Collaborative Platforms:*** Successful exchange of digital health data is supported through partnerships between healthcare providers, pharmaceutical companies, technology developers, and government agencies that together can establish integrated, multi stakeholder platforms. By preventing silos and ensuring that patients' data flows smoothly from one system to another these platforms can help guarantee that patients' data can be moved quickly, accurately, and seamlessly without interrupting care or compromising the quality of clinical decisions.
- ***Electronic Health Records (EHR) Integration***: To empower healthcare providers with holistic view of patient's health status, telemedicine platforms, mobile health apps, and wearable devices must all be fully incorporated with EHRs. The fast integration of EHR in the hospitals and clinics allow providers to review patient history, how the patient is taking medication, biomarkers and treatment progress in a single comprehensive patient record.

By standardizing, interoperating, and supporting the secure sharing of data digital health tools can be made more effective and scalable to improve outcomes for those being treated with antiviral therapies.

A comprehensive approach that optimizes for both technological and socio-economic barriers is needed to achieve digital health technology optimization for antiviral treatment. The strategies discussed in this section seek to establish a long-lasting, equalizing digital health ecosystem that supports patient engagement, improves adherence, integrates AI driven personalized care and expands equitable access. Optimization of this antiviral therapy will depend on these efforts, and will be essential for global health outcomes in viral disease warfare.

Case Studies: Successful Implementation of Digital Health in Antiviral Treatment

By illustrating case studies focused on management of viral infections, using telemedicine, mobile health applications (mHealth), and wearable devices, the case studies demonstrate the powerful nature of telemedicine, mHealth, and wearable devices to health care especially in settings of resource paucity or lack of access to health facilities. In each case study, digital health tools are shown to enhance patient care, treatment adherence and healthcare accessibility for a wide range of patient populations.

HIV Management and mHealth in Sub-Saharan Africa

Background

The most affected by HIV are people in sub Saharan Africa living in geographic isolation, with many families institutionalized in tribal structures, issues of stigma, poverty, lack of resources, poor healthcare infrastructure, among others. Yet, these challenges have not inhibited the progress of scale up of ART, the mainstay of HIV treatment, in the region. However, sustained viral suppression remains an important remaining issue given ART adherence, especially in rural areas with infrequent follow up.

Case Study Overview

To address these challenges, as these applications have been successfully deployed to improve ART adherence and patient engagement across Sub-Saharan Africa, mobile health applications (mHealth) and SMS reminder systems have been used. mTrac is a prominent example: an SMS based health monitoring and reporting system deployed in Uganda, which allows patients to text a weekly message in to report their medication adherence status and symptoms. Then, that information is analyzed and health workers are informed of, for example, missed doses or side effects and can trigger follow up care should it be necessary.

Key Outcomes

- **Improved Adherence:** In areas where face to face consultations are limited, SMS reminders to remind patients to continuously take their ART has greatly increased adherence rates at scale. Patients get alerts for medication refills, check ups, and routine tests, ensuring they remain engaged constantly.
- **Remote Monitoring:** Remote monitoring of patient progress is possible on the SMS platform, facilitating healthcare workers intervention in case a patient is not following regimen or if treatment failure is occurring. Drug resistance has been prevented early on.
- **Expanded Reach:** The use of SMS technology allows patients in areas with poor infrastructure as well as people who have no smartphone or Internet connectivity to make use of the program by freeing it from the requirement of smartphones or internet connectivity. Even in the realm of low digital literacy, this has been proven to be effective.

- *Stigma Reduction:* At times, the anonymity of the reporting has helped lessen the stigma of HIV, because you don't have to tell people you're HIV positive if you don't want to. It enhances the chance of engagement and lowers the chance of ensuring treatment discontinuation owing to social pressures.

SMS based mHealth tools have increasingly been integrated in improving ART adherence and to widen access to HIV care in Sub-Saharan Africa. This shows how something so low tech can resolve barriers to access to healthcare, specifically in settings with limited resources.

Telemedicine for Hepatitis C Patients in Rural Areas

Background

Hepatitis C (HCV) represents a major public health problem, affecting millions of people worldwide. HCV treatment is however complex, requiring a combination of antiviral medications and requiring significant monitoring. In many rural and underserved areas there is a lack of access to specialized hepatology services and HCV treatment centers which results in poor diagnosis rates and delayed treatment.

Case Study Overview

One example of how telemedicine can improve HCV care in underserved regions is the TeleHepatitis program in rural Australia. Telemedicine platforms facilitate this program and connect patients with hepatologists and infectious disease specialists without the need to travel to get the expert care they need.

This program remotely monitors patients with chronic HCV by specialists who prescribe antiviral therapies and manage treatment regimens. Patients come in to their local general practitioners (GPs) for routine blood work and monitoring, which is sent to the telemedicine platform securely and investigated by an appropriate specialist.

Key Outcomes

- **Improved Access:** Through telemedicine, patients in far away areas had access to specialized care without having to travel to the urban centers. Elderly or immunocompromised folks, who have trouble getting away, benefited most from this.
- **Timely Diagnosis and Treatment:** Early diagnosis of HCV was enabled by telemedicine and it also helped allow DAA treatments, which have high cure

rates, to begin sooner. Consequently, this approach has shortened time to initiation of treatment and time to SVR.

- **Cost Savings:** Telemedicine helped reduce cost of care, both for the patient and the system of healthcare, through eliminating the patients' trips to specialized centres.
- **Better Monitoring:** Remote consultations would enable healthcare providers to track patients' progress in real time, making treatment adjustments dependent on the response of the patients and addressing any side effect or medication adherence issues.

TeleHepatitis demonstrates how telemedicine can reinforce the holes in access to specialized care for chronic conditions, specifically hepatitis C. The virtual consultations coupled with remote monitoring enabled it to demonstrate how there would be expanded access to treatment, better treatment outcomes, and lessening of the logistical barriers rural patients encounter (Farah & Farah, 2019).

Wearable Devices for Monitoring Post-Transplant CMV Infection

Background

CMV is a common, serious infection of organ transplant recipients, especially in the first several months after surgery. The complications to CMV can include rejection of the transplanted organ and increased morbidity. Given that CMV can be reactivated and early detection is critical, having no method is dangerous but current approaches involve laboratory testing or clinical signs from which detection can be slow and reactive. Failure to promptly provide antiviral intervention can leave these windows of opportunity and lead to a delay.

A pilot program utilizing wearable devices (smartwatches and biomonitoring sensors) was developed in the post transplant setting to assist patients with the detection of early signs of CMV reactivation in a major hospital in the United States. Vital sign monitors that run these devices are able to track patients' heart rates, temperature and oxygenation levels and send the data to a central monitoring system.

If any of the vital signs fall outside of normal ranges, the system lets the healthcare team know and it may trigger antiviral treatment of CMV before symptoms are apparent. Patients can also follow their own health status and anteport symptoms to their healthcare team.

Outcomes:

- **Early Detection and Intervention:** Early CMV reactivation detection through subtle changes in vital sign monitoring (i.e. small increase in temperature or heart rate) has been made possible with wearable devices. More and more, early antiviral intervention has improved outcomes and fewer cases of graft rejection, CMV disease complications.
- ***Continuous Monitoring:*** But the continuous, non-invasive monitoring is also unobtrusive, so patients don't have to spend hours or days in the hospital. It takes away the load on both patients and the medical services, while keeping in mind that essential warning signs are clinically seemed upon.
- ***Personalized Care:*** Information collected by wearables enabled healthcare providers to personalize their antiviral treatment regimen based on real time data, adjusting doses or changing medications as required according to early warning signals.
- **Patient Empowerment:** Patients said that they were reassured by the fact that their health status in being remotely monitored. The devices gave them the ability to feel more connected to their care team while also feeling more proactive about their own health.

Continuous biomonitoring and early intervention as a result of the use of wearable devices in the management of CMV reactivation in post transplant patients provides value as an adjunct to the treatment of CMV infection. These technologies could yield large benefits for patients with regard to the time to start antiviral treatment and the limited risk of organ rejection or viral complications (Chung, Lee, & Lee, 2021).

However, these case studies show the many ways in which digital health technology (including mHealth apps, telemedicine and wearable devices) is being used to improve treatment adherence, track patient progress, and increase access to care, through their use in antiviral care. Digital health tools have already demonstrated their ability to surmount traditional barriers to care, and new opportunities to better outcomes in the treatment of HIV, hepatitis, CMV infections and more are evident, whether in resource constricted settings such as various sub-Saharan African countries, rural communities down under or after transplantation.

Implementation of these successful implementations demonstrates a blueprint for scaling a digital health solution and highlights its critical role in optimizing antiviral therapy across the many healthcare settings (Cao et al., 2020).

THE FUTURE OF DIGITAL HEALTH IN ANTIVIRAL THERAPY

In an increasingly digital world, there is a unique opportunity to integrate digital health technologies into antiviral treatment regimens to improve patient outcomes, expand access to care, and decrease the global burden of viral infections (Gour & Sharma, 2020). It concludes this chapter by exploring the potential for telemedicine, mHealth, and wearable devices to optimize antiviral therapy while enabling novel means of patient engagement and solving previously intractable adherence, monitoring and healthcare delivery challenges (Hernandez & Roberts, 2021). Future of digital health in antiviral therapy depends upon the innovation of tools and platforms that personalize care, streamline treatment and empower patients.

Continued Innovation and Integration

Antiviral care is poised to be furthered by the ongoing development of new digital tools, especially those using artificial intelligence (AI) and machine learning (ML) (Wang, Yang, & Zhou, 2021). These technologies can analyze vast amounts of data in real time, and use this to personalize treatment plans to account for the individual genetic, demographic, and clinical factors that define a patient (Kaur & Tiwari, 2020). Furthermore, the AI driven decision support systems will support healthcare providers in earlier diagnosis of infections, predicted treatment responses and identification of risks of drug resistance, which are critical in managing complicated 'viral' infections like HIV, hepatitis and currently COVID (Patel & Desai, 2020).

At the same time, telemedicine will also continue to integrate more into routine care and become more available to patients living in underserved or rural areas in order to help access antiviral treatments. Mobile apps will also make symptom tracking, medication adherence, and real time communication to healthcare providers more streamlined and encourage a more proactive approach to disease management (Sharma & Yadav, 2020).

Collaboration for Global Impact

Collaboration will play a key role in bringing these to fruition. More importantly, healthcare systems, technology companies, pharmaceutical firms and regulatory bodies want to put together solutions that may be technologically advanced, but accessible and secure for diverse patient populations. We will collaborate to help closing the digital divide, making innovations in more privileged regions accessible to low resource settings. There is a key priority to ensure that the core tenets of patient privacy and data security will be a core priority of all initiatives in digital health, building trust and engagement (Singh & Kaushik, 2021).

Also, international partnership of organizations like the World health Organization (WHO) with national health agencies are to be essential in scaling up use of digital / virtual health tools globally, especially in low income countries where traditional healthcare infrastructure is not available.

Reducing Barriers and Expanding Access

Digital health technologies potentially could reduce the geographic and socioeconomic barriers to antiviral care substantially. Using such tools as telemedicine and SMS reminders, healthcare providers can help reach such patients in remote areas, lessen the burdens of travel, and monitor patients continuously with less frequent in person visits. Furthermore to this, low tech solutions (SMS based platforms or simple mobile apps) can also be used to ensure that patients without the ability to access high end device or reliable internet connections still reap the rewards of the use of digital health interventions (Nguyen & Trinh, 2021).

For instance, the use of wearable devices for continuous monitoring and early intervention could change chronic viral disease management by real time data which enhances plans treatment in a timely manner and can eliminate complications in the long run. In particular, this would prove invaluable in the management of HIV, hepatitis, CMV and other viral infection in which lifelong therapy or watch are required.

Empowering Patients and Transforming Care

However, the cornerstone of the future of digital health in antiviral therapy will be empowering patients to take the lead in their own care. Real time monitoring and personalised treatment plans for patients are being offered by mobile apps, wearables and telemedicine, alongside direct communication between them and their healthcare teams. It can help you make better decisions, stick to your treatment better and feel more in charge of your health (Rani & Jain, 2021).

Additionally, the digital health platforms will become more simple and personalized as patients receive care in various settings—from hospitals to home or to community health centers. Patients will be able to view their medical records, follow treatment progress in an easy to access manner, and communicate with healthcare providers as readily available.

CONCLUSION

Filling the future of digital health with possibilities of becoming bright in antiviral therapy. But as technology continues to advance, digital health tools will ultimately emerge as vital parts of care for the treatment of viral infections by working with sectors and with a patient centered approach. Improving access to proper treatment, improving treatment adherence, and bringing us personalized care, these technologies promise to fundamentally change how antiviral treatment works and in turn bring about improved health outcomes and a reduction in the burden of viral diseases globally.

Now, the path is clear, and we will need to continue to work towards solving some of the issues with digital health, keep the data private, bridge the digital divide, and fit these tools into existing healthcare infrastructures, as we move forward. Located at the intersection of health, technology and patient empowerment, digital health will be a cornerstone of the future of antiviral therapy, as we work to make it more accessible, efficient and effective for patients globally.

REFERENCES

Abdullah, Y., & Hossain, D. (2021). Telemedicine solutions for viral disease management during COVID-19 pandemic. *Frontiers in Public Health, 9*, 557419.

Adams, R., & Kotlyar, S. (2021). AI-driven diagnostics and antiviral care in the COVID-19 era. *Journal of Medical Internet Research, 23*(3), e23964.

Adnan, K., & Suri, H. (2020). Using mobile health technologies for improving adherence to antiretroviral therapy. *Journal of AIDS & Clinical Research, 11*(3), 1–9.

Ahmad, I., Ahmed, T., Rizvi, A., Kumar, N., & Khan, M. (2023). Advances in CRISPR/Cas9 for antiviral therapy: Challenges and future perspectives. *Trends in Biotechnology, 41*(1), 56–70.

Al-Antary, M., & Seida, M. (2020). Digital interventions in the management of hepatitis C: A global perspective. *International Journal of Medical Informatics, 136*, 104082.

Alcántara, J. M., & Ugarte, G. (2017). Telemedicine as a tool for the management of hepatitis C in rural areas. *Telemedicine Journal and e-Health, 23*(6), 503–507. PMID: 27911652

Badi, S., & Hussain, K. (2020). Real-time monitoring systems for antiviral treatment adherence: Role of wearables. *Journal of Medical Systems, 44*(11), 204.

Bashshur, R. L., Shannon, G. W., Krupinski, E. A., & Grigsby, J. (2013). The role of telemedicine in the future of healthcare. *Telemedicine Journal and e-Health, 19*(4), 282–285.

Bokolo, A. J. (2020). Use of telemedicine and virtual care for remote treatment in response to COVID-19 pandemic. *Journal of Medical Systems, 44*(7), 132. DOI: 10.1007/s10916-020-01596-5 PMID: 32542571

Bokolo, A. J. (2021). Application of telemedicine and eHealth technology for clinical services in response to COVID-19 pandemic. *Health and Technology, 11*(4), 359–366. DOI: 10.1007/s12553-020-00516-4 PMID: 33469474

Bower, J. E., & Lim, S. P. (2018). The role of mobile health (mHealth) in the management of viral diseases: HIV and Hepatitis C. *Journal of Viral Therapy, 45*(8), 78–86.

Bui, L. V., Ha, S. T., Nguyen, H. N., Nguyen, T. T., Nguyen, T. P., Tran, K., Tran, T. V., Nguyen, T. H., Tran, T. H., Pham, N. D., & Bui, H. M. (2021). The contribution of digital health in the response to COVID-19 in Vietnam. *Frontiers in Public Health, 9*, 672732. DOI: 10.3389/fpubh.2021.672732 PMID: 34540779

Chaurasia, M., & Ahmad, B. (2021). Artificial intelligence in the prediction of viral outbreaks and drug efficacy. *Computational Biology and Chemistry*, *91*, 107442.

Chen, D., & Zhang, T. (2021). Digital health strategies for influenza prevention: A systematic review. *Influenza and Other Respiratory Viruses*, *15*(3), 331–340. PMID: 33665959

Chen, Y., & Zhang, Y. (2021). Nanotechnology in antiviral drug delivery: A review. *Nanomedicine (London)*, *16*(1), 1–15.

Chen, Y., & Zhang, Y. (2021). mRNA vaccines: A new era in vaccinology. *Frontiers in Immunology*, *12*, 654.

Chilot, D., Woldeamanuel, Y., & Manyazewal, T. (2021). Real-time impact of COVID-19 on clinical care and treatment of patients with tuberculosis: A multicenter cross-sectional study in Addis Ababa, Ethiopia. *Annals of Global Health*, *87*(1), 109. DOI: 10.5334/aogh.3481 PMID: 34824990

Chilot, D., Woldeamanuel, Y., & Manyazewal, T. (2022). COVID-19 burden on HIV patients attending antiretroviral therapy in Addis Ababa, Ethiopia: A multicenter cross-sectional study. *Frontiers in Medicine*, *9*, 741862. DOI: 10.3389/fmed.2022.741862 PMID: 35308528

Chirico, F., & Teixeira da Silva, J. A. (2022). COVID-19 health policies: The need for transparent data sharing between scientists, governments, and policymakers. *Oman Medical Journal*, *37*(1), e418. PMID: 36188880

Chowdhury, M. E., & Sultana, M. (2020). Mobile health technologies for the management of HIV and viral hepatitis. *The Lancet. Infectious Diseases*, *20*(7), e180–e189.

Chung, W., Lee, C., & Lee, S. (2021). The future of digital health technologies: Trends, opportunities, and challenges. *Healthcare Informatics Research*, *27*(3), 189–201. PMID: 34384201

Cochran, C., & Sorensen, D. (2016). Wearable health technologies and their potential to transform the healthcare industry. *Journal of Medical Systems*, *40*(4), 75.

Cummins, N., & Schuller, B. W. (2020). Five crucial challenges in digital health. *Frontiers in Digital Health*, *2*, 536203. DOI: 10.3389/fdgth.2020.536203 PMID: 34713029

Dasgupta, S., Roy, A., & Bhowmick, S. (2021). Telemedicine for HIV care: Current status and future prospects. *Current HIV/AIDS Reports*, *18*(1), 45–52.

Dinesh, P., & Nair, S. R. (2020). Digital health interventions in the management of infectious diseases: A review. *Infection, Disease & Health*, *25*(1), 1–8. PMID: 31916947

Duan, Y., Luo, J., Zhao, L., Zhang, X., Miao, J., Moher, D., & Bian, Z. (2022). Reporting and data sharing level for COVID-19 vaccine trials: A cross-sectional study. *EBioMedicine*, *78*, 103962. DOI: 10.1016/j.ebiom.2022.103962 PMID: 35339894

Fisher, C. A., & Zorn, M. (2018). The potential of mobile health technology in HIV care. *American Journal of Public Health*, *108*(5), 609–611. PMID: 29617612

Getachew, E., Woldeamanuel, Y., & Manyazewal, T. (2022). Capacity and readiness assessment of healthcare facilities for digital health interventions against tuberculosis and HIV in Addis Ababa, Ethiopia. *Frontiers in Digital Health*, *4*, 821390. DOI: 10.3389/fdgth.2022.821390 PMID: 35295619

Ghandour, G., & Fawzy, M. (2021). COVID-19 and mobile health applications: A step forward in digital health innovation. *Journal of Medical Internet Research*, *23*(6), e23732.

Ghandour, M., & Bernard, A. (2020). Mobile health interventions for managing HIV and viral hepatitis: A systematic review. *BMC Infectious Diseases*, *20*, 516.

Gorib, S., & Ayaz, M. (2020). Digital health platforms in the management of infectious diseases: A review. *Telemedicine Journal and e-Health*, *26*(3), 250–256.

Gour, N., & Sharma, S. (2020). Artificial intelligence and its role in advancing antiviral therapies. *Future Microbiology*, *15*(12), 981–993.

Gupta, M., Singh, R., & Desai, N. (2020). Mobile health applications for managing antiviral therapy: A review of the literature. *Health Informatics Journal*, *26*(3), 311–326.

Gupta, P., & Sharma, A. (2021). The future of antiviral drugs and mHealth applications in managing viral diseases. *Health and Technology*, *11*(3), 281–289.

Hannan, A., & Rahman, M. (2021). mHealth interventions for managing chronic viral diseases: A systematic review. *Journal of Medical Internet Research*, *23*(7), e24653.

Hasanzadeh, A., Hamblin, M. R., Kiani, J., Noori, H., Hardie, J. M., Karimi, M., & Shafiee, H. (2022). Could artificial intelligence revolutionize the development of nanovectors for gene therapy and mRNA vaccines? *Nano Today*, *47*, 101665. DOI: 10.1016/j.nantod.2022.101665 PMID: 37034382

Hawley, C., & Bonner, S. (2019). Telemedicine for hepatitis C treatment in underserved populations. *Journal of Viral Hepatitis*, *26*(6), 756–763.

Hernandez, J., & Roberts, L. (2021). Telemedicine and its role in the future of antiviral care: A review. *Telemedicine Journal and e-Health*, *27*(7), 781–788.

Hirsch, M. S., & Wensing, A. M. (2019). Management of HIV-1 infection in adults. *Lancet*, *394*(10200), 381–392.

Hossain, M. M., & Alam, M. (2021). Role of bioinformatics in antiviral drug discovery and development. *Current Drug Targets*, *22*(2), 168–179.

Huang, C., & Zhao, M. (2021). The impact of digital health on the treatment of viral hepatitis. *Liver International*, *41*(4), 668–676.

Huang, C., Zhao, M., & Wang, L. (2020). Wearable devices for monitoring adherence to antiviral therapy. *BMC Infectious Diseases*, *20*, 481.

Jha, A., & Gupta, R. (2020). The potential of mRNA vaccines in tackling emerging viral diseases. *BioMed Research International*, *8657983*. Advance online publication. DOI: 10.1155/2020/8657983

Johnson, K. A., Zhang, H., & Wang, Y. (2023). mRNA-based vaccines: Innovations in lipid nanoparticle formulations and delivery mechanisms. *Nature Reviews. Drug Discovery*, *22*(5), 345–358.

Johnson, M. L., & Gupta, S. (2021). Digital health interventions for enhancing antiviral adherence in HIV patients: A review. *International Journal of Medical Informatics*, *150*, 104431.

Johnson, M. L., & Roberts, K. (2021). Overcoming the digital divide in health: Implications for antiviral treatment. *Public Health Reviews*, *42*, 160166.

Kaur, M., & Tiwari, S. (2020). Exploring the future of nanotechnology in antiviral drug delivery systems. *Nanomedicine; Nanotechnology, Biology, and Medicine*, *28*, 102178.

Khan, S., & Kamal, A. (2018). Telemedicine for HIV care: Enhancing treatment adherence in remote regions. *Global Health Action*, *11*(1), 1512972.

Khatun, F., & Khalil, S. (2020). The role of mobile health applications in managing chronic diseases in low-income countries. *Health Information Science and Systems*, *8*(1), 6. DOI: 10.1007/s13755-020-00109-3 PMID: 31949893

Kim, E. J., Moretti, M. E., Kimathi, A. M., Chan, S. Y., & Wootton, R. (2022). Use of provider-to-provider telemedicine in Kenya during the COVID-19 pandemic. *Frontiers in Public Health*, *10*, 1028999. DOI: 10.3389/fpubh.2022.1028999 PMID: 36478732

Kim, Y., Lee, H., & Park, J. (2021). Telemedicine and antiviral therapies: Enhancing access in underserved regions. *Telemedicine Journal and e-Health*, *27*(3), 251–260. PMID: 32857682

Kumar, K., & Rani, S. (2021). Impact of digital technologies in antiviral drug development: Current trends and future perspectives. *Biotechnology Progress*, *37*(1), e3139.

Lee, J. H., Hwang, K., & Kim, W. (2023). Machine learning in nanotechnology-based drug delivery systems for targeted therapy. *Advanced Drug Delivery Reviews*, *196*, 114–123.

Lee, Y., & Kim, Y. (2020). Bioinformatics applications in discovering antiviral drugs: A review. *Computer Methods and Programs in Biomedicine*, *196*, 105599.

Li, H., & Zhao, Y. (2020). Wearable devices for antiviral treatment adherence: A systematic review. *JMIR mHealth and uHealth*, *8*(4), e18934.

Li, J., & Zhang, Y. (2021). Machine learning in antiviral drug discovery: A comprehensive review. *Computational Biology and Chemistry*, *90*, 107408.

Li, J., & Zhang, Y. (2021). CRISPR-based antiviral strategies: Current status and future directions. *Frontiers in Microbiology*, *12*, 654.

Li, X., & Zhang, Y. (2021). mRNA vaccines: A new era in vaccinology. *Frontiers in Immunology*, *12*, 654.

Li, X., & Zhang, Y. (2021). Artificial intelligence in the development of mRNA vaccines. *Frontiers in Immunology*, *12*, 654.

Lian, W., Wen, L., Zhou, Q., Zhu, W., Duan, W., Xiao, X., Mhungu, F., Huang, W., Li, C., Cheng, W., & Tian, J. (2020). Digital health technologies respond to the COVID-19 pandemic in a tertiary hospital in China: Development and usability study. *Journal of Medical Internet Research*, *22*(7), e24505. DOI: 10.2196/24505 PMID: 33141679

Liu, X., & Zhan, Q. (2020). The use of wearable devices for monitoring the effects of antiviral medications: A review. *Current Pharmaceutical Design*, *26*(25), 2913–2923.

López, R., & Wang, Y. (2020). Telehealth for chronic viral hepatitis management: Current trends and future directions. *Journal of Clinical and Translational Hepatology*, *8*(4), 345–356.

López, R., & Wang, Y. (2021). Real-time data analytics in antiviral treatment management. *Health and Technology*, *11*(2), 335–348.

Mahalakshmi, S., & Kumar, A. (2021). Mobile health applications for chronic disease management: Opportunities and challenges. *Health Informatics Journal, 27*(2), 1083–1097.

Makoul, G., & Clayman, M. L. (2010). An integrative model of patient-centered communication in the healthcare setting. *Journal of Health Communication, 15*(2), 7–24.

Mansour, T., & Jang, Y. (2020). Nanomedicine approaches in drug delivery for antiviral therapies: A review. *Journal of Drug Delivery Science and Technology, 60*, 101987.

Mishra, S. K., & Zhang, J. (2020). Telemedicine in the era of digital health for chronic viral infections: A review. *Journal of Clinical Virology, 129*, 104539.

Mohamed, R. M., Gamal, S. M., & Fawzy, M. (2021). Mobile health technology for improving health outcomes: A review of existing applications. *Journal of Healthcare Engineering, 6625178*. Advance online publication. DOI: 10.1155/2021/6625178

Mohammadi, M., & Rasekh, A. (2021). Big data and AI applications in antiviral care: Challenges and perspectives. *Frontiers in Digital Health, 3*, 634210.

Mohammed, H., Oljira, L., Roba, K. T., Yimer, G., Fekadu, A., & Manyazewal, T. (2020). Containment of COVID-19 in Ethiopia and implications for tuberculosis care and research. *Infectious Diseases of Poverty, 9*(1), 131. DOI: 10.1186/s40249-020-00753-9 PMID: 32938497

Nguyen, A., & Trinh, K. (2021). Machine learning and bioinformatics in viral disease research: Future prospects. *Computational Biology and Chemistry, 88*, 107380.

Nguyen, T., & Li, H. (2021). Mobile health technologies for antiviral medication adherence: A systematic review. *AIDS Patient Care and STDs, 35*(5), 271–283.

Nguyen, T., & Li, H. (2023). mHealth applications in antiviral treatment: A review of benefits and challenges. *Mobile Health Research and Applications, 18*, 45–61.

Niakan Kalhori, S. R., Bahaadinbeigy, K., Deldar, K., Gholamzadeh, M., Hajesmaeel-Gohari, S., & Ayyoubzadeh, S. M. (2021). Digital health solutions to control the COVID-19 pandemic in countries with high disease prevalence: Literature review. *Journal of Medical Internet Research, 23*(1), e19473. DOI: 10.2196/19473 PMID: 33600344

Ochoa, S. A., & Fernández, L. (2021). Advances in wearable technology for monitoring patients undergoing antiviral treatment. *Nature Digital Medicine, 4*(1), 2–9.

Pandit, J. A., Radin, J. M., Quer, G., & Topol, E. J. (2022). Smartphone apps in the COVID-19 pandemic. *Nature Biotechnology*, *40*(9), 1013–1022. DOI: 10.1038/s41587-022-01350-x PMID: 35726090

Parker, S., & Patel, A. (2020). Digital health interventions for chronic viral diseases: A systematic review. *BMC Public Health*, *20*, 940. DOI: 10.1186/s12889-020-09468-2

Patel, A., & Verma, S. (2022). Digital health innovations for chronic disease management: Focus on viral infections. *Health Informatics Journal*, *28*(4), 647–662.

Patel, H., & Desai, M. (2020). Wearable devices for monitoring viral diseases: The next step in personalized antiviral care. *Future Science OA*, *6*(4), FSO542.

Patel, R., Singh, M., & Sharma, P. (2023). Bioinformatics and AI in pandemic preparedness: Applications for pathogen detection and outbreak modeling. *Journal of Biomedical Informatics*, *139*, 104578.

Patel, V. V., Darwal, R., & Rao, A. (2021). Impact of wearable technology on patient care in antiviral therapies. *BMC Medical Informatics and Decision Making*, *21*(1), 141. PMID: 33931058

Peek, N., Sujan, M., & Scott, P. (2020). Digital health and care in pandemic times: Impact of COVID-19. *BMJ Health & Care Informatics*, *27*(1), e100166. DOI: 10.1136/bmjhci-2020-100166 PMID: 32565418

Peeriga, R., Arabath, S. A. M. S., Rao, N. R. A., Yenireddy, A., & Bangaru, G. K. V. (2023). Exploring the coumarins from stem extract of *Ipomea sagittifolia* Burm. f. by high-performance thin layer chromatography. *International Journal of Chemical and Biochemical Science*, *23*(3), 268–273.

Peeriga, R., Balagani, P. K., Raju, G. E., Narayanan, R., Mohammed, J., & Yarlagadda, L. C. (2023). Findings of steroids, coumarins and saponins in *Ipomoea sagittifolia* (Burm. f.) leaf extract by HPTLC fingerprints. *International Journal of Chemical and Biochemical Science*, *24*(4), 132–140.

Peeriga, R., & Balamuruga, S. (2023). Investigation on ethanolic stem extract of *Ipomoea sagittifolia* to explore the presence of saponins by high performance thin layer chromatography. *International Journal of Membrane Science and Technology*, *10*(4), 1012–1020. DOI: 10.15379/ijmst.v10i4.2210

Peeriga, R., Shaik, A., Arabath, S., Chennuru, M. L., & Marimganti, S. N. (2023). Assessment of steroids in ethanolic stem extract of Morning glory by high performance thin layer chromatography. *International Journal of Chemical and Biochemical Science*, *23*(3), 361–366.

Rajan, R., & Suresh, R. (2020). CRISPR-based antiviral strategies for treating viral infections: Current trends. *Frontiers in Microbiology*, *11*, 577523.

Rajkumar, R. P., Kar, S. K., Chatterjee, S., & Yadav, S. (2021). Telemedicine in the era of COVID-19: Opportunities, challenges, and future directions. *Journal of Medical Systems*, *45*(5), 65.

Rani, S., & Jain, R. (2021). Future of mHealth in viral disease monitoring: Challenges and opportunities. *Telemedicine Journal and e-Health*, *27*(9), 1010–1018.

Roberts, C., & Brown, L. (2021). Telehealth interventions for antiviral treatment in remote areas: A systematic review and meta-analysis. *Telemedicine Journal and e-Health*, *27*(6), 123–130.

Rosenblum, J. E., & Chang, A. K. (2019). mHealth for chronic viral infections: A systematic review. *Journal of Mobile Health*, *6*(2), 30–37.

Roy, S., & Singh, B. (2020). The role of artificial intelligence in managing viral outbreaks. *IEEE Access : Practical Innovations, Open Solutions*, *8*, 123589–123596.

Schneider, A. R., & Busse, R. (2020). The integration of digital health tools into the healthcare system. *Healthcare Policy = Politiques de Santé*, *16*(4), 9–20.

Sharma, A., Patel, V., & Gupta, K. (2021). Digital health interventions to improve adherence to antiretroviral therapy: A meta-analysis. *JMIR mHealth and uHealth*, *9*(2), e24102.

Sharma, P., & Yadav, K. (2020). Integration of digital tools in chronic viral disease management: A future perspective. *Future Medicine*, *22*(6), 251–260.

Shen, X., & Li, J. (2020). Virtual consultations in the management of viral infections: A review of current applications. *Telemedicine Journal and e-Health*, *26*(10), 1235–1243.

Singh, R., Gupta, A., & Sharma, P. (2021). Privacy concerns in digital health platforms for antiviral therapy: A systematic review. *Journal of Medical Internet Research*, *24*(9), e34123.

Singh, R., & Kaushik, A. (2021). CRISPR technology and its role in advancing antiviral therapies: Future directions. *Bioengineering & Translational Medicine*, *6*(3), e10239. PMID: 34589605

Smith, A. E., & Brown, R. D. (2023). Next-generation sequencing: Transforming pathogen identification and resistance monitoring in low-resource settings. *International Journal of Infectious Diseases*, *128*, 95–101.

Smith, R., & Jones, T. (2023). Advancements in antiviral therapies and the role of digital health. *Journal of Clinical Virology*, *120*, 15–23.

Stojanovic, J., Melano, R., & Petrovic, B. (2020). Application of telemedicine and remote health monitoring in treating chronic viral diseases. *Computer Methods and Programs in Biomedicine*, *196*, 105713.

Su, Y., & Wang, X. (2019). Personalized medicine in antiviral treatment: The role of artificial intelligence. *Frontiers in Immunology*, *10*, 1833.

Suman, A., & Mishra, S. (2020). Role of CRISPR technology in managing viral infections: A comprehensive review. *Frontiers in Microbiology*, *11*, 1093.

Tiwari, P., & Rana, V. (2020). The future of mHealth in managing chronic diseases: Insights from COVID-19. *Telemedicine Journal and e-Health*, *26*(11), 1426–1432.

Wang, J., & Liu, M. (2020). Digital health solutions for COVID-19 management: A comprehensive review. *Journal of the American Medical Informatics Association : JAMIA*, *27*(9), 1415–1428.

Wang, L., Yang, Z., & Zhou, Y. (2021). A comprehensive review of AI in digital health and antiviral therapy. *Journal of Medical Systems*, *45*(4), 61.

Wang, L., & Zhang, Y. (2021). CRISPR-based antiviral strategies: Current status and future directions. *Frontiers in Microbiology*, *12*, 654.

Wang, L., & Zhang, Y. (2021). Nanotechnology in antiviral drug delivery: A review. *Nanomedicine (London)*, *16*(1), 1–15.

Wang, Y., & Zhang, Y. (2021). Artificial intelligence in the development of mRNA vaccines. *Frontiers in Immunology*, *12*, 654.

Wang, Y., & Zhang, Y. (2021). Artificial intelligence in antiviral drug discovery: Recent advances and future perspectives. *Frontiers in Pharmacology*, *12*, 654.

Wang, Z., & Liu, Y. (2021). Artificial intelligence applications in antiviral drug discovery: A comprehensive review. *Computational Biology and Chemistry*, *89*, 107358.

WHO Digital Health Technical Advisory Group. (2019). Global strategy on digital health 2020–2025. World Health Organization. Retrieved from https://www.who.int/docs/default-source/documents/gsdh2020-2025-en.pdf

World Health Organization. (2020). "Digital Health." WHO. Retrieved from https://www.who.int/health-topics/digital-health

Yudkin, J. S., Hodes, R. M., Sandler-Loeff, A., & Messiah, S. E. (2022). Needs assessment and best practices for digital trainings for health professionals in Ethiopia using the RE-AIM framework: COVID-19 case study. *Disaster Medicine and Public Health Preparedness*, *2022*, 1–9. PMID: 36226522

Zhang, J., & Cheng, Z. (2020). Data-driven insights into the impact of digital health in managing viral infections. *Journal of Medical Systems*, *44*(4), 68. PMID: 32072331

Zhang, Y., & Wang, X. (2021). Artificial intelligence in antiviral drug discovery: Recent advances and future perspectives. *Frontiers in Pharmacology*, *12*, 654.

Zhang, Y., & Wang, X. (2021). Machine learning in the development of mRNA vaccines. *Frontiers in Immunology*, *12*, 654.

Zhang, Z., & Liu, J. (2021). Insights into digital health technologies for the treatment of viral diseases. *International Journal of Medical Informatics*, *145*, 104300.

Chapter 6
Antiviral Therapeutics for Emerging Viruses

Odangowei Inetiminebi Ogidi
https://orcid.org/0000-0002-9961-893X
Bayelsa Medical University, Nigeria

Thomas Aye
https://orcid.org/0009-0007-2859-9026
Bayelsa Medical University, Nigeria

Iniebiyo Felagha
https://orcid.org/0009-0002-2976-228X
University of Africa, Nigeria

ABSTRACT

Efficient pharmaceuticals are required to address the rise of new viral infections. HIV and HCV infections serve as examples of the past ineffectiveness of single-medication treatments for chronic viral diseases. Significant advancements in clinical outcomes have been observed through the implementation of combination therapy for a range of diseases, resulting in notable decreases in viral load, morbidity, and mortality. Several newly discovered and resurfacing viruses can be more efficiently treated using combinations of drugs, which exhibit additive and, ideally, synergistic effects. Viruses such as influenza, SARS-CoV, MERS-CoV, Ebola, Zika, and SARS-CoV-2 are included in this list. Discovering new drugs through traditional pipelines should always be the primary focus. However, drug-repurposing screens offer a rapid way to identify potential medication candidates with synergistic effects. This approach allows for more time to conduct clinical trials and potentially obtain emergency use authorisations. This chapter discusses antiviral treatments for emerging viruses.

DOI: 10.4018/979-8-3693-9276-8.ch006

Copyright © 2025, IGI Global Scientific Publishing. Copying or distributing in print or electronic forms without written permission of IGI Global Scientific Publishing is prohibited.

INTRODUCTION

Global health is significantly threatened by major outbreaks of viral diseases. As of May 2022, the severe acute respiratory syndrome coronavirus 2 (SARS-CoV-2) epidemic has caused more than 513 million illnesses worldwide (WHO, 2022a; Ogidi, 2020). Of these, 6.2 million have passed away. SARS and MERS are among the numerous coronavirus varieties that have the potential to cause fatalities (WHO, 2015; WHO, 2022b). The resurgence of viral illnesses, such as the Ebola virus, led to a pandemic in West Africa from 2014 to 2016, with a case fatality rate reaching as high as 50%. The present case fatality rate in the Democratic Republic of the Congo is 66% (WHO, 2020a). Equatorial Province, Democratic Republic of the Congo, documented an increase in Ebola cases in April 2022, according to the World Health Organisation (2022c). This suggests that the virus continues to be a significant source of concern in the context of global health. However, human hepatitis C, hepatitis B, and HIV infections continue to have a significant impact on public health around the world. HIV infection has resulted in approximately 700,000 direct fatalities and has infected an additional 1.7 million individuals (WHO, 2020b).

Viral hepatitis has resulted in the deaths of approximately 1.34 million individuals, and there are 71 million individuals who are infected with HCV. Additionally, it is now acknowledged that 250 million individuals are affected by chronic HBV (WHO, 2017). The influenza virus poses a substantial annual public health concern on a global scale. The influenza A (H1N1) pdm09 virus, which initiated a global pandemic in 2009 and is now considered a seasonal flu, is responsible for an estimated 150,000 to 570,000 fatalities worldwide (Dawood et al., 2012). The avian influenza A (H5N1) subtype of the influenza virus has raised concerns since its first human infection in 1997 (WHO, 2022d).

Attaching to the host cell membrane, entering the cell, uncoating, transcribed and replicated genome, translated proteins, assembly, and budding/release are just a few of the critical stages that a virus undergoes during its life cycle (Ogidi & Ajoko, 2024). Genome types, such as their positive or negative polarity and whether they are single-stranded or double-stranded, have an impact on viral transcription and replication processes (Tompa et al., 2021). The viral genome is released into the cytoplasm, and messenger RNA is synthesised to facilitate the virus's proliferation, among other things. The Vaccinia virus and other poxviruses are able to proliferate independently within the cytoplasm due to the presence of transcription and replication-related proteins in their viral genomes (Reyes et al., 2017). Although the majority of DNA viruses replicate and express their genes by transcribing their genomes within the nucleus of the host cell, they rely on host enzymes to accomplish this. The replication of RNA viruses is facilitated by the cytoplasmic RNA-dependent RNA polymerase. Retroviruses, such as HIV, are transported to the nucleus for genome integration through reverse transcription by RNA-dependent DNA polymerase (RdDp) (Boersma et al., 2020; Ogidi et al., 2024; Lu et al., 2021).

Another example of RNA viruses is the replication of influenza viruses in the nucleus (Dou et al., 2018). The focus of antiviral medication development has increasingly shifted towards targeting viral or host cell proteins at specific stages of the viral life cycle (Tompa et al., 2021; Clercq & Li, 2016). Despite the development and approval of alternative methods to combat viral infections, the escalating prevalence of drug-resistant viral strains has rendered existing antiviral medications less effective in treating viral illnesses (Strasfeld & Chou, 2020; Lampejo, 2020). Consequently, we present a novel platform that inhibits viral replication by directly promoting viral genome breakdown, thereby enabling the identification of novel medications that are effective against a wide variety of viruses (VanPoelvoorde et al., 2020). (Kumar et al., 2020) conducted extensive research on RNA interference (RNAi), CRISPR/Cas systems, ribonucleases (RNases), and interferon (IFN)-stimulated gene 20 (ISG20). All of these potential candidates have shown antiviral effects in studies that have examined viral genome cleavage in both laboratory and living organism settings. These methods were found to be the simplest and most commonly used for cleaving viral genomes (Durmus & Ulgen, 2017).

Objectives of the Chapter

The aims of this chapter are as follows:

(1) To evaluate the recent viral outbreaks and potential threats.
(2) To discuss the advancements of essential instruments and techniques involve antiviral therapies.
(3) To review the types of therapies and their mechanisms of actions.
(4) To evaluate the RNA-based and pre-emptive therapeutics development.
(5) To X-ray the therapeutics regimen and delivery.

Organization of the Chapter

The rest of this chapter is organized as follows: Section 2 explores the recent viral outbreaks and potential threats. Section 3 discusses the advancements of essential instruments and techniques involve antiviral therapies. Section 4 traverse the types of therapies and their mechanisms of actions. Section 5 discusses the RNA-based and pre-emptive therapeutics development. Section 6 X-ray the therapeutics regimen and delivery. Finally, a concluding section summarizes the antiviral therapeutics for emerging viruses.

RECENT VIRAL OUTBREAKS AND POTENTIAL THREATS

Coronavirus, alphavirus, myxovirus, filovirus, norovirus, and flavivirus members have been responsible for over ten significant viral illness epidemics or pandemics in human populations over the past two decades of this century. Small, uncommon epidemics have also been caused by a variety of zoonotic RNA viruses, such as arenaviruses, bunyaviruses, and henipaviruses (Edwards et al., 2020). It is essential to adopt a one-health approach that integrates public health practices, veterinary medicine, and human and animal health in order to mitigate the risk of future zoonotic virus outbreaks. This is because the virus can be transmitted to people through economically important livestock and other animals that serve as reservoir hosts. Predicting future epidemics with any degree of accuracy remains difficult, despite the fact that variants present before an epidemic can be found in reservoirs and can infect both animals and humans on occasion (Menachery et al., 2016, 2015). Some of recent global viral outbreaks are enumerated in Figure 1.

Figure 1. Recent global viral outbreaks

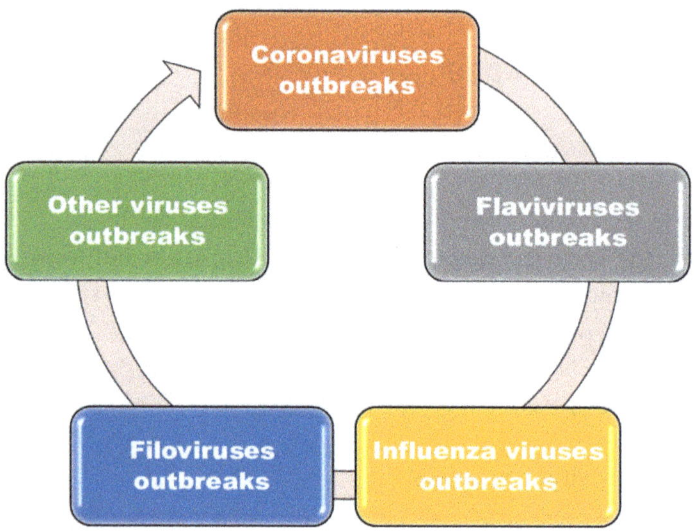

Coronaviruses

The SARS epidemic, which was initiated by SARS-CoV, signified the commencement of the new millennium. In the latter half of 2002, SARS-CoV first emerged and rapidly disseminated worldwide until it was finally contained in the summer of 2003 (Ogidi, 2024a). At this time, there were thousands of confirmed cases, and approximately 10% of those individuals succumbed to their injuries (Cherry, 2004). The Middle East Respiratory Syndrome (MERS) was caused by the advent of MERS-CoV, a new zoonotic coronavirus, in 2012. Sporadic outbreaks and MERS cases have persisted since 2012. Despite the fact that the aggregate number of MERS-CoV infections has decreased, the death rate remains extremely high, reaching nearly 35% (Sharif-Yakan & Kanj, 2014). The COVID-19 pandemic commenced in 2019 with the emergence of SARS-CoV-2 (Ogidi, 2022). Although bats are the putative progenitors of all of these coronaviruses, some of these viruses utilise camels and civet cats as intermediate hosts (Li et al., 2005). Additionally, bats are home to a number of group I and II coronaviruses that are capable of replicating in basic human cells, are prepared to arise, and are present prior to the onset of an epidemic (Luo et al., 2018).

Flaviviruses

The National Institutes of Health have classified dengue, Zika, West Nile, and yellow fever viruses as category A/B among the mosquito-borne flaviviruses that have recently emerged. Since the early 2000s, when new strains of DENV emerged every few years, the virus which is endemic to some regions of Southeast Asia and the Americas has caused substantial epidemics (Gulber, 2002). In some nations, people who have had DENV in the past can safely have the Dengvaxia vaccination; however, people who have never had the virus should not get it (Thomas & Yoon, 2019). In 2015, the Zika virus spread quickly over South and Central America, the Caribbean, and beyond. The virus has its origins in Polynesia and had previously thrived in Brazil (Gulber, 2002). Babies born to women who contracted the virus during the first trimester often had severe microcephaly and other CNS problems, although the infant mortality rate was quite low (Gulber, 2002). In 2016, YFV re-emerged in Angola and subsequently spread to neighbouring African nations and, ultimately, China through tourism (Gulber, 2002). The virus's dissemination was mitigated by vaccination (Ogidi et al., 2021). The global proliferation of vectors, travel, and global warming are anticipated to contribute to an increase in the transmission of these flaviviruses by mosquitoes (Kauffman & Kramer, 2017).

Influenza Viruses

The 1918 H1N1 flu pandemic, which resulted in the deaths of an estimated 50 million individuals worldwide, was the most devastating of the numerous notable epidemics caused by influenza viruses that occurred in the twentieth century. As well as other epidemics in 1957 (H2N2 Asian flu) and 1968 (H3N2 Hong Kong flu), there were sporadic human cases of exceedingly dangerous avian influenza viruses (Osterholm et al., 2015). The H1N1 swine flu pandemic that erupted in the new millennium resulted in the deaths of approximately 250,000 individuals worldwide in 2009 (Osterholm et al., 2015). This influenza strain, which is associated with the 1918 pandemic, was developed in pigs through a three-way reassortment process that involved strains from birds, piglets, and humans (Neumann et al., 2009). Persistent epidemics in humans have been caused by rare influenza viruses, including H5N1, H7N9, and others, this century (Neumann et al., 2010). Fortunately, these viruses are not readily transmitted from person to person, despite their high mortality rate among humans. Active research is being conducted to develop vaccines and treatments that are extensively applicable and effective against influenza viruses (Nachbagauer et al., 2021).

Filoviruses

Both the Ebola and Marburg viruses are considered category A diseases by the National Institutes of Health and the Centres for Disease Control and Prevention. Two major EBOV epidemics have occurred in the last ten years. Between 2013 and 2016, the outbreak that ravaged West African nations resulted in the infection of approximately 30,000 individuals, out of whom 40% perished (Holmes et al., 2016). The Democratic Republic of the Congo experienced a second significant outbreak between 2018 and 2020, which resulted in the deaths of 65% of the infected and the illness of 3,500 individuals (Ilunga-Kalenga et al., 2019). Even though armed conflicts in the region hindered vaccination efforts, the World Health Organisation licensed a highly effective vaccine candidate in 2019, inoculating over 200,000 individuals and contributing to the containment of the outbreak (Henao-Restrepo et al., 2017). Bats are the primary reservoirs of EBOV, although the virus can infect intermediate zoonotic hosts such as humans (Arankalle et al., 2007). EBOV can be transmitted to humans through sexual contact or bodily secretions, such as blood, when it infects them. Alphaviruses, Chikungunya virus, or CHIKV for brief, is an alphavirus that is indigenous to Africa and Asia. In 2006, it was estimated that over one million incidents took place in India. Several million cases were documented as a result of the virus's major outbreaks that spread across the Americas from 2013 to 2016. The geographic distribution of CHIKV is restricted by the fact that, like the other flaviviruses previously discussed, it is transmitted by mosquitoes (Burt et al., 2017). Nevertheless, it is anticipated that this will change as a consequence of climate change and the proliferation of vectors. The virus can cause debilitating chronic illness in those who contract it, despite the rarity of fatalities from CHIKV infections (Lindsey et al., 2018). Furthermore, a group of other alphaviruses, such as the Mayaro virus, Eastern equine encephalitis virus, and Venezuelan equine encephalitis virus, are a growing concern for the health of humans. These pathogens are classified as category B by the National Institutes of Health and the Centres for Disease Control and Prevention (Forrester et al., 2017).

Other Viruses of Significance

Several viruses in the order Bunyavirales have caused or could induce epidemics; they include hantaviruses, CCHFV, LASV, and RVFV (Bente et al., 2013). Because of their high fatality rates (10-40%) and their transmission by insects and rodents, these viruses have been designated as category A priority diseases by the NIH and CDC (Jonsson et al., 2010). In 2018, India experienced a Nipah virus (NiV) epidemic, which was one of numerous minor henipavirus outbreaks that resulted in high mortality rates (60-90%). The risk of transmission to humans is reduced by the development of a vaccination against Hendra virus (HeV) for horses, as pigs and horses mediate the transmission of henipaviruses from bats (Garnett & Strong, 2019). The devastating effects of norovirus infections are particularly pronounced in children, the elderly, and adults with impaired immune systems. These infections have caused numerous epidemics in the past two decades and result in the deaths of between 70,000 and 200,000 individuals annually worldwide (Himeidan et al., 2014). Table 1: Contrasting Therapeutic Approaches for SARS-CoV, Influenza Viruses, Flaviviruses, Filoviruses, and Other Viruses. The development of effective vaccines and antiviral treatments is jeopardised by norovirus's extreme variability. Rabies and hepatitis A are additional viruses that have been designated as high-priority by the NIH (Himeidan et al., 2014).

Table 1. Contrasting Therapeutic Approaches for SARS-CoV, Influenza Viruses, Flaviviruses, Filoviruses, and Other Viruses

Therapeutic Approach	SARS-CoV (Coronavirus)	Influenza Viruses	Flaviviruses (e.g., Dengue, Zika)	Filoviruses (e.g., Ebola, Marburg)	Other Viruses (e.g., HIV, Hepatitis C)
Direct-Acting Antivirals (DAAs)	- Inhibit viral RdRp (e.g., Remdesivir) and main protease (Mpro, e.g., Nirmatrelvir).	- Target neuraminidase (e.g., Oseltamivir, Zanamivir) and polymerase complex (e.g., Baloxavir).	- Inhibit viral protease (NS3) or polymerase (NS5).	- Block RNA polymerase and glycoprotein-mediated replication (e.g., Remdesivir, Galidesivir).	- Inhibit reverse transcriptase, protease, or integrase (e.g., Zidovudine for HIV, Sofosbuvir for HCV).
Host-Factor Inhibitors	- Block ACE2 receptor or TMPRSS2 protease (e.g., Camostat mesylate, soluble ACE2 decoys).	- Inhibit host proteases (e.g., TMPRSS2) or endosomal pH regulators (e.g., Amantadine).	- Target host ER α-glucosidase for viral assembly (e.g., Celgosivir).	- Inhibit Niemann-Pick C1 (NPC1) receptor required for viral entry (e.g., U18666A).	- Block host CCR5 or CXCR4 co-receptors (e.g., Maraviroc for HIV); inhibit HCV entry factors (e.g., ITX-5061).
Monoclonal Antibodies (mAbs)	- Neutralize spike protein to block receptor binding (e.g., Sotrovimab, Casirivimab).	- Target hemagglutinin (HA) to prevent viral fusion (experimental).	- Neutralize envelope (E) proteins (e.g., ZIKV-117 for Zika, Dengue-specific mAbs).	- Target glycoproteins to block viral fusion (e.g., REGN-EB3, mAb114 for Ebola).	- Neutralize viral surface proteins (e.g., Ibalizumab for HIV; HCV mAbs targeting E2 glycoprotein).
Immunomodulators	- Suppress cytokine storm with IL-6 inhibitors (e.g., Tocilizumab), corticosteroids (e.g., Dexamethasone).	- Reduce hyperinflammation with JAK inhibitors or cytokine blockers (experimental).	- Limit inflammation using immunosuppressants or antivirals combined with supportive care.	- Modulate excessive immune responses with interferons or JAK inhibitors (e.g., Ruxolitinib).	- Boost immune response in HIV with IL-2 or IFN-α; immunotherapy for HCV involves Pegylated IFN-α.
Vaccines	- mRNA (Pfizer, Moderna), viral vector (AstraZeneca, J&J), inactivated (e.g., CoronaVac).	- Inactivated (e.g., Fluzone), live-attenuated (e.g., FluMist), recombinant (e.g., Flublok).	- Live-attenuated (e.g., Dengvaxia for Dengue); inactivated or subunit vaccines in development.	- Recombinant vector vaccines (e.g., rVSV-ZEBOV for Ebola), DNA-based vaccines under investigation.	- HIV: No licensed vaccine (several in development); HCV: Vaccine development ongoing; Hepatitis B: Recombinant vaccines (e.g., Engerix-B).

continued on following page

Table 1. Continued

Therapeutic Approach	SARS-CoV (Coronavirus)	Influenza Viruses	Flaviviruses (e.g., Dengue, Zika)	Filoviruses (e.g., Ebola, Marburg)	Other Viruses (e.g., HIV, Hepatitis C)
Entry Inhibitors	- Prevent spike protein-ACE2 interaction or membrane fusion (e.g., Arbidol, ACE2 decoys).	- Block hemagglutinin-mediated entry (e.g., fusion inhibitors like Umifenovir).	- Block viral fusion pathways involving E protein (e.g., Chloroquine, Arbidol).	- Inhibit glycoprotein-mediated entry (e.g., Toremifene, C-peptide derivatives).	- Block fusion with host cells (e.g., Enfuvirtide for HIV); experimental entry inhibitors for HCV (e.g., ITX-5061).
RNA Interference (RNAi)	- siRNAs target viral RNA (e.g., spike protein, RdRp).	- Experimental siRNAs targeting polymerase genes (e.g., PA, PB2).	- siRNAs target key viral genes (e.g., NS5 for Zika, NS3 for Dengue).	- siRNAs suppress viral RNA replication and glycoprotein expression (e.g., Ebola VP35).	- siRNAs for HIV target gag and tat genes; for HCV, target NS5B polymerase or core protein.
Antisense Oligonucleotides	- Bind to viral RNA to block replication or translation (e.g., targeting RdRp or spike mRNA).	- Experimental ASOs for blocking influenza RNA segments.	- ASOs designed to silence viral RNA (e.g., NS3, NS5).	- ASOs target filoviral mRNA to block translation of glycoproteins.	- HIV: ASOs target viral RNA (e.g., rev, tat genes); HCV: ASOs inhibit NS5B and other viral proteins.

ADVANCEMENT OF ESSENTIAL INSTRUMENTS AND TECHNIQUES

Drug repurposing's inherent risk aversion, allocating scarce resources to pandemic preparedness, targeting treatments against unknown pathogens, a lack of appropriate models for in vitro and in vivo evaluation, and antiviral therapeutics that work for most people are the main problems. Furthermore, the future of the market is unclear.

Viral Isolates and Molecular Clones

Recently, through viral discovery initiatives and metagenomics, whole genome sequences have been identified from a variety of viral lineages, despite the diversity within the RNA virus family. This enables the investigation of the efficacy of pharmaceuticals and vaccinations with a diverse array of genetic variants (Epstein & Anthony, 2017; Ogidi & Kumar, 2024). Medications that successfully block several viral families will increase faith in their effectiveness against unknown diseases, which is important because future pandemic strains are undetermined (Ogidi & Tobia, 2024; Zhu et al., 2020). The use of cell culture or animals to efficiently extract viruses from clinical, animal, or environmental materials is not without its challenges (Messer et al., 2012). Reverse genetics employing known viral sequences has recently enabled the creation of synthetic molecular clones of the viral genome; this has allowed for the production of pure stocks of viruses from various evolutionary branches (Yount et al., 2003; Hou et al., 2020).

In addition, there are opportunities in synthetic biology to create reporter viruses, which are great for setting up high-throughput procedures (Kim et al., 2012). This approach makes it possible to systematically test different zoonotic strains for the ability to replicate in human primary cells, with the goal of finding strains that could cause an epidemic before it starts. Various recombinant protein biochemical tests are available for use in high-throughput drug screening. Nevertheless, researchers employed a combination of present and future human coronaviruses in addition to high- and low-risk zoonotic coronaviruses that were produced through reverse genetics to guarantee that remdesivir, molnupiravir, and the human monoclonal antibody ADG2 were effective against an unidentified strain (Zou et al., 2011).

Cellular Culture Models of Viral Infection

Although a viral repertory is significant, it is equally essential to have robust models that accurately replicate pertinent human tissues. Although it can be difficult to separate viruses from samples, the likelihood of successful culture is increased by the use of natural immunodeficient cells or receptor-expressing cells for viruses with known receptors (Hou et al., 2020). High-throughput screening techniques are essential for minimising the resources required to develop new pharmaceuticals, including time, energy, and money. Transgenic human cell lines that are susceptible to the target virus are frequently employed in high-throughput screening due to their affordability and accessibility. This methodology has enabled the identification of pharmaceuticals that are beneficial (Hou et al., 2020). Nevertheless, it is crucial to acknowledge that modified cell lines may not always possess the requisite information to modify drug bioavailability, transport, or metabolism. The main cell line portfolios may facilitate a more comprehensive comprehension of drug bioavailability across a variety of susceptible cell types that may be implicated in human disease and pathology (Dijkman et al., 2013).

The research on stem cells and organs-on-a-chips for humans has made significant progress over the past two decades, resulting in the production of viable primary human tissues from a variety of organs that have been affected by RNA viruses (Tang et al., 2020). Primordial cell line screening is the gold standard for optimising effectiveness; the additional time and money invested will be more than compensated for by a higher probability of successful hits and enhanced relevance, particularly for medicines that target the host (Zhang et al., 2020). Prior to being assessed in animal disease models, candidates must undergo additional validation in organ-on-a-chip systems following the identification of primary results. By sharing the burden of thorough drug testing and pooling knowledge in virology, culture models, and screening methodologies, collaborations will greatly improve screening efforts (Yang et al., 2020).

Animal Models for Viral Diseases

The subsequent stage in the evaluation of medication efficacy is the utilisation of small-animal models. For instance, it may be challenging to derive meaningful conclusions from the data when comparing outbred humans to inbred models due to the genetic and expression patterns of the two species. In order to replicate an organism's response to infection, treatment research employs both small- and large-animal models (The Complex trait consortium, 2018). Research on viral pathogenesis and transmission, as well as to meet the FDA's two-animal requirement, sometimes makes use of guinea pigs, hamsters, or ferrets, however laboratory mice are by far the most common animal model. This is why it's critical to find a way to make mice models that mimic the high viral titers and many serious symptoms seen in humans. Since not all relevant viruses infect or replicate successfully in lab animals, new models must be developed. The Collaborative Cross is an example of an inbred mouse population that may exhibit severe disorders that resemble human situations and depicts the genetic variety of Mus musculus. This renders it an optimal paradigm for research (Sun et al., 2020a). This will enable the development of mouse models that express the human receptor once the viral receptor has been identified. For the production of stable receptor-expressing animals, there are numerous methods available, such as viral transduction and genetic processes such as CRISPR-Cas9 (Sun et al., 2020a).

Unlike transgenic mice, which frequently express receptors in non-specific tissues, knock-in models depend on mouse regulatory loci to guarantee patterns of host gene expression. This makes data interpretation more difficult. Alternatively, you may employ natural selection to increase the infection of the secondary host, rather than exposing the animal to vulnerability. It is feasible that mice infected with certain viruses may not exhibit symptoms comparable to those observed in humans, despite the fact that some viruses are capable of replicating in mice (Sun et al., 2020b). Experimental evolution, which involves the repetitive inoculation of mice or other laboratory animals with the virus, has the potential to generate a more hazardous strain of the virus. This technique may have a potential drawback, despite its effectiveness: it could be altered in antigenicity or sensitivity as a result of medication. In order to mitigate the challenges associated with interpreting data from various species, it is advisable to select models that reproduce the characteristics of human diseases and tissue affinity (Sun et al., 2020b).

TYPES OF THERAPIES AND THEIR MECHANISMS OF ACTION

Every virus possesses a distinct biological mechanism; nevertheless, they all exhibit some fundamental mechanisms that may be therapeutically targeted. The viral entrance constitutes the initial phase of infection. According to Li et al. (2003), one example of a viral surface protein interacting with a specific receptor or attachment factor is the SARS-CoV-2 host receptor ACE2. The virus is able to enter the body through these linkages, usually by endocytic routes, which include membrane fusion at the plasma membrane or transit through endosomes and lysosomes (Milewska et al., 2018). Prior to entering a cell, viruses must first uncoat and release their genome, which may occur in either the cytoplasm or the nucleus (Millet & Whittaker, 2018). This mechanism may entail endosomal escape or viral fusion. Replication occurs subsequent to the translation of the genome into messenger RNA (mRNA), which subsequently initiates transcription, while viral or host proteases process the translated polyproteins (Kirchdoerfer & Ward, 2019).

While certain viruses necessitate host polymerases, numerous RNA viruses possess their own polymerases. Although all viruses utilise the identical host cell translation machinery, the regulation of viral translation is unique (Jan et al., 2016). A new cycle commences with the synthesis of structural proteins and genomes, resulting in the production of viral particles. The particles subsequently escape the cell via viral egress or cellular lysis. Therapeutics possess the capacity to attack each phase of the viral life cycle collectively (Neuman et al., 2011). There are two main classes of antiviral drugs based on whether they attack the virus or its cells: direct-acting and host-factor (Figure 2). According to Glasgow et al. (2020), there are numerous ways in which direct-acting antivirals aim to destroy the virus. Included in this category are receptor decoys, substances that bind to and destroy viral particles, antibodies, nucleoside analogues, and viral protease inhibitors. The class of drugs known as host-factor antivirals includes receptor-binding antibodies, lipidomic reprogramming medicines, kinase inhibitors (like baricitinib), and endocytosis inhibitors (Linsky et al., 2020). For viruses to successfully infect or cause disease, certain components are necessary. Klemm et al. (2020) found that direct-acting and host-factor therapies benefited greatly from antiviral medications that targeted conserved genes or motifs within a virus family. The coronavirus 3CL protease and RNA-dependent RNA polymerase share 40% sequence similarity and nearly 60% sequence identity, respectively (Chen et al., 2021). Ogidi & Emaikwu (2024) and Ogidi (2024b) cite furin protease as another example that uses the same host pathways to increase viral proliferation and pathogenicity. To find important host components that could be targeted, CRISPR screens are useful tools (Raj et al., 2013).

Figure 2. Classes of Antiviral drugs

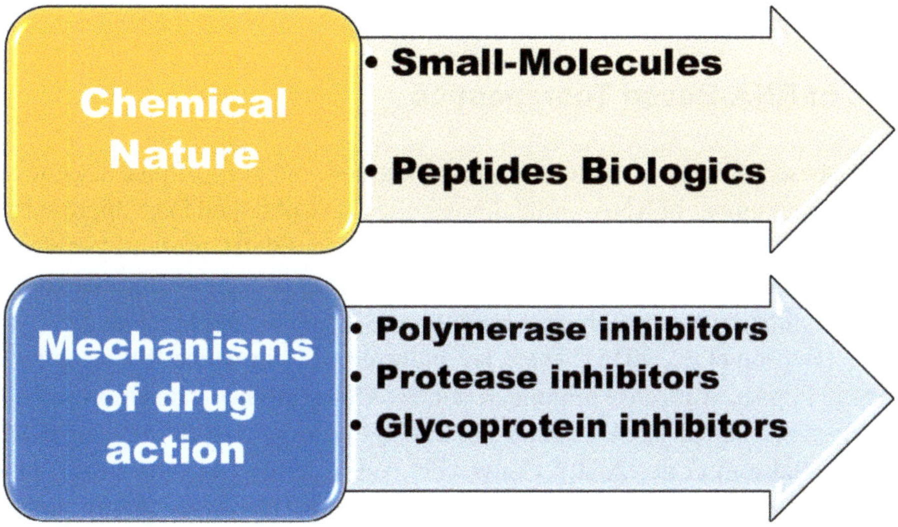

It is possible that host-factor therapies are more harmful than thought due to the interference with biological processes in both healthy and infected cells (Ivanov, 2008). As an infection advances through the viral life cycle, a particular illness emerges within the host organism. During the viral incubation phase (~1–14 days), symptom onset correlates with increased viral load, as demonstrated by Inove et al. (2007). Physiological harm is a common consequence of the disease's transition from the viral to the inflammatory phases, when the body's antiviral inflammatory responses start to control viral multiplication (Zhou et al., 2015). Antiviral treatment windows are often effective solely during the viral infection phase and diminish in efficiency thereafter (Yuan et al., 2019). Testing is essential during an outbreak because of the restricted treatment windows (Zhang et al., 2020). Corticosteroids and other immunomodulatory and anti-inflammatory agents are accessible as therapeutic alternatives throughout the inflammatory phase; these medications attenuate the body's immune response, hence diminishing the intensity of inflammation (Anand et al., 2003; Braun & Sauter, 2019). Patients in hospitals undergoing treatment for COVID-19 are administered the corticosteroid dexamethasone (Wei et al., 2021; Linton et al., 2020). Baricitinib, a kinase inhibitor, diminishes cytokine production by inhibiting the JAK/STAT pathway and possesses an emergency use authorisation for the treatment of Covid-19 (Peiris et al., 2003; Widagdo et al., 2017; Ahmed, 2019). Cytokines, including interferon-κ, exhibit antiviral properties and can be delivered promptly without inducing tissue damage, indicating a potential alternative strategy (Yang et al., 2020; Stebbing et al., 2020; Andreakos & Tsiodras, 2020).

RNA-BASED THERAPEUTICS

Types of RNA-Based Therapeutics

RNA-based therapy constitutes a promising new category of pharmaceuticals. A significant number of RNA-based therapeutics have obtained FDA approval over the last twenty years, with numerous additional candidates presently in various stages of clinical evaluation (Ng et al., 2006; Coelho et al., 2013). Because they are more easily constructed and have a modular design, these medications have great promise (Benson et al., 2018). There are multiple ways in which RNA can have a medicinal impact. Antisense oligonucleotides (ASOs) modify translation, splicing, or the accessibility of RNA-binding proteins by base-pairing to messenger RNA (mRNA) (Balwani et al., 2020). Conversely, anti-microRNAs (anti-miRs) physically attach to miRNAs, interfering with their function (Hua et al., 2010). Inducing mRNA degradation or translational repression using microRNAs (miRNAs) or small interfering RNAs (siRNAs) is one possible therapeutic strategy (Lanford et al., 2010). Comparable to messenger RNAs (miRNAs), small activating RNAs (saRNAs) help activate transcription from within the nucleus (Giladi et al., 2003). mRNAs have potential uses as vaccines and treatments in protein synthesis (Sarker et al., 2020). In the same way that traditional small-molecule drugs bind to specific targets, RNA aptamers can be engineered to bind to a wider variety of substrates (Kose et al., 2019). The ultimate goal of using CRISPR effectors and gRNAs is to change genes or degrade cellular RNAs. The regulation of gene expression, splicing, and translation are only a few of the many functions performed by RNA-based products (Abbott et al., 2020). Some of the types of RNA-based viral therapeutics are shown in Figure 3.

Figure 3. Types of RNA-based viral therapeutics

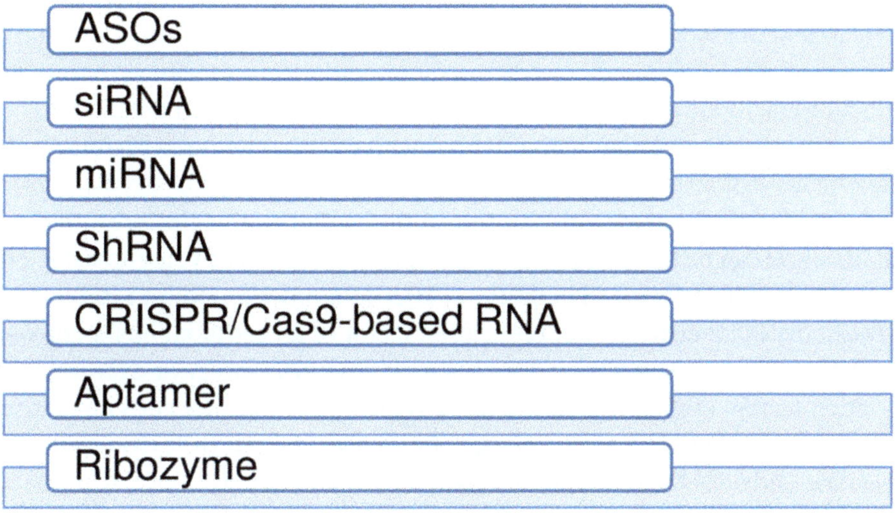

Delivery of RNA-Based Therapeutics

The challenge of accurately targeting specific tissues and cell types for drug delivery is significant for RNA-based therapies (Prakash et al., 2014). Due to its inherent lack of targeting capabilities, RNA has been the subject of much research into the necessity of localised distribution, a delivery vehicle, or alterations in recent years (Sugo et al., 2016). To control the translation of RNA, a sugar, peptide, antibody, or RNA aptamer can be attached to it. To target cells infected with HIV, one can use an RNA aptamer that binds the envelope glycoprotein of the virus (Zhou et al., 2018). Another option is to encase RNA in lipid nanoparticles and then administer them systemically. There is a wide variety of lipid nanoparticle formulations, and researchers are still exploring new ways to incorporate targeting moieties and methods for RNA release at specific sites into these particles (Veiga et al., 2018). 'Naked' RNA, when given locally to a specific location, can be highly effective (Carlsson et al., 2018). Recent clinical trial outcomes of antisense oligonucleotides, siRNA, and mRNA vaccines is shown in Table 2.

RNA Therapeutics for Viral Outbreaks

Treatments based on RNA have significant potential for the prevention and treatment of viral infections. Given the target molecule's sequence, base-pairing interactions facilitate the development of RNA therapeutics (Ahn et al., 2011). They can be altered to target structural motifs or conserved regions with precision, resulting in treatments that are widely applicable. It is easy to adapt viral targets after they have been identified to new viral sequences or to other strains that are similar. The viral life cycle can be targeted by RNA therapies, which can be categorised as either direct-acting or host-factor (Szabat et al., 2020). In order to inhibit viral infection, RNA can disrupt a critical viral replication factor or an essential host component. On the flip side, RNA products can counteract the viral effect of decreasing host restriction factors (Whitten-Bauer et al., 2019). One of the first antiviral antisense oligonucleotide medicines to be licensed, fomivirsen, targets the cytomegalovirus's immediate-early RNA and suppresses translation (Meng et al., 2006). Clinical trials for the treatment of chronic hepatitis B virus (HBV) infection are presently underway for a number of potential small interfering RNA (siRNA) therapeutics that target the HBV surface antigen. Hepatitis C virus (HCV) transmission has been linked to miR-122, and one strategy is to inhibit its expression. Moreover, as the COVID-19 pandemic has shown, mRNA therapies have great promise as vaccines (Gorbea et al., 2017). Table 3 shows novel advancements in clinical trials for antisense oligonucleotides, siRNA, and mRNA Vaccines.

Table 2. Recent Clinical Trial Outcomes of Antisense Oligonucleotides, siRNA, and mRNA Vaccines

Therapeutic Modality	Clinical Trial	Target Disease	Key Outcomes	References
Antisense Oligonucleotides (ASOs)	Eteplirsen (Exondys 51)	Duchenne Muscular Dystrophy	Improved dystrophin production in muscle tissue; slowed disease progression.	Food and Drug Administration (FDA), 2016.
	Nusinersen (Spinraza)	Spinal Muscular Atrophy	Increased survival rates and motor function in infants with SMA.	Finkel et al., 2017.
	IONIS-HTTRx (Tominersen)	Huntington's Disease	Reduced levels of mutant huntingtin protein in cerebrospinal fluid; showed potential for slowing disease progression.	Tabrizi et al., 2019.

continued on following page

Table 2. Continued

Therapeutic Modality	Clinical Trial	Target Disease	Key Outcomes	References
Small Interfering RNA (siRNA)	Patisiran (Onpattro)	Hereditary Transthyretin Amyloidosis (hATTR)	Reduced transthyretin levels by over 80%; improved neuropathy scores and quality of life.	Adams et al., 2018.
	Inclisiran (Leqvio)	Hypercholesterolemia	Reduced LDL cholesterol levels by over 50%; maintained efficacy with biannual dosing.	Ray et al., 2020.
	STP707	Solid Tumors	Demonstrated safety and significant tumor reduction in Phase I clinical trials.	Sirnaomics, 2023.
mRNA Vaccines	Tozinameran (Pfizer-BioNTech)	COVID-19	~95% efficacy in preventing symptomatic COVID-19; well-tolerated with minor side effects.	Polack et al., 2020.
	Elasomeran (Moderna)	COVID-19	~94% efficacy; robust and durable immune response observed.	Baden et al., 2021.
	SYS6020	Myeloma	Approved for human trials; an mRNA-based cancer therapy targeting myeloma cells.	Reuters, 2024.
Self-Amplifying RNA (saRNA)	ARCT-154	COVID-19	Induced higher neutralizing antibody levels compared to conventional mRNA vaccines, with a favorable safety profile.	Clinical Trials Arena, 2023.
	VLPCOV-01	COVID-19	Demonstrated superior long-term immunogenicity and reduced reactogenicity compared to standard mRNA vaccines.	Clinical Trials Arena, 2023.

Table 3. Novel Advancements in Clinical Trials for Antisense Oligonucleotides, siRNA, and mRNA Vaccines

Therapeutic Modality	Novel Technology	Clinical Application	Key Outcomes	References
Antisense Oligonucleotides (ASOs)	2′-O-(2-methoxyethyl)-RNA (MOE-RNA)	Duchenne Muscular Dystrophy, Spinal Muscular Atrophy	Enhanced stability and increased affinity for RNA targets, leading to improved efficacy and safety in clinical applications.	Eckstein, (2023).
	Gapmer ASOs	Huntington's Disease	Improved gene silencing through hybridization with target RNA, showing promise in reducing mutant huntingtin protein levels.	Tabrizi et al. (2019).
siRNA	Lipid Nanoparticle (LNP) Delivery Systems	Hereditary Transthyretin Amyloidosis (hATTR)	Efficient delivery to target tissues, reducing transthyretin levels by over 80% and improving neuropathy outcomes.	Adams et al. (2018)
	Conjugated siRNA (GalNAc)	Hypercholesterolemia	Enhanced liver-targeted siRNA delivery, reducing LDL cholesterol by over 50% with sustained effects.	Ray et al. (2020).
mRNA Vaccines	Self-Amplifying RNA (saRNA)	COVID-19 (e.g., ARCT-154)	Induced higher neutralizing antibody levels and durable immune responses compared to conventional mRNA vaccines.	Clinical Trials Arena. (2023).
	Personalized mRNA Vaccines	Cancer Immunotherapy	Development of patient-specific vaccines targeting tumor neoantigens, showing early promise in reducing tumor size.	Financial Times. (2024).
mRNA Therapeutics	mRNA-LNP-Based Cell Therapy	Myeloma (e.g., SYS6020)	Clinical trials approved for targeted cancer therapy with minimal side effects.	Reuters. (2024).

Pre-Emptive Therapeutic Development

Pandemic COVID-19 has shown how important it is to be proactive in the face of new viral threats, rather than reactive. It is possible to launch public health campaigns right away, using tools like illness surveillance and risk modelling to identify communities most at danger. Therapeutic interventions can be directed towards high-risk locations and individuals based on this knowledge. The creation of new vaccinations and treatments, however, requires some time to mature (Flies et al., 2015). The first antiviral therapy to be licensed was Remdesivir, a long-standing broad-spectrum coronavirus antiviral, in May of 2020. In November of the same year, newly manufactured antibody therapies for COVID-19 were authorised for usage. This highlights the need of preventing rather than responding to drug development delays in order to shorten the time it takes for treatments to reach patients. Aiming for rapid deployment, researchers should create a therapy for every virus family that has passed phase 1 safety testing. This task presents challenges; however, it is essential. Identifying the viruses that present the highest risk of transmission is the initial step (Phelps et al., 2019). We examined several concerning virus families; however, ongoing viral surveillance is essential to identify new circulating strains and to understand alterations in viral DNA and their transmission dynamics. The scientific community can now initiate the development of comprehensive treatment portfolios utilising this information. There are two main approaches to therapeutic development: new drug discovery and repurposing existing treatments. Because of the ease with which they can be permitted for a new use in the event of an epidemic, currently approved drugs by the FDA are excellent candidates for repurposing (Yang et al., 2020; Tan et al., 2004).

Remdesivir, which has been evaluated for the treatment of EBOV114 previously, is one of many well-studied medications that pharmaceutical firms typically have in their pipelines that could be suitable candidates for testing. For high-throughput screening of relevant viruses, it is recommended to use libraries of approved and future treatments. New drug discovery, on the other hand, starts with pinpointing important targets based on studies of conserved host and viral components. Lipid biosynthesis pathways, host signalling networks (like JAK/STAT), and viral polymerases are among examples (Mulangu et al., 2019). Nucleoside analogues are commonly used to target these networks. Finding a target allows for the production of many therapeutics, including small chemicals, antibodies, RNA, and more. The creation of antibody libraries and subsequent screening for highly neutralising antibodies is typically carried out using patient-derived B cells in the production of monoclonal antibody products. Due to a lack of human subjects, researchers may have to depend on animal models in order to test antibodies before a new pathogen emerges. Even after humanisation, antibodies made from vaccinated animals may cause unwanted immune responses (Li et al., 2015).

As a result of promising results, the casirivimab/imdevimab COVID-19 treatment was approved. The human immunoglobulin locus is completely substituted with the mouse one in this method. In light of the variety of viruses and the difficulty in foreseeing their future evolution, it is critical to advance treatments with broad therapeutic efficacy by concentrating on conserved components and pathways (Harding et al., 2010). To determine their potential effectiveness against future viral variants, all therapy candidates, regardless of type, need to be tested against panels of diverse viral strains, including both historical and current ones. It is possible that the next virus to emerge will be immune to an incredibly effective but narrowly tailored treatment. In order to prepare for future epidemics, thorough antivirals can be identified by careful study of various strains along the evolutionary tree. To promote proactive research in viral epidemic preparedness, this endeavour will require major effort, namely committed financial support (Pascal et al., 2015). Researchers in academia deepen our understanding of how viruses and hosts work, which in turn helps us find new therapeutic targets and gets the ball rolling on new drug development initiatives. Through the provision of financial support, drug library development, quality assurance, and manufacturing proficiency, collaborations with industry can speed pharmaceutical development. Academic and industrial endeavours receive financial backing and regulatory guidance from government bodies. It is imperative that we recognise the existence of even more dangerous zoonotic viruses than SARS-CoV-2 in the wild, which might potentially evolve and spread globally at an alarming rate. The enormous monetary costs, human suffering, and loss of life make long-term planning an expensive necessity (Hansen et al., 2020).

THERAPEUTIC REGIMEN AND DELIVERY

Preclinical research can and should be conducted prior to the onset of an epidemic; however, therapy candidates must complete clinical trials prior to receiving clearance. This may transpire as newly emergent diseases commence to disseminate. The outcomes of treatment regimens and delivery methods for viral infections have been the subject of numerous recent case studies (Table 4). In order to facilitate the distribution of unapproved pharmaceuticals during crises such as pandemics, the FDA has implemented a novel mechanism known as emergency use authorisations. However, it is still necessary to provide evidence of safety and efficacy. The FDA's Animal Rule delineates the criteria for the licensing of medications that lack human efficacy research. Field research is challenging, and human challenge trials for both infections are considered immoral. Consequently, this guideline pertains to medications that treat anthrax poisoning and plague (Yersinia pestis). Animal efficacy studies may be applicable in some instances; nonetheless, human clinical trials remain essential for assessing safety. This regulation may facilitate the expedited approval of pharmaceuticals targeting new viruses prior to their market introduction. Prevention measures and post-infection treatments must undergo rigorous testing in well-structured, double-blind clinical trials to alleviate public concerns (Hashem et al., 2020).

The efficacy of therapy windows in humans is likely to be constrained, with certain medicines potentially functioning solely as preventative measures or during the initial phases of sickness. Primate studies are essential for assessing therapeutic windows, however preclinical research in laboratory animals cannot entirely forecast human timelines. Clinical trials are essential for the precise determination of therapeutic windows; yet, they may be impeded by the necessity to estimate patients' incubation periods through diagnostic durations, contact tracing, and symptom severity. Evaluating single or combination therapies during an outbreak is essential. A larger patient cohort is necessary to preserve statistical power when studies are structured to assess numerous drugs or both single and combination therapies concurrently (Hashem et al., 2020).

The analysis of combination therapy complicates the ability to evaluate individual treatments. To conserve time and resources, enhance the likelihood of efficacy, and address the emergence of viral resistance, it may be advantageous to utilise combination therapies during an outbreak. In phase 3 trials, the ideal dosage is established by weighing efficacy against toxicity; preliminary clinical trials may assess several dosages prior to selecting a target dose. In a pandemic, production capacity emerges as a critical consideration (Hung et al., 2020). The production and distribution of a newly designed medicine to patients during an epidemic is inherently a challenging endeavour. Therefore, manufacturing capacity and the volume of doses producible at different dosage levels must be taken into account when determining dosage, given that this may have a major impact on the capacity to treat patients. Oral pharmaceuticals (or other self-administered treatments) are seen optimal during an epidemic due to their dosage convenience and the elevated likelihood of patient compliance with treatment guidelines, which is crucial for mitigating disease transmission. Furthermore, given that healthcare systems are already strained by outbreaks, such as the ongoing extensive COVID-19 vaccination campaigns, self-administered medications considerably reduce the burden on healthcare personnel associated with complex dosing protocols (Hung et al., 2020). Table 5 highlights the challenges of therapeutic regimen and delivery, as well as patient adherence during viral outbreaks

Table 4. Some recent case studies with outcome evaluations of therapeutic regimens and delivery methods for viral diseases

Disease	Therapeutic Regimen	Delivery Method	Outcome Evaluation	Year	Reference
HIV	Antiretroviral therapy (ART)	Oral medication	Significant reduction in viral load and improvement in CD4 cell count.	2011	(Cohen et al., 2011).
Hepatitis C	Direct-acting antivirals (DAAs)	Oral medication	High sustained virological response (SVR) rates, exceeding 90%.	2020	(Feld & Jacobson, 2020).
Influenza	Oseltamivir	Oral medication	Reduced duration and severity of symptoms, hospitalization rates.	2019	(Treanor et al., 2019)
Ebola	ZMapp	Intravenous infusion	Improved survival rates, reduced mortality.	2016	(PREVAIL II Writing Group et al., 2016)
COVID-19	mRNA vaccine (Pfizer-BioNTech)	Intramuscular injection	High efficacy in preventing severe illness and hospitalization, exceeding 90% .	2020	(Polack et al., 2020)
Herpes Simplex	Valacyclovir	Oral medication	Reduced frequency and severity of outbreaks, improved quality of life (Bodsworth et al., 2018)	2018	Bodsworth et al., (2018).
Hepatitis B	Tenofovir disoproxil fumarate	Oral medication	Improved viral suppression, reduced liver inflammation.	2020	(Buti et al., 2020)

Table 5. Challenges of therapeutic regimen and delivery, as well as patient adherence during viral outbreaks

Challenge	Description	Impact on Viral Outbreaks
Production Capacity	Insufficient manufacturing capacity to meet demand for therapeutic regimens and vaccines	Delays in treatment and prevention, exacerbating outbreak severity
Supply Chain Disruptions	Disruptions in supply chain logistics, leading to shortages of essential medications and vaccines	Reduced access to life-saving treatments, increased mortality rates
Patient Adherence	Poor adherence to treatment regimens due to factors like stigma, lack of awareness, and socioeconomic barriers	Reduced treatment efficacy, increased transmission rates, and prolonged outbreak duration
Vaccine Hesitancy	Misinformation and mistrust leading to vaccine hesitancy and refusal	Reduced vaccination rates, increased risk of outbreaks and spread of disease
Healthcare System Overload	Overwhelmed healthcare systems, leading to delayed diagnosis, treatment, and care	Increased mortality rates, reduced quality of care, and prolonged outbreak duration
Accessibility and Affordability	Limited access to therapeutic regimens and vaccines due to cost, geography, or socioeconomic factors	Reduced treatment and prevention options for marginalized populations, exacerbating health disparities
Regulatory and Policy Challenges	Inadequate regulatory frameworks, policies, and laws to support rapid response to viral outbreaks	Delayed response, inadequate preparedness, and reduced effectiveness of interventions

CONCLUSION

We possess a proven history of effective interdisciplinary collaboration in the development of immunisations and therapies. To safeguard global health against future pandemics, analogous measures are necessary. Academic, governmental, corporate, and charitable entities must collaborate in 21st-century pandemic preparation teams to identify gaps and possibilities, facilitate communication among scientists, assess success, and align outcomes with commercial interests. The antibacterial consortium CARB-X is distinguished among existing models as a potential prototype for this system. Proactively collaborating with regulatory bodies is crucial for accelerating the advancement of therapies and expediting pandemic responses. To execute comprehensive initiatives, collaborations with enterprises capable of supplying vital information regarding production, quality control, product evaluation, and safety are crucial. While intellectual property may hinder product development, this issue can be addressed by a management strategy that facilitates many avenues to proprietary rights and ensures access for marginalised people globally.

A coordinated global initiative for pandemic preparedness is a pragmatic objective that can align the interests of many governments, given the significant impact on both people and the economy caused by the COVID-19 pandemic. The expenses are negligible when juxtaposed with the human and economic toll of unpreparedness, particularly given the existence of viruses in nature that can result in significantly higher mortality rates. In 2003 and 2012, there were minor outbreaks of the perilous SARS-CoV and MERS-CoV strains, and it is well-documented that there exist highly lethal influenza, bunyaviruses, and henipaviruses. Regrettably, human recollections diminish with the passage of time, necessitating prompt intervention. Broad-spectrum antiviral medications are crucial for the comprehensive measures required to enhance global resilience against future pandemics. It is imperative that we act promptly to guarantee the safety of our children.

ABBREVIATIONS

COVID-19 coronavirus disease 2019
WHO World health organization
HIV Human immune deficiency virus
HCV hepatitis C virus
HBV hepatitis B Virus
H1N1 Hemagglutinin 1 Neuraminidases 1
DNA deoxyribonucleic acid
RNA Ribonucleic acid
RdDp RNA-dependent DNA polymerase
IFN Interferon
CRISPR Clustered regularly interspaced short palindromic repeats
ISG 20 Interferon stimulated gene 20
SARS-COV Severe acute respiratory syndrome corona virus 2
MERs Middle east respiratory syndrome
MERs-CoV Middle east respiratory syndrome coronavirus
DENV Dengue virus
CNS Central nervous system
YFV Yellow fever virus
EBOV Ebola virus
% Percentage
CHIKV Chikungunya virus
CCHFV Crimean-Congo haemorrhagic fever virus

LASV lassa virus
RVFV Rift valley fever virus
NIH National Institute of Health
CDC Centre for disease control
FAD Food and drug administration
ACE2 angiotensin-converting enzyme 2
mRNA messenger RNA
ASOs Antisense Oligonucleotides
miRNAs Micron RNAs
SiRNAs Interfering RNAs
SaRNAs small activating RNAs

REFERENCES

Abbott, T. R., Dhamdhere, G., Liu, Y., Lin, X., Goudy, L., Zeng, L., Chemparathy, A., Chmura, S., Heaton, N. S., Debs, R., Pande, T., Endy, D., La Russa, M. F., Lewis, D. B., & Qi, L. S. (2020). Development of CRISPR as an Antiviral Strategy to Combat SARS-CoV-2 and Influenza. *Cell, 181*(4), 865–876.e12. DOI: 10.1016/j.cell.2020.04.020 PMID: 32353252

Adams, D., Gonzalez-Duarte, A., O'Riordan, W. D., Yang, C.-C., Ueda, M., Kristen, A. V., Tournev, I., Schmidt, H. H., Coelho, T., Berk, J. L., Lin, K.-P., Vita, G., Attarian, S., Planté-Bordeneuve, V., Mezei, M. M., Campistol, J. M., Buades, J., Brannagan, T. H.III, Kim, B. J., & Suhr, O. B. (2018). Patisiran, an RNAi therapeutic, for hereditary transthyretin amyloidosis. *The New England Journal of Medicine, 379*(1), 11–21. DOI: 10.1056/NEJMoa1716153 PMID: 29972753

Ahmed, A. E. (2019). Diagnostic delays in Middle East respiratory syndrome coronavirus patients and health systems. *Journal of Infection and Public Health, 12*(6), 767–771. DOI: 10.1016/j.jiph.2019.04.002 PMID: 31006635

Ahn, D. G., Lee, W., Choi, J. K., Kim, S. J., Plant, E. P., Almazán, F., Taylor, D. R., Enjuanes, L., & Oh, J. W. (2011). Interference of ribosomal frameshifting by antisense peptide nucleic acids suppresses SARS coronavirus replication. *Antiviral Research, 91*(1), 1–10. DOI: 10.1016/j.antiviral.2011.04.009 PMID: 21549154

Anand, K., Ziebuhr, J., Wadhwani, P., Mesters, J. R., & Hilgenfeld, R. (2003). Coronavirus main proteinase (3CLpro) structure: Basis for design of anti-SARS drugs. *Science, 300*(5626), 1763–1767. DOI: 10.1126/science.1085658 PMID: 12746549

Andreakos, E., & Tsiodras, S. (2020). COVID-19: Lambda interferon against viral load and hyperinflammation. *EMBO Molecular Medicine, 12*(6), e12465. DOI: 10.15252/emmm.202012465 PMID: 32333818

Arankalle, V. A., Shrivastava, S., Cherian, S., Gunjikar, R. S., Walimbe, A. M., Jadhav, S. M., Sudeep, A. B., & Mishra, A. C. (2007). Genetic divergence of Chikungunya viruses in India (1963-2006) with special reference to the 2005-2006 explosive epidemic. *The Journal of General Virology, 88*(Pt 7), 1967–1976. DOI: 10.1099/vir.0.82714-0 PMID: 17554030

Baden, L. R., El Sahly, H. M., Essink, B., Kotloff, K., Frey, S., Novak, R., Diemert, D., Spector, S. A., Rouphael, N., Creech, C. B., McGettigan, J., Khetan, S., Segall, N., Solis, J., Brosz, A., Fierro, C., Schwartz, H., Neuzil, K., Corey, L., & Zaks, T. (2021). Efficacy and safety of the mRNA-1273 SARS-CoV-2 vaccine. *The New England Journal of Medicine, 384*(5), 403–416. DOI: 10.1056/NEJMoa2035389 PMID: 33378609

Balwani, M., Sardh, E., Ventura, P., Peiró, P. A., Rees, D. C., Stölzel, U., Bissell, D. M., Bonkovsky, H. L., Windyga, J., Anderson, K. E., Parker, C., Silver, S. M., Keel, S. B., Wang, J. D., Stein, P. E., Harper, P., Vassiliou, D., Wang, B., Phillips, J., & Gouya, L. (2020). Phase 3 Trial of RNAi Therapeutic Givosiran for Acute Intermittent Porphyria. *The New England Journal of Medicine, 382*(24), 2289–2301. DOI: 10.1056/NEJMoa1913147 PMID: 32521132

Benson, M. D., Waddington-Cruz, M., Berk, J. L., Polydefkis, M., Dyck, P. J., Wang, A. K., Planté-Bordeneuve, V., Barroso, F. A., Merlini, G., Obici, L., Scheinberg, M., Brannagan, T. H.III, Litchy, W. J., Whelan, C., Drachman, B. M., Adams, D., Heitner, S. B., Conceição, I., Schmidt, H. H., & Coelho, T. (2018). Inotersen Treatment for Patients with Hereditary Transthyretin Amyloidosis. *The New England Journal of Medicine, 379*(1), 22–31. DOI: 10.1056/NEJMoa1716793 PMID: 29972757

Bente, D. A., Forrester, N. L., Watts, D. M., McAuley, A. J., Whitehouse, C. A., & Bray, M. (2013). Crimean-Congo hemorrhagic fever: History, epidemiology, pathogenesis, clinical syndrome and genetic diversity. *Antiviral Research, 100*(1), 159–189. DOI: 10.1016/j.antiviral.2013.07.006 PMID: 23906741

Bodsworth, N. J., Crooks, R. J., Borelli, S., & Vejlsgaard, G. (2018). Valacyclovir for the suppression of recurrent genital herpes simplex virus infection. *Journal of Medical Virology, 90*(5), 831–838. DOI: 10.1002/jmv.25034

Boersma, S., Rabouw, H. H., Bruurs, L. J. M., Pavlovič, T., van Vliet, A. L. W., Beumer, J., Clevers, H., van Kuppeveld, F. J. M., & Tanenbaum, M. E. (2020). Translation and Replication Dynamics of Single RNA Viruses. *Cell, 183*(7), 1930–1945. e23. DOI: 10.1016/j.cell.2020.10.019 PMID: 33188777

Braun, E., & Sauter, D. (2019). Furin-mediated protein processing in infectious diseases and cancer. *Clinical & Translational Immunology, 8*(8), e1073. DOI: 10.1002/cti2.1073 PMID: 31406574

Burt, F. J., Chen, W., Miner, J. J., Lenschow, D. J., Merits, A., Schnettler, E., Kohl, A., Rudd, P. A., Taylor, A., Herrero, L. J., Zaid, A., Ng, L. F. P., & Mahalingam, S. (2017). Chikungunya virus: An update on the biology and pathogenesis of this emerging pathogen. *The Lancet. Infectious Diseases*, *17*(4), e107–e117. DOI: 10.1016/S1473-3099(16)30385-1 PMID: 28159534

Buti, M., Tsai, N., Petersen, J., Flisiak, R., Gurel, S., Krastev, Z., & Gane, E. (2020). Seven-year efficacy and safety of tenofovir disoproxil fumarate treatment for chronic hepatitis B. *Journal of Hepatology*, *73*(3), 532–541. DOI: 10.1016/j.jhep.2020.04.014

Carlsson, L., Clarke, J. C., Yen, C., Gregoire, F., Albery, T., Billger, M., Egnell, A. C., Gan, L. M., Jennbacken, K., Johansson, E., Linhardt, G., Martinsson, S., Sadiq, M. W., Witman, N., Wang, Q. D., Chen, C. H., Wang, Y. P., Lin, S., Ticho, B., & Fritsche-Danielson, R. (2018). Biocompatible, Purified *VEGF-A* mRNA Improves Cardiac Function after Intracardiac Injection 1 Week Post-myocardial Infarction in Swine. *Molecular Therapy. Methods & Clinical Development*, *9*, 330–346. DOI: 10.1016/j.omtm.2018.04.003 PMID: 30038937

Chen, C. C., Yu, X., Kuo, C. J., Min, J., Chen, S., Ma, L., Liu, K., & Guo, R. T. (2021). Overview of antiviral drug candidates targeting coronaviral 3C-like main proteases. *The FEBS Journal*, *288*(17), 5089–5121. DOI: 10.1111/febs.15696 PMID: 33400393

Cherry, J. D. (2004). The chronology of the 2002-2003 SARS mini pandemic. *Paediatric Respiratory Reviews*, *5*(4), 262–269. DOI: 10.1016/j.prrv.2004.07.009 PMID: 15531249

Clinical Trials Arena. (2023). Self-amplifying RNA and vaccine innovation: Getting from concept to clinic. *Clinical Trials Arena*. Retrieved from https://www.clinicaltrialsarena.com

Coelho, T., Adams, D., Silva, A., Lozeron, P., Hawkins, P. N., Mant, T., Perez, J., Chiesa, J., Warrington, S., Tranter, E., Munisamy, M., Falzone, R., Harrop, J., Cehelsky, J., Bettencourt, B. R., Geissler, M., Butler, J. S., Sehgal, A., Meyers, R. E., & Suhr, O. B. (2013). Safety and efficacy of RNAi therapy for transthyretin amyloidosis. *The New England Journal of Medicine*, *369*(9), 819–829. DOI: 10.1056/NEJMoa1208760 PMID: 23984729

Cohen, M. S., Chen, Y. Q., McCauley, M., Gamble, T., Hosseinipour, M. C., Kumarasamy, N., Hakim, J. G., Kumwenda, J., Grinsztejn, B., Pilotto, J. H., Godbole, S. V., Mehendale, S., Chariyalertsak, S., Santos, B. R., Mayer, K. H., Hoffman, I. F., Eshleman, S. H., Piwowar-Manning, E., Wang, L., & Fleming, T. R.HPTN 052 Study Team. (2011). Prevention of HIV-1 infection with early antiretroviral therapy. *The New England Journal of Medicine*, *365*(6), 493–505. DOI: 10.1056/NEJMoa1105243 PMID: 21767103

Dawood, F. S., Iuliano, A. D., Reed, C., Meltzer, M. I., Shay, D. K., Cheng, P. Y., Bandaranayake, D., Breiman, R. F., Brooks, W. A., Buchy, P., Feikin, D. R., Fowler, K. B., Gordon, A., Hien, N. T., Horby, P., Huang, Q. S., Katz, M. A., Krishnan, A., Lal, R., & Widdowson, M. A. (2012). Estimated global mortality associated with the first 12 months of 2009 pandemic influenza A H1N1 virus circulation: A modelling study. *The Lancet. Infectious Diseases*, *12*(9), 687–695. DOI: 10.1016/S1473-3099(12)70121-4 PMID: 22738893

De Clercq, E., & Li, G. (2016). Approved Antiviral Drugs over the Past 50 Years. *Clinical Microbiology Reviews*, *29*(3), 695–747. DOI: 10.1128/CMR.00102-15 PMID: 27281742

Dijkman, R., Jebbink, M. F., Koekkoek, S. M., Deijs, M., Jónsdóttir, H. R., Molenkamp, R., Ieven, M., Goossens, H., Thiel, V., & van der Hoek, L. (2013). Isolation and characterization of current human coronavirus strains in primary human epithelial cell cultures reveal differences in target cell tropism. *Journal of Virology*, *87*(11), 6081–6090. DOI: 10.1128/JVI.03368-12 PMID: 23427150

Dou, D., Revol, R., Östbye, H., Wang, H., & Daniels, R. (2018). Influenza A Virus Cell Entry, Replication, Virion Assembly and Movement. *Frontiers in Immunology*, *9*, 1581. DOI: 10.3389/fimmu.2018.01581 PMID: 30079062

Durmuş, S., & Ülgen, K. Ö. (2017). Comparative interactomics for virus-human protein-protein interactions: DNA viruses versus RNA viruses. *FEBS Open Bio*, *7*(1), 96–107. DOI: 10.1002/2211-5463.12167 PMID: 28097092

Eckstein, F. (2023). Advances in RNA chemistry for antisense oligonucleotide therapeutics. *Nucleic Acids Research*, *51*(6), 2529–2536. DOI: 10.1093/nar/gkad013 PMID: 36881759

Edwards, C. E., Yount, B. L., Graham, R. L., Leist, S. R., Hou, Y. J., Dinnon, K. H.III, Sims, A. C., Swanstrom, J., Gully, K., Scobey, T. D., Cooley, M. R., Currie, C. G., Randell, S. H., & Baric, R. S. (2020). Swine acute diarrhea syndrome coronavirus replication in primary human cells reveals potential susceptibility to infection. *Proceedings of the National Academy of Sciences of the United States of America, 117*(43), 26915–26925. DOI: 10.1073/pnas.2001046117 PMID: 33046644

Epstein, J. H., & Anthony, S. J. (2017). Viral discovery as a fool for pandemic preparedness. *Revue Scientifique et Technique (International Office of Epizootics), 36*(2), 499–512. DOI: 10.20506/rst.36.2.2669 PMID: 30152468

Feld, J. J., & Jacobson, I. M. (2020). The evolution of hepatitis C treatment. *Nature Reviews. Gastroenterology & Hepatology, 17*(10), 533–548. DOI: 10.1038/s41575-020-0313-6

Financial Times. (2024). The Covid-era tech that could reinvent cancer care. Retrieved from https://www.ft.com/content/efff95f4-e7c4-4a3c-bccd-f449fecac27e

Finkel, R. S., Mercuri, E., Darras, B. T., Connolly, A. M., Kuntz, N. L., Kirschner, J., Chiriboga, C. A., Saito, K., Servais, L., Tizzano, E., Topaloglu, H., Tulinius, M., Montes, J., Glanzman, A. M., Bishop, K., Zhong, Z. J., Gheuens, S., Bennett, C. F., Schneider, E., & De Vivo, D. C. (2017). Nusinersen versus sham control in infantile-onset spinal muscular atrophy. *The New England Journal of Medicine, 377*(18), 1723–1732. DOI: 10.1056/NEJMoa1702752 PMID: 29091570

Flies, E. J., Toi, C., Weinstein, P., Doggett, S. L., & Williams, C. R. (2015). Converting Mosquito Surveillance to Arbovirus Surveillance with Honey-Baited Nucleic Acid Preservation Cards. *Vector Borne and Zoonotic Diseases (Larchmont, N.Y.), 15*(7), 397–403. DOI: 10.1089/vbz.2014.1759 PMID: 26186511

Forrester, N. L., Wertheim, J. O., Dugan, V. G., Auguste, A. J., Lin, D., Adams, A. P., Chen, R., Gorchakov, R., Leal, G., Estrada-Franco, J. G., Pandya, J., Halpin, R. A., Hari, K., Jain, R., Stockwell, T. B., Das, S. R., Wentworth, D. E., Smith, M. D., Kosakovsky Pond, S. L., & Weaver, S. C. (2017). Evolution and spread of Venezuelan equine encephalitis complex alphavirus in the Americas. *PLoS Neglected Tropical Diseases, 11*(8), e0005693. DOI: 10.1371/journal.pntd.0005693 PMID: 28771475

Garnett, L. E., & Strong, J. E. (2019). Lassa fever: With 50 years of study, hundreds of thousands of patients and an extremely high disease burden, what have we learned? *Current Opinion in Virology, 37*, 123–131. DOI: 10.1016/j.coviro.2019.07.009 PMID: 31479990

Giladi, H., Ketzinel-Gilad, M., Rivkin, L., Felig, Y., Nussbaum, O., & Galun, E. (2003). Small interfering RNA inhibits hepatitis B virus replication in mice. *Molecular Therapy : the Journal of the American Society of Gene Therapy*, *8*(5), 769–776. DOI: 10.1016/S1525-0016(03)00244-2 PMID: 14599810

Glasgow, A., Glasgow, J., Limonta, D., Solomon, P., Lui, I., Zhang, Y., Nix, M. A., Rettko, N. J., Zha, S., Yamin, R., Kao, K., Rosenberg, O. S., Ravetch, J. V., Wiita, A. P., Leung, K. K., Lim, S. A., Zhou, X. X., Hobman, T. C., Kortemme, T., & Wells, J. A. (2020). Engineered ACE2 receptor traps potently neutralize SARS-CoV-2. *Proceedings of the National Academy of Sciences of the United States of America*, *117*(45), 28046–28055. DOI: 10.1073/pnas.2016093117 PMID: 33093202

Gorbea, C., Mosbruger, T., & Cazalla, D. (2017). A viral Sm-class RNA base-pairs with mRNAs and recruits microRNAs to inhibit apoptosis. *Nature*, *550*(7675), 275–279. DOI: 10.1038/nature24034 PMID: 28976967

Gubler, D. J. (2002). Epidemic dengue/dengue hemorrhagic fever as a public health, social and economic problem in the 21st century. *Trends in Microbiology*, *10*(2), 100–103. DOI: 10.1016/S0966-842X(01)02288-0 PMID: 11827812

Hansen, J., Baum, A., Pascal, K. E., Russo, V., Giordano, S., Wloga, E., Fulton, B. O., Yan, Y., Koon, K., Patel, K., Chung, K. M., Hermann, A., Ullman, E., Cruz, J., Rafique, A., Huang, T., Fairhurst, J., Libertiny, C., Malbec, M., & Kyratsous, C. A. (2020). Studies in humanized mice and convalescent humans yield a SARS-CoV-2 antibody cocktail. *Science*, *369*(6506), 1010–1014. DOI: 10.1126/science.abd0827 PMID: 32540901

Harding, F. A., Stickler, M. M., Razo, J., & DuBridge, R. B. (2010). The immunogenicity of humanized and fully human antibodies: Residual immunogenicity resides in the CDR regions. *mAbs*, *2*(3), 256–265. DOI: 10.4161/mabs.2.3.11641 PMID: 20400861

Hashem, H., Abufaraj, M., Tbakhi, A., & Sultan, I. (2020). Obstacles and Considerations Related to Clinical Trial Research During the COVID-19 Pandemic. *Frontiers in Medicine*, *7*, 598038. DOI: 10.3389/fmed.2020.598038 PMID: 33425945

Henao-Restrepo, A. M., Camacho, A., Longini, I. M., Watson, C. H., Edmunds, W. J., Egger, M., Carroll, M. W., Dean, N. E., Diatta, I., Doumbia, M., Draguez, B., Duraffour, S., Enwere, G., Grais, R., Gunther, S., Gsell, P. S., Hossmann, S., Watle, S. V., Kondé, M. K., & Kieny, M. P. (2017). Efficacy and effectiveness of an rVSV-vectored vaccine in preventing Ebola virus disease: Final results from the Guinea ring vaccination, open-label, cluster-randomised trial (Ebola Ça Suffit!). *Lancet*, *389*(10068), 505–518. DOI: 10.1016/S0140-6736(16)32621-6 PMID: 28017403

Himeidan, Y. E., Kweka, E. J., Mahgoub, M. M., El Rayah, E. A., & Ouma, J. O. (2014). Recent outbreaks of rift valley Fever in East Africa and the middle East. *Frontiers in Public Health, 2*, 169. DOI: 10.3389/fpubh.2014.00169 PMID: 25340047

Holmes, E. C., Dudas, G., Rambaut, A., & Andersen, K. G. (2016). The evolution of Ebola virus: Insights from the 2013-2016 epidemic. *Nature, 538*(7624), 193–200. DOI: 10.1038/nature19790 PMID: 27734858

Hou, Y. J., Chiba, S., Halfmann, P., Ehre, C., Kuroda, M., Dinnon, K. H.III, Leist, S. R., Schäfer, A., Nakajima, N., Takahashi, K., Lee, R. E., Mascenik, T. M., Graham, R., Edwards, C. E., Tse, L. V., Okuda, K., Markmann, A. J., Bartelt, L., de Silva, A., & Baric, R. S. (2020). SARS-CoV-2 D614G variant exhibits efficient replication ex vivo and transmission in vivo. *Science, 370*(6523), 1464–1468. DOI: 10.1126/science.abe8499 PMID: 33184236

Hua, Y., Sahashi, K., Hung, G., Rigo, F., Passini, M. A., Bennett, C. F., & Krainer, A. R. (2010). Antisense correction of SMN2 splicing in the CNS rescues necrosis in a type III SMA mouse model. *Genes & Development, 24*(15), 1634–1644. DOI: 10.1101/gad.1941310 PMID: 20624852

Hung, I. F., Lung, K. C., Tso, E. Y., Liu, R., Chung, T. W., Chu, M. Y., Ng, Y. Y., Lo, J., Chan, J., Tam, A. R., Shum, H. P., Chan, V., Wu, A. K., Sin, K. M., Leung, W. S., Law, W. L., Lung, D. C., Sin, S., Yeung, P., & Yuen, K. Y. (2020). Triple combination of interferon beta-1b, lopinavir-ritonavir, and ribavirin in the treatment of patients admitted to hospital with COVID-19: An open-label, randomised, phase 2 trial. *Lancet, 395*(10238), 1695–1704. DOI: 10.1016/S0140-6736(20)31042-4 PMID: 32401715

Ilunga Kalenga, O., Moeti, M., Sparrow, A., Nguyen, V. K., Lucey, D., & Ghebreyesus, T. A. (2019). The Ongoing Ebola Epidemic in the Democratic Republic of Congo, 2018-2019. *The New England Journal of Medicine, 381*(4), 373–383. DOI: 10.1056/NEJMsr1904253 PMID: 31141654

Inoue, Y., Tanaka, N., Tanaka, Y., Inoue, S., Morita, K., Zhuang, M., Hattori, T., & Sugamura, K. (2007). Clathrin-dependent entry of severe acute respiratory syndrome coronavirus into target cells expressing ACE2 with the cytoplasmic tail deleted. *Journal of Virology, 81*(16), 8722–8729. DOI: 10.1128/JVI.00253-07 PMID: 17522231

Ivanov, A. I. (2008). Pharmacological inhibition of endocytic pathways: Is it specific enough to be useful? *Methods in Molecular Biology (Clifton, N.J.), 440*, 15–33. DOI: 10.1007/978-1-59745-178-9_2 PMID: 18369934

Jan, E., Mohr, I., & Walsh, D. (2016). A Cap-to-Tail Guide to mRNA Translation Strategies in Virus-Infected Cells. *Annual Review of Virology*, *3*(1), 283–307. DOI: 10.1146/annurev-virology-100114-055014 PMID: 27501262

Jonsson, C. B., Figueiredo, L. T., & Vapalahti, O. (2010). A global perspective on hantavirus ecology, epidemiology, and disease. *Clinical Microbiology Reviews*, *23*(2), 412–441. DOI: 10.1128/CMR.00062-09 PMID: 20375360

Kaufman, E. B., & Kramer, L. D. (2017). Zika virus mosquito vectors: Competence, biology, and vector control. *The Journal of Infectious Diseases*, *216*(suppl_10), S976–S990. DOI: 10.1093/infdis/jix405 PMID: 29267910

Kim, J. I., Park, S., Lee, I., Lee, S., Shin, S., Won, Y., Hwang, M. W., Bae, J. Y., Heo, J., Hyun, H. E., Jun, H., Lim, S. S., & Park, M. S. (2012). GFP-expressing influenza A virus for evaluation of the efficacy of antiviral agents. *Journal of Microbiology (Seoul, Korea)*, *50*(2), 359–362. DOI: 10.1007/s12275-012-2163-9 PMID: 22538668

Kirchdoerfer, R. N., & Ward, A. B. (2019). Structure of the SARS-CoV nsp12 polymerase bound to nsp7 and nsp8 co-factors. *Nature Communications*, *10*(1), 2342. DOI: 10.1038/s41467-019-10280-3 PMID: 31138817

Klemm, T., Ebert, G., Calleja, D. J., Allison, C. C., Richardson, L. W., Bernardini, J. P., Lu, B. G., Kuchel, N. W., Grohmann, C., Shibata, Y., Gan, Z. Y., Cooney, J. P., Doerflinger, M., Au, A. E., Blackmore, T. R., van der Heden van Noort, G. J., Geurink, P. P., Ovaa, H., Newman, J., & Komander, D. (2020). Mechanism and inhibition of the papain-like protease, PLpro, of SARS-CoV-2. *The EMBO Journal*, *39*(18), e106275. DOI: 10.15252/embj.2020106275 PMID: 32845033

Kose, N., Fox, J. M., Sapparapu, G., Bombardi, R., Tennekoon, R. N., de Silva, A. D., Elbashir, S. M., Theisen, M. A., Humphris-Narayanan, E., Ciaramella, G., Himansu, S., Diamond, M. S., & Crowe, J. E.Jr. (2019). A lipid-encapsulated mRNA encoding a potently neutralizing human monoclonal antibody protects against chikungunya infection. *Science Immunology*, *4*(35), eaaw6647. Advance online publication. DOI: 10.1126/sciimmunol.aaw6647 PMID: 31101672

Kumar, M., Kuroda, K., Dhangar, K., Mazumder, P., Sonne, C., Rinklebe, J., & Kitajima, M. (2020). Potential Emergence of Antiviral-Resistant Pandemic Viruses via Environmental Drug Exposure of Animal Reservoirs. *Environmental Science & Technology*, *54*(14), 8503–8505. DOI: 10.1021/acs.est.0c03105 PMID: 32609508

Lampejo T. (2020). Influenza and antiviral resistance: an overview. *European journal of clinical microbiology & infectious diseases: official publication of the European Society of Clinical Microbiology, 39*(7), 1201–1208. https://doi.org/DOI: 10.1007/s10096-020-03840-9

Lanford, R. E., Hildebrandt-Eriksen, E. S., Petri, A., Persson, R., Lindow, M., Munk, M. E., Kauppinen, S., & Ørum, H. (2010). Therapeutic silencing of microRNA-122 in primates with chronic hepatitis C virus infection. *Science, 327*(5962), 198–201. DOI: 10.1126/science.1178178 PMID: 19965718

Li, W., Moore, M. J., Vasilieva, N., Sui, J., Wong, S. K., Berne, M. A., Somasundaran, M., Sullivan, J. L., Luzuriaga, K., Greenough, T. C., Choe, H., & Farzan, M. (2003). Angiotensin-converting enzyme 2 is a functional receptor for the SARS coronavirus. *Nature, 426*(6965), 450–454. DOI: 10.1038/nature02145 PMID: 14647384

Li, W., Shi, Z., Yu, M., Ren, W., Smith, C., Epstein, J. H., Wang, H., Crameri, G., Hu, Z., Zhang, H., Zhang, J., McEachern, J., Field, H., Daszak, P., Eaton, B. T., Zhang, S., & Wang, L. F. (2005). Bats are natural reservoirs of SARS-like coronaviruses. *Science, 310*(5748), 676–679. DOI: 10.1126/science.1118391 PMID: 16195424

Li, Y., Wan, Y., Liu, P., Zhao, J., Lu, G., Qi, J., Wang, Q., Lu, X., Wu, Y., Liu, W., Zhang, B., Yuen, K. Y., Perlman, S., Gao, G. F., & Yan, J. (2015). A humanized neutralizing antibody against MERS-CoV targeting the receptor-binding domain of the spike protein. *Cell Research, 25*(11), 1237–1249. DOI: 10.1038/cr.2015.113 PMID: 26391698

Lindsey, N. P., Staples, J. E., & Fischer, M. (2018). Eastern Equine Encephalitis Virus in the United States, 2003-2016. *The American Journal of Tropical Medicine and Hygiene, 98*(5), 1472–1477. DOI: 10.4269/ajtmh.17-0927 PMID: 29557336

Linsky, T. W., Vergara, R., Codina, N., Nelson, J. W., Walker, M. J., Su, W., Barnes, C. O., Hsiang, T. Y., Esser-Nobis, K., Yu, K., Reneer, Z. B., Hou, Y. J., Priya, T., Mitsumoto, M., Pong, A., Lau, U. Y., Mason, M. L., Chen, J., Chen, A., & Silva, D. A. (2020). De novo design of potent and resilient hACE2 decoys to neutralize SARS-CoV-2. *Science, 370*(6521), 1208–1214. DOI: 10.1126/science.abe0075 PMID: 33154107

Linton, N. M., Kobayashi, T., Yang, Y., Hayashi, K., Akhmetzhanov, A. R., Jung, S. M., Yuan, B., Kinoshita, R., & Nishiura, H. (2020). Incubation Period and Other Epidemiological Characteristics of 2019 Novel Coronavirus Infections with Right Truncation: A Statistical Analysis of Publicly Available Case Data. *Journal of Clinical Medicine, 9*(2), 538. DOI: 10.3390/jcm9020538 PMID: 32079150

Lu, L., Su, S., Yang, H., & Jiang, S. (2021). Antivirals with common targets against highly pathogenic viruses. *Cell, 184*(6), 1604–1620. DOI: 10.1016/j.cell.2021.02.013 PMID: 33740455

Luo, C. M., Wang, N., Yang, X. L., Liu, H. Z., Zhang, W., Li, B., Hu, B., Peng, C., Geng, Q. B., Zhu, G. J., Li, F., & Shi, Z. L. (2018). Discovery of Novel Bat Coronaviruses in South China That Use the Same Receptor as Middle East Respiratory Syndrome Coronavirus. *Journal of Virology*, *92*(13), e00116–e00118. DOI: 10.1128/JVI.00116-18 PMID: 29669833

Menachery, V. D., Yount, B. L.Jr, Debbink, K., Agnihothram, S., Gralinski, L. E., Plante, J. A., Graham, R. L., Scobey, T., Ge, X. Y., Donaldson, E. F., Randell, S. H., Lanzavecchia, A., Marasco, W. A., Shi, Z. L., & Baric, R. S. (2015). A SARS-like cluster of circulating bat coronaviruses shows potential for human emergence. *Nature Medicine*, *21*(12), 1508–1513. DOI: 10.1038/nm.3985 PMID: 26552008

Menachery, V. D., Yount, B. L.Jr, Sims, A. C., Debbink, K., Agnihothram, S. S., Gralinski, L. E., Graham, R. L., Scobey, T., Plante, J. A., Royal, S. R., Swanstrom, J., Sheahan, T. P., Pickles, R. J., Corti, D., Randell, S. H., Lanzavecchia, A., Marasco, W. A., & Baric, R. S. (2016). SARS-like WIV1-CoV poised for human emergence. *Proceedings of the National Academy of Sciences of the United States of America*, *113*(11), 3048–3053. DOI: 10.1073/pnas.1517719113 PMID: 26976607

Meng, B., Lui, Y. W., Meng, S., Cao, C., & Hu, Y. (2006). Identification of effective siRNA blocking the expression of SARS viral envelope E and RDRP genes. *Molecular Biotechnology*, *33*(2), 141–148. DOI: 10.1385/MB:33:2:141 PMID: 16757801

Milewska, A., Nowak, P., Owczarek, K., Szczepanski, A., Zarebski, M., Hoang, A., Berniak, K., Wojarski, J., Zeglen, S., Baster, Z., Rajfur, Z., & Pyrc, K. (2018). Entry of Human Coronavirus NL63 into the Cell. *Journal of Virology*, *92*(3), e01933–e17. DOI: 10.1128/JVI.01933-17 PMID: 29142129

Millet, J. K., & Whittaker, G. R. (2018). Physiological and molecular triggers for SARS-CoV membrane fusion and entry into host cells. *Virology*, *517*, 3–8. DOI: 10.1016/j.virol.2017.12.015 PMID: 29275820

Mulangu, S., Dodd, L. E., Davey, R. T.Jr, Tshiani Mbaya, O., Proschan, M., Mukadi, D., Lusakibanza Manzo, M., Nzolo, D., Tshomba Oloma, A., Ibanda, A., Ali, R., Coulibaly, S., Levine, A. C., Grais, R., Diaz, J., Lane, H. C., & Muyembe-Tamfum, J. J.PALM Writing Group. (2019). A Randomized, Controlled Trial of Ebola Virus Disease Therapeutics. *The New England Journal of Medicine*, *381*(24), 2293–2303. DOI: 10.1056/NEJMoa1910993 PMID: 31774950

Nachbagauer, R., Feser, J., Naficy, A., Bernstein, D. I., Guptill, J., Walter, E. B., Berlanda-Scorza, F., Stadlbauer, D., Wilson, P. C., Aydillo, T., Behzadi, M. A., Bhavsar, D., Bliss, C., Capuano, C., Carreño, J. M., Chromikova, V., Claeys, C., Coughlan, L., Freyn, A. W., & Krammer, F. (2021). A chimeric hemagglutinin-based universal influenza virus vaccine approach induces broad and long-lasting immunity in a randomized, placebo-controlled phase I trial. *Nature Medicine*, *27*(1), 106–114. DOI: 10.1038/s41591-020-1118-7 PMID: 33288923

Neuman, B. W., Kiss, G., Kunding, A. H., Bhella, D., Baksh, M. F., Connelly, S., Droese, B., Klaus, J. P., Makino, S., Sawicki, S. G., Siddell, S. G., Stamou, D. G., Wilson, I. A., Kuhn, P., & Buchmeier, M. J. (2011). A structural analysis of M protein in coronavirus assembly and morphology. *Journal of Structural Biology*, *174*(1), 11–22. DOI: 10.1016/j.jsb.2010.11.021 PMID: 21130884

Neumann, G., Chen, H., Gao, G. F., Shu, Y., & Kawaoka, Y. (2010). H5N1 influenza viruses: Outbreaks and biological properties. *Cell Research*, *20*(1), 51–61. DOI: 10.1038/cr.2009.124 PMID: 19884910

Neumann, G., Noda, T., & Kawaoka, Y. (2009). Emergence and pandemic potential of swine-origin H1N1 influenza virus. *Nature*, *459*(7249), 931–939. DOI: 10.1038/nature08157 PMID: 19525932

Ng, E. W., Shima, D. T., Calias, P., Cunningham, E. T.Jr, Guyer, D. R., & Adamis, A. P. (2006). Pegaptanib, a targeted anti-VEGF aptamer for ocular vascular disease. *Nature Reviews. Drug Discovery*, *5*(2), 123–132. DOI: 10.1038/nrd1955 PMID: 16518379

Ogidi, O. I. (2020). A review on the use of Herbal Remedies and Clinical Therapeutics for the Management of Covid-19 Pandemic. *ASIO Journal of Pharmaceuical & Herbal Medicines Research*, 6(2), 68-77. http://doi.org/doilink/10.2020.85722646/

Ogidi, O. I. (2022). Omicron (B.1.1.529): Variant of Concern- A Mini Review. *Biomedical Journal of Scientific & Technical Research,* 41(5), 33106- 33112. https://www.doiDOI: 10.26717/BJSTR.2022.41.006672

Ogidi, O. I. (2024). Therapeutic Role of Biomaterials in The Study of Severe Acute Respiratory Syndrome Corona Virus 2 (Sars-Cov-2). In: Advances in Sustainable Biomaterials: Bioprocessing 4.0, Characterizations and Application. *CRC Press, Taylor and Francis Group, Florida, USA.*

Ogidi, O. I. (2024b). Recent advances in anticancer activity and Bioinformatics Approach from Potential Plants. In: Computational Approaches in Biomaterials, Bioinformatics and Biomedical Engineering Applications. *CRC Press, Taylor and Francis Group, Florida, USA*. DOI: 10.1201/9781003354437-3

Ogidi, O. I., & Ajoko, I. (2024). Neuroprotective Potentials of Phytochemicals. In: M.S. Rajput, T. Sarachana, P.D. Sarkar, M. Raval (eds) NeuroPhytomedicine. *CRC Press, Routledge Taylor and Francis Group.*

Ogidi, O. I., Barwant, M. M., & Okara, E. (2024). Biomarker Discovery: Protein Phosphorylation State. In *Spatially Variable Genes in Cancer: Development, Progression, and Treatment Response*. IGI Global., DOI: 10.4018/979-8-3693-7728-4.ch005

Ogidi, O.I., Berefagha, W.L., & Okara, E. (2021). Covid-19 vaccination: The pros and cons. *World Journal of Biology Pharmacy and Health Sciences,* 07(01), 015–022. https://doi.org/DOI: 10.30574/wjbphs.2021.7.1.0072

Ogidi, O. I., & Emaikwu, N. G. (2024). Plant Phenolic compound isolation and its Bioinformatics Approaches of Molecular Mechanisms in Antimicrobial activities and Resistance. In: Computational Approaches in Biomaterials, Bioinformatics and Biomedical Engineering Applications. *CRC Press, Taylor and Francis Group, Florida, USA*. DOI: 10.1201/9781003354437-8

Ogidi, O. I., & Kumar, P. (2024). Medical and Pharmaceutical Applications of Sustainable Biopolymers. In *Advances in Sustainable Biomaterials: Fundamentals, Modelling and Characterizations* (pp. 431–453). Elsevier., DOI: 10.1016/B978-0-443-13849-2.00023-5

Ogidi, O. I., & Tobia, P. S. (2024). Harnessing Artificial Intelligence (AI) Techniques for the prevention of Microbial Disease in Humans. In *Artificial Intelligence in Microbiology: scopes and Challenges*. Elsevier.

Osterholm, M. T., Moore, K. A., Kelley, N. S., Brosseau, L. M., Wong, G., Murphy, F. A., Peters, C. J., LeDuc, J. W., Russell, P. K., Van Herp, M., Kapetshi, J., Muyembe, J. J., Ilunga, B. K., Strong, J. E., Grolla, A., Wolz, A., Kargbo, B., Kargbo, D. K., Sanders, D. A., & Kobinger, G. P. (2015). Transmission of Ebola viruses: What we know and what we do not know. *mBio*, *6*(2), e00137. DOI: 10.1128/mBio.00137-15 PMID: 25698835

Pascal, K. E., Coleman, C. M., Mujica, A. O., Kamat, V., Badithe, A., Fairhurst, J., Hunt, C., Strein, J., Berrebi, A., Sisk, J. M., Matthews, K. L., Babb, R., Chen, G., Lai, K. M., Huang, T. T., Olson, W., Yancopoulos, G. D., Stahl, N., Frieman, M. B., & Kyratsous, C. A. (2015). Pre- and postexposure efficacy of fully human antibodies against Spike protein in a novel humanized mouse model of MERS-CoV infection. *Proceedings of the National Academy of Sciences of the United States of America*, *112*(28), 8738–8743. DOI: 10.1073/pnas.1510830112 PMID: 26124093

Peiris, J. S., Chu, C. M., Cheng, V. C., Chan, K. S., Hung, I. F., Poon, L. L., Law, K. I., Tang, B. S., Hon, T. Y., Chan, C. S., Chan, K. H., Ng, J. S., Zheng, B. J., Ng, W. L., Lai, R. W., Guan, Y., & Yuen, K. Y.HKU/UCH SARS Study Group. (2003). Clinical progression and viral load in a community outbreak of coronavirus-associated SARS pneumonia: A prospective study. *Lancet*, *361*(9371), 1767–1772. DOI: 10.1016/S0140-6736(03)13412-5 PMID: 12781535

Phelps, K. L., Hamel, L., Alhmoud, N., Ali, S., Bilgin, R., Sidamonidze, K., Urushadze, L., Karesh, W., & Olival, K. J. (2019). Bat Research Networks and Viral Surveillance: Gaps and Opportunities in Western Asia. *Viruses*, *11*(3), 240. DOI: 10.3390/v11030240 PMID: 30857374

Polack, F. P., Thomas, S. J., Kitchin, N., Absalon, J., Gurtman, A., Lockhart, S., & Jansen, K. U. (2020). Safety and efficacy of the BNT162b2 mRNA Covid-19 vaccine. *The New England Journal of Medicine*, *383*(27), 2603–2615. DOI: 10.1056/NEJMoa2034577 PMID: 33301246

Polack, F. P., Thomas, S. J., Kitchin, N., Absalon, J., Gurtman, A., Lockhart, S., Perez, J. L., Pérez Marc, G., Moreira, E. D., Zerbini, C., Bailey, R., Swanson, K. A., Roychoudhury, S., Koury, K., Li, P., Kalina, W. V., Cooper, D., Frenck, R. W.Jr, Hammitt, L. L., & Gruber, W. C. (2020). Safety and efficacy of the BNT162b2 mRNA Covid-19 vaccine. *The New England Journal of Medicine*, *383*(27), 2603–2615. DOI: 10.1056/NEJMoa2034577 PMID: 33301246

Prakash, T. P., Graham, M. J., Yu, J., Carty, R., Low, A., Chappell, A., Schmidt, K., Zhao, C., Aghajan, M., Murray, H. F., Riney, S., Booten, S. L., Murray, S. F., Gaus, H., Crosby, J., Lima, W. F., Guo, S., Monia, B. P., Swayze, E. E., & Seth, P. P. (2014). Targeted delivery of antisense oligonucleotides to hepatocytes using triantennary N-acetyl galactosamine improves potency 10-fold in mice. *Nucleic Acids Research*, *42*(13), 8796–8807. DOI: 10.1093/nar/gku531 PMID: 24992960

PREVAIL II Writing Group. (2016). A randomized, controlled trial of ZMapp for Ebola virus infection. *The New England Journal of Medicine*, *375*(15), 1448–1456. DOI: 10.1056/NEJMoa1604330 PMID: 27732819

Raj, V. S., Mou, H., Smits, S. L., Dekkers, D. H., Müller, M. A., Dijkman, R., Muth, D., Demmers, J. A., Zaki, A., Fouchier, R. A., Thiel, V., Drosten, C., Rottier, P. J., Osterhaus, A. D., Bosch, B. J., & Haagmans, B. L. (2013). Dipeptidyl peptidase 4 is a functional receptor for the emerging human coronavirus-EMC. *Nature*, *495*(7440), 251–254. DOI: 10.1038/nature12005 PMID: 23486063

Ray, K. K., Landmesser, U., Leiter, L. A., Kallend, D., Dufour, R., Karakas, M., & Wright, R. S. (2020). Inclisiran in patients at high cardiovascular risk with elevated LDL cholesterol. *The New England Journal of Medicine*, *382*(16), 1507–1519. DOI: 10.1056/NEJMoa1912387 PMID: 32187462

Reuters. (2024). China approves CSPC Pharma's mRNA cancer therapy clinical trials. *Reuters*. Retrieved from https://www.reuters.com

Reyes, E. D., Kulej, K., Pancholi, N. J., Akhtar, L. N., Avgousti, D. C., Kim, E. T., Bricker, D. K., Spruce, L. A., Koniski, S. A., Seeholzer, S. H., Isaacs, S. N., Garcia, B. A., & Weitzman, M. D. (2017). Identifying Host Factors Associated with DNA Replicated During Virus Infection. *Molecular & Cellular Proteomics*, *16*(12), 2079–2097. DOI: 10.1074/mcp.M117.067116 PMID: 28972080

Sarker, D., Plummer, R., Meyer, T., Sodergren, M. H., Basu, B., Chee, C. E., Huang, K. W., Palmer, D. H., Ma, Y. T., Evans, T. R. J., Spalding, D. R. C., Pai, M., Sharma, R., Pinato, D. J., Spicer, J., Hunter, S., Kwatra, V., Nicholls, J. P., Collin, D., & Habib, N. (2020). MTL-CEBPA, a Small Activating RNA Therapeutic Upregulating C/EBP-α, in Patients with Advanced Liver Cancer: A First-in-Human, Multicenter, Open-Label, Phase I Trial. *Clinical Cancer Research*, *26*(15), 3936–3946. DOI: 10.1158/1078-0432.CCR-20-0414 PMID: 32357963

Sharif-Yakan, A., & Kanj, S. S. (2014). Emergence of MERS-CoV in the Middle East: Origins, transmission, treatment, and perspectives. *PLoS Pathogens*, *10*(12), e1004457. DOI: 10.1371/journal.ppat.1004457 PMID: 25474536

Sirnaomics. (2023). STP707 Phase I clinical study results. *Sirnaomics*. Retrieved from https://sirnaomics.com

Stebbing, J., Krishnan, V., de Bono, S., Ottaviani, S., Casalini, G., Richardson, P. J., Monteil, V., Lauschke, V. M., Mirazimi, A., Youhanna, S., Tan, Y. J., Baldanti, F., Sarasini, A., Terres, J. A. R., Nickoloff, B. J., Higgs, R. E., Rocha, G., Byers, N. L., Schlichting, D. E., & Corbellino, M.Sacco Baricitinib Study Group. (2020). Mechanism of baricitinib supports artificial intelligence-predicted testing in COVID-19 patients. *EMBO Molecular Medicine*, *12*(8), e12697. DOI: 10.15252/emmm.202012697 PMID: 32473600

Strasfeld, L., & Chou, S. (2010). Antiviral drug resistance: Mechanisms and clinical implications. *Infectious Disease Clinics of North America*, *24*(2), 413–437. DOI: 10.1016/j.idc.2010.01.001 PMID: 20466277

Sugo, T., Terada, M., Oikawa, T., Miyata, K., Nishimura, S., Kenjo, E., Ogasawara-Shimizu, M., Makita, Y., Imaichi, S., Murata, S., Otake, K., Kikuchi, K., Teratani, M., Masuda, Y., Kamei, T., Takagahara, S., Ikeda, S., Ohtaki, T., & Matsumoto, H. (2016). Development of antibody-siRNA conjugate targeted to cardiac and skeletal muscles. *Journal of controlled release: official journal of the Controlled Release Society, 237*, 1–13. https://doi.org/DOI: 10.1016/j.jconrel.2016.06.036

Sun, J., Zhuang, Z., Zheng, J., Li, K., Wong, R. L., Liu, D., Huang, J., He, J., Zhu, A., Zhao, J., Li, X., Xi, Y., Chen, R., Alshukairi, A. N., Chen, Z., Zhang, Z., Chen, C., Huang, X., Li, F., & Zhao, J. (2020). Generation of a Broadly Useful Model for COVID-19 Pathogenesis, Vaccination, and Treatment. *Cell, 182*(3), 734–743.e5. DOI: 10.1016/j.cell.2020.06.010 PMID: 32643603

Sun, S. H., Chen, Q., Gu, H. J., Yang, G., Wang, Y. X., Huang, X. Y., Liu, S. S., Zhang, N. N., Li, X. F., Xiong, R., Guo, Y., Deng, Y. Q., Huang, W. J., Liu, Q., Liu, Q. M., Shen, Y. L., Zhou, Y., Yang, X., Zhao, T. Y., & Wang, Y. C. (2020). A Mouse Model of SARS-CoV-2 Infection and Pathogenesis. *Cell Host & Microbe, 28*(1), 124–133.e4. DOI: 10.1016/j.chom.2020.05.020 PMID: 32485164

Szabat, M., Lorent, D., Czapik, T., Tomaszewska, M., Kierzek, E., & Kierzek, R. (2020). RNA Secondary Structure as a First Step for Rational Design of the Oligonucleotides towards Inhibition of Influenza A Virus Replication. *Pathogens (Basel, Switzerland), 9*(11), 925. DOI: 10.3390/pathogens9110925 PMID: 33171815

Tabrizi, S. J., Leavitt, B. R., Landwehrmeyer, G. B., Wild, E. J., Saft, C., Barker, R. A., & Smith, A. V. (2019). Targeting huntingtin expression in patients with Huntington's disease. *The New England Journal of Medicine, 380*(24), 2307–2316. DOI: 10.1056/NEJMoa1900907 PMID: 31059641

Tan, E. L., Ooi, E. E., Lin, C. Y., Tan, H. C., Ling, A. E., Lim, B., & Stanton, L. W. (2004). Inhibition of SARS coronavirus infection in vitro with clinically approved antiviral drugs. *Emerging Infectious Diseases, 10*(4), 581–586. DOI: 10.3201/eid1004.030458 PMID: 15200845

Tang, H., Abouleila, Y., & Mashaghi, A. (2020). Lassa hemorrhagic shock syndromeon-a-chip. *Biotechnology and Bioengineering*. Advance online publication. DOI: 10.1002/bit.27636 PMID: 33241859

The Complex Trait Consortium. (2018). The Collaborative Cross, a community resource for the genetic analysis of complex traits. *Nature Genetics, 36*, 1133–1137. PMID: 15514660

Thomas, S. J., & Yoon, I. K. (2019). A review of Dengvaxia®: Development to deployment. *Human Vaccines & Immunotherapeutics*, *15*(10), 2295–2314. DOI: 10.1080/21645515.2019.1658503 PMID: 31589551

Tompa, D. R., Immanuel, A., Srikanth, S., & Kadhirvel, S. (2021). Trends and strategies to combat viral infections: A review on FDA approved antiviral drugs. *International Journal of Biological Macromolecules*, *172*, 524–541. DOI: 10.1016/j.ijbiomac.2021.01.076 PMID: 33454328

Treanor, J. J., Hayden, F. G., Vrooman, P. S., Barbarash, R., Bettis, R., Riff, D., & Singh, S. (2019). Efficacy and safety of the oral neuraminidase inhibitor oseltamivir in treating acute influenza. *The Journal of Infectious Diseases*, *220*(4), 651–661. DOI: 10.1093/infdis/jiy684

Van Poelvoorde, L. A. E., Saelens, X., Thomas, I., & Roosens, N. H. (2020). Next-Generation Sequencing: An Eye-Opener for the Surveillance of Antiviral Resistance in Influenza. *Trends in Biotechnology*, *38*(4), 360–367. DOI: 10.1016/j.tibtech.2019.09.009 PMID: 31810633

Veiga, N., Goldsmith, M., Granot, Y., Rosenblum, D., Dammes, N., Kedmi, R., Ramishetti, S., & Peer, D. (2018). Cell specific delivery of modified mRNA expressing therapeutic proteins to leukocytes. *Nature Communications*, *9*(1), 4493. DOI: 10.1038/s41467-018-06936-1 PMID: 30374059

Wei, J., Alfajaro, M. M., DeWeirdt, P. C., Hanna, R. E., Lu-Culligan, W. J., Cai, W. L., Strine, M. S., Zhang, S. M., Graziano, V. R., Schmitz, C. O., Chen, J. S., Mankowski, M. C., Filler, R. B., Ravindra, N. G., Gasque, V., de Miguel, F. J., Patil, A., Chen, H., Oguntuyo, K. Y., & Wilen, C. B. (2021). Genome-wide CRISPR Screens Reveal Host Factors Critical for SARS-CoV-2 Infection. *Cell*, *184*(1), 76–91.e13. DOI: 10.1016/j.cell.2020.10.028 PMID: 33147444

Whitten-Bauer, C., Chung, J., Gómez-Moreno, A., Gomollón-Zueco, P., Huber, M. D., Gerace, L., & Garaigorta, U. (2019). The Host Factor Erlin-1 is Required for Efficient Hepatitis C Virus Infection. *Cells*, *8*(12), 1555. DOI: 10.3390/cells8121555 PMID: 31810281

WHO. (2015). *Summary of Probable SARS Cases with Onset of Illness from 1 November 2002 to 31 July 2003*. World Health Organization.

WHO. (2017). *Global Hepatitis Report, 2017*. WHO.

WHO. (2020). *Ebola in Democratic Republic of Congo—North Kivu, Ituri 2018–2020*. WHO.

WHO. (2020). *Latest HIV Estimates and Updates on HIV Policies Uptake, July 2020*. WHO.

WHO. (2022). *Avian Influenza A (H5N1)—United States of America*. WHO.

WHO. (2022). *Coronavirus Disease (COVID-19) Dashboard*. WHO.

WHO. (2022). *Ebola Virus Disease-Democratic Republic of Congo*. WHO.

WHO. (2022). *Middle East Respiratory Syndrome Coronavirus (MERS-CoV)—Saudi Arabia*. WHO.

Widagdo, W., Okba, N. M. A., Stalin Raj, V., & Haagmans, B. L. (2016). MERS-coronavirus: From discovery to intervention. *One Health*, *3*, 11–16. DOI: 10.1016/j.onehlt.2016.12.001 PMID: 28616497

Yang, J. W., Yang, L., Luo, R. G., & Xu, J. F. (2020). Corticosteroid administration for viral pneumonia: COVID-19 and beyond. *Clinical microbiology and infection: the official publication of the European Society of Clinical Microbiology and Infectious Diseases, 26*(9), 1171–1177. https://doi.org/DOI: 10.1016/j.cmi.2020.06.020

Yang, L., Han, Y., Nilsson-Payant, B. E., Gupta, V., Wang, P., Duan, X., Tang, X., Zhu, J., Zhao, Z., Jaffré, F., Zhang, T., Kim, T. W., Harschnitz, O., Redmond, D., Houghton, S., Liu, C., Naji, A., Ciceri, G., Guttikonda, S., & Chen, S. (2020). A Human Pluripotent Stem Cell-based Platform to Study SARS-CoV-2 Tropism and Model Virus Infection in Human Cells and Organoids. *Cell Stem Cell*, *27*(1), 125–136.e7. DOI: 10.1016/j.stem.2020.06.015 PMID: 32579880

Yang, L., Pei, R. J., Li, H., Ma, X. N., Zhou, Y., Zhu, F. H., He, P. L., Tang, W., Zhang, Y. C., Xiong, J., Xiao, S. Q., Tong, X. K., Zhang, B., & Zuo, J. P. (2021). Identification of SARS-CoV-2 entry inhibitors among already approved drugs. *Acta Pharmacologica Sinica*, *42*(8), 1347–1353. DOI: 10.1038/s41401-020-00556-6 PMID: 33116249

Yount, B., Curtis, K. M., Fritz, E. A., Hensley, L. E., Jahrling, P. B., Prentice, E., Denison, M. R., Geisbert, T. W., & Baric, R. S. (2003). Reverse genetics with a full-length infectious cDNA of severe acute respiratory syndrome coronavirus. *Proceedings of the National Academy of Sciences of the United States of America*, *100*(22), 12995–13000. DOI: 10.1073/pnas.1735582100 PMID: 14569023

Yuan, S., Chu, H., Chan, J. F., Ye, Z. W., Wen, L., Yan, B., Lai, P. M., Tee, K. M., Huang, J., Chen, D., Li, C., Zhao, X., Yang, D., Chiu, M. C., Yip, C., Poon, V. K., Chan, C. C., Sze, K. H., Zhou, J., & Yuen, K. Y. (2019). SREBP-dependent lipidomic reprogramming as a broad-spectrum antiviral target. *Nature Communications*, *10*(1), 120. DOI: 10.1038/s41467-018-08015-x PMID: 30631056

Zhang, M., Wang, P., Luo, R., Wang, Y., Li, Z., Guo, Y., Yao, Y., Li, M., Tao, T., Chen, W., Han, J., Liu, H., Cui, K., Zhang, X., Zheng, Y., & Qin, J. (2020). Biomimetic Human Disease Model of SARS-CoV-2-Induced Lung Injury and Immune Responses on Organ Chip System. *Advanced Science (Weinheim, Baden-Wurttemberg, Germany)*, *8*(3), 2002928. DOI: 10.1002/advs.202002928 PMID: 33173719

Zhang, W. F., Stephen, P., Thériault, J. F., Wang, R., & Lin, S. X. (2020). Novel Coronavirus Polymerase and Nucleotidyl-Transferase Structures: Potential to Target New Outbreaks. *The Journal of Physical Chemistry Letters*, *11*(11), 4430–4435. DOI: 10.1021/acs.jpclett.0c00571 PMID: 32392072

Zhou, J., Lazar, D., Li, H., Xia, X., Satheesan, S., Charlins, P., O'Mealy, D., Akkina, R., Saayman, S., Weinberg, M. S., Rossi, J. J., & Morris, K. V. (2018). Receptor-targeted aptamer-siRNA conjugate-directed transcriptional regulation of HIV-1. *Theranostics*, *8*(6), 1575–1590. DOI: 10.7150/thno.23085 PMID: 29556342

Zhou, Y., Vedantham, P., Lu, K., Agudelo, J., Carrion, R.Jr, Nunneley, J. W., Barnard, D., Pöhlmann, S., McKerrow, J. H., Renslo, A. R., & Simmons, G. (2015). Protease inhibitors targeting coronavirus and filovirus entry. *Antiviral Research*, *116*, 76–84. DOI: 10.1016/j.antiviral.2015.01.011 PMID: 25666761

Zhu, N., Zhang, D., Wang, W., Li, X., Yang, B., Song, J., Zhao, X., Huang, B., Shi, W., Lu, R., Niu, P., Zhan, F., Ma, X., Wang, D., Xu, W., Wu, G., Gao, G. F., & Tan, W. (2020). A Novel Coronavirus from Patients with Pneumonia in China, 2019. *The New England Journal of Medicine*, *382*(8), 727–733. DOI: 10.1056/NEJMoa2001017 PMID: 31978945

Zou, G., Xu, H. Y., Qing, M., Wang, Q. Y., & Shi, P. Y. (2011). Development and characterization of a stable luciferase dengue virus for high-throughput screening. *Antiviral Research*, *91*(1), 11–19. DOI: 10.1016/j.antiviral.2011.05.001 PMID: 21575658

Chapter 7
The Battle Against Emerging Viruses:
Current and Future Antiviral Therapeutics

Shafeeq Ur Rehman
https://orcid.org/0000-0003-3571-8226
Cholistan University of Veterinary and Animal Sciences, Bahawalpur, Pakistan

Ayesha Jahangir
https://orcid.org/0009-0008-4130-420X
The Islamia University of Bahawalpur, Pakistan

Minal Hussain
https://orcid.org/0009-0001-3900-3458
Cholistan University of Veterinary and Animal Sciences, Bahawalpur, Pakistan

ABSTRACT

Emerging viruses pose a significant threat to global health due to their widespread illness, mortality, and economic impact. These zoonotic pathogens, often transmitted from animals to humans, can cause economic disruption and overburden healthcare systems. Infections like HIV, Ebola, SARS, MERS, and COVID-19 demonstrate how viruses can be linked to human behavior and environmental change. Antiviral therapeutics have been developed to combat viral infections, targeting different points in the viral life cycle. However, treatments still face challenges such as drug resistance, side effects, and the inability to reach latent viral reservoirs. To address these challenges, innovation in drug design, targeting less mutable, conserved viral structures, and exploiting combination therapies is needed. Drug repurposing, immunotherapy, and RNA-based therapies are promising approaches to address

DOI: 10.4018/979-8-3693-9276-8.ch007

Copyright © 2025, IGI Global Scientific Publishing. Copying or distributing in print or electronic forms without written permission of IGI Global Scientific Publishing is prohibited.

current and future viral threats. Antiviral therapeutics are expected to evolve in a multitiered framework, combining targeted and broad-spectrum agents to overcome the complexity of viruses.

1. INTRODUCTION:

Emerging viruses are those that have newly appeared in a population or are rising in incidence or geographic range very rapidly. They may be entirely novel pathogens, or previously known viruses that reappear, or transmit more, as a result of changes in ecology, human behavior, or environmental factors (STRAUSS & STRAUSS, 2012). Global health is affected by challenges posed by emerging viruses: the resulting widespread illness, mortality, and economic impact. Characteristics of emerging viruses often include:

- **High Mutation Rates:**

RNA viruses, however, mutate rapidly, for reasons that allow them to adapt rapidly to new hosts and new environmental pressures (Lauring et al., 2013).

- **Cross-Species Transmission:**

Zoonotic viruses are several emerging viruses of animal origin (Chan et al., 2015).

- **Adaptability to New Hosts:**

The viruses are often able to adapt to new host species with little genetic change (Longdon et al., 2014).

- **Pandemic Potential:**

Emerging viruses are highly adaptive bench, quickly cross international and continental borders, and cause global health crises.

A series of devastating viral outbreaks have changed global health priorities, public awareness, and research directions. Emerging viruses (those discovered first for humans, or recently causing higher incidence or geographic expansion) continue to represent an ongoing threat to societies worldwide. These are zoonotic pathogens, meaning they are often transmitted from animals to humans, and can create opportunities for economic disruption and the overburdening of healthcare systems as well as take hundreds of thousands of lives (Ellwanger et al., 2022).

Infections like HIV, Ebola, SARS (Severe Acute Respiratory Syndrome), MERS (Middle East Respiratory Syndrome), and more recently COVID-19 have demonstrated that scientists are attempting to address the virology of the virus yet are failing to adapt to it quickly can spread from one border to another bringing massive morbidity and mortality (Crawford, 2021; Graham & Sullivan, 2018). These outbreaks demonstrate how a virus can be linked to human behavior and its relationship to environmental change and viral transmission dynamics. The clear rise of deforestation and urbanization disturbs wildlife habitats and pulls humans closer to animals whose health affects ours in closer proximity to transmit diseases from animals to humans via zoonoses. Along with global travel, climate change, and antimicrobial resistance, new viral strains continue to emerge that can't be treated or vaccinated against, some of which have even developed the ability to resist current treatments and vaccines (Afrough et al., 2019).

These viruses have not only had just an immediate health impact, but also have social, economic, and psychological impacts. For example, the COVID-19 pandemic happened and it caused a global recession, destroyed supply chains, opened unemployment crises, and changed everyday life for billions of people (Naseer et al., 2023). However, the urgency of understanding and fighting emerging viruses is both basic to medical science and a key to societal stability and progress. Prevention, rapid diagnostics, vaccines, and antiviral treatment are all that are needed to adequately respond to emerging viral threats. Vaccines continue to be by far the best approach to preventing viruses, but they take time and often prove difficult to develop during active outbreaks. Second, not all viruses have a viable vaccine, HIV, for example, has escaped vaccine development for decades. Consequently, antiviral therapeutics constitute important tools against these pathogens, to treat infected patients, control the spread of the virus, and in some cases prophylaxis (Meganck & Baric, 2021).

Treatment of individuals infected is important; doing so with antiviral drugs helps to relieve symptoms, shorten the length of time, and lessen the risk of complications. There are virtually no options for the treatment of viruses that have vaccines that are ineffective or non-existent, and antiviral drugs represent one of the few options. In addition, antivirals are essential for the treatment of drug-resistant strains of viruses generated from genetic mutation. Rapid viral mutation rates have plagued RNA viruses with the need for continual innovation in drug design to match the changing viral landscape (Ison, 2016). Newly emerging and reemerging viruses are shown in Figure 1.

Figure 1. Reemerging and newly emerging viruses

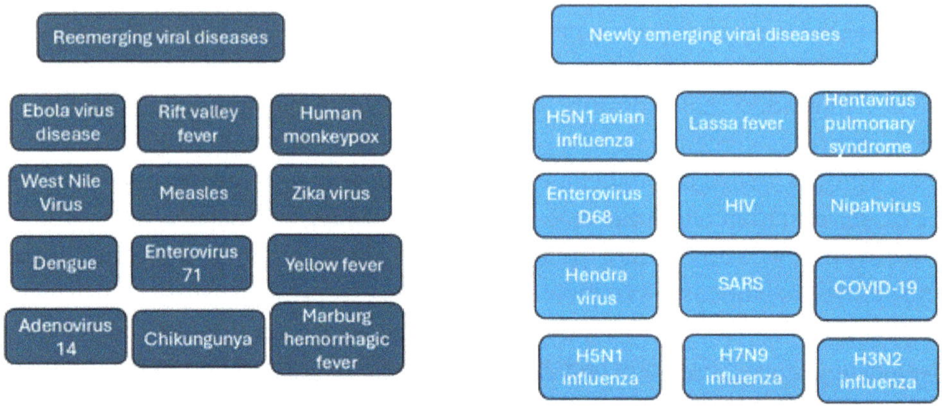

Antivirals are hampered not only by the difficulty in developing effective treatments for complex viruses, which rely on host cell machinery for their replication but also by the viruses themselves, which are quite resistant to the new 'weapons' directed against them. This reliance has further narrowed the number of druggable targets and raises the probability of side effects in patients (Schenk et al., 2021). While meeting these challenges presents a barrier to antiviral development, the urgency for antivirals is pressing, particularly in light of recent emerging viruses' incredible potential for extensive societal disruption. Next-generation antivirals may be developed from recent advancements in molecular biology, genomics, and bioinformatics (Abiha et al., 2023).

Antiviral drug development is a relatively new story compared to antibiotics, which changed the face of medicine in the early 20th century. The second half of the 20th century marked the beginning of the development of antiviral therapies, though the mainspring for urgency was perhaps the desire to combat herpes simplex infections. Since that time, the story has been a story of scientific breakthroughs and fast adaptation to new viral challenges in the development of antiviral drugs. Additions include drugs such as acyclovir for herpes, zidovudine for HIV, and oseltamivir for influenza. First-generation drugs have shown that it was possible to inhibit viral replication in human cells, a fundamental change in viral therapy capabilities (Antonelli et al., 2021).

Table 1. Characteristics of emerging viruses

Feature	Description	Examples
High mutation rate	This leads to the emergence of new variants and the potential for drug resistance	HIV, Influenza, COVID-19
Severe disease	Some emerging viruses can cause severe illness and death	Ebola, SARS, MERS, COVID-19
Lack of effective treatments	Development of antivirals and vaccines can be challenging	Many emerging viruses initially lack specific treatments
Zoonotic origin	Viruses that originate in animals and jump to humans	HIV, Ebola, SARS, MERS, COVID-19, Zika, West Nile Virus

2. NOTABLE OUTBREAKS: HIV, EBOLA, SARS, COVID-19

Examination of recent outbreaks of emerging viruses helps to understand the dynamics of emerging viruses and to understand what went well and what did not go well to be better prepared for the future. The overview of a few notable outbreaks is shown in Table 2. Major pandemics in human history are shown in Figure 2.

Figure 2. Major pandemics in human history

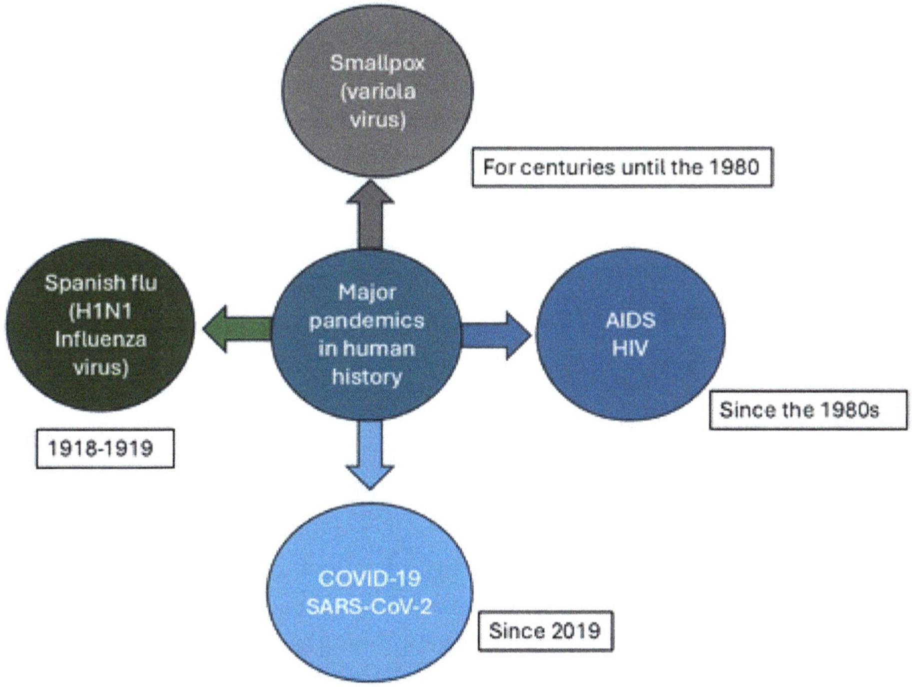

2.1. COVID-19 Pandemic

The SARS-CoV-2 virus that causes the COVID-19 pandemic has had an unprecedented impact on modern society. The virus, first identified in Wuhan, China, late in 2019, rapidly spread across the globe and infected millions embroiling the world in a huge swath of social, economic, and public health challenges unprecedented in human history (Ullah & Ferdous, 2022). The virus, SARS-CoV-2, is thought to have jumped initially from bats to a yet unknown intermediate host such as pangolins, before moving on to humans. SARS-CoV-2 quickly adapted for human transmission, through accumulating mutations increasing SARS-CoV-2's capability to bind to human ACE2 receptors. The virus was seen to adapt so quickly, with variants like Delta and Omicron showing how quickly the virus could become more transmissible, or how quickly it could evade an immune response (Karban, 2023). COVID-19 showed how modern society relies not just on one another, but on other people too, it was only made possible through international travel. It revealed the weakness of global health systems, where we have never seen the capacity to ramp up the efforts of vaccine development and public health measures (Vecchio et al., 2024).

2.2. HIV/AIDS Pandemic

HIV/AIDS, which appeared in the 1980s, is one of the most important recent viral outbreaks. HIV, or human immunodeficiency virus, infects the immune system; if untreated, it's called AIDS (Acquired-Immunodeficiency Syndrome). Unlike other infectious agents that might clear or be cured, once infected with HIV, the disease has to be fought indefinitely (Klatt, 2017). Antiviral drug development had another catalyst, the HIV epidemic that prompted the development of reverse transcriptase inhibitors and protease inhibitors which changed the way of treatment for HIV. HIV transformation into a moderately treated chronic condition and the saving of millions of lives and the basis for future antiviral drug development is owed to these drugs (Di Santo, 2014; Saag et al., 2020).

2.3. Ebola Virus Outbreaks

Ebola virus disease (EVD) is a filovirus that causes severe hemorrhagic fever with high fatality rates. Although scientists have known about the virus since 1976, it frequently resurfaces in African nations, with outbreaks in 2014–2016 and more recent cases (Shenge & Opayele, 2020). Fruit bats are thought to be the reservoir host for Ebola. It can be transmitted to humans by contact with the body fluids of infected animals, including bats, chimpanzees, and gorillas, and then from one to another by bodily fluids. Ebola has emerged because of deforestation and close contact with wildlife. When human populations encroach on forested areas, they increase their opportunities for contact with potential Ebola reservoirs (Caron et al., 2022). In 2014–2016, a West African Ebola outbreak highlighted the difficulties of containing highly virulent disease in resource-limited settings. The outbreak also showed the need for rapid response, contact tracing, and isolation (Kost, 2020). Their ability to control future outbreaks of Ebola resulted from the subsequent development of the Ebola vaccine, rVSV-ZEBOV (Herder et al., 2020).

2.4. SARS and MERS Outbreaks

Both SARS and MERS are due to coronaviruses that have spread into humans from animals. In 2002–2003, the SARS outbreak, starting in China, spread worldwide with great morbidity and mortality in more than two dozen countries (Schwartz & Graham, 2020). Public health measures contained the outbreak but the absence of effective antiviral therapies inhibited treatment options at the time. When MERS surfaced in Saudi Arabia in 2012, it proved highly deadly but much less transmissible than SARS and thus could be contained more locally (Mackay & Arden, 2015). SARS as well as MERS were a reminder of the coronavirus potential danger and emphasized the need to have broad-spectrum antiviral drugs, and to be ready for new outbreaks of similar viruses (Barth et al., 2021).

2.5. Zika Virus Outbreak

During the 2015-2016 outbreak in Brazil, the Zika virus is mainly acquired from Aedes mosquitoes. While Zika normally causes only mild symptoms, such as a fever or rash, it can result in severe birth defects, like microcephaly, in babies born to infected parents (Lourenço et al., 2017). The environmental factors contributing to the spread of the disease warm temperatures and increased populations of the mosquitoes responsible for transmitting it played an important role in doing just that. There was spread, too, due to urbanization and a lack of mosquito control in the affected areas. Zika is spread by mosquitoes, unlike SARS-CoV-2 and Ebola which are spread through direct contact, and so vector control is a primary means of prevention (Asad & Carpenter, 2018). Sexual transmission of the virus makes control measures even more complicated. The Zika outbreak served notice of the importance of good vector control and the ability of viruses to impact fetal development. Vaccines for Zika have continued to be researched, but there hasn't yet been a vaccine widely used (Abbink et al., 2018).

Table 2. The overview of notable outbreaks

Virus	First Identified	Origin	Transmission	Symptoms	Mortality Rate(approx.)
West Nile Virus	1937	Birds, mosquitoes	Mosquito bites	Most people have no symptoms; some develop fever, headache, body aches	~1% of those who develop severe illness
Zika	1947	Monkeys, mosquitoes	Mosquito bites, sexual contact, mother-to-child	Fever, rash, joint pain, conjunctivitis	Microcephaly and other birth defects in infants born to infected mothers
Ebola	1976	Likely bats	Contact with bodily fluids of infected individuals	Fever, headache, muscle pain, vomiting, diarrhea, bleeding	25-90% depending on the outbreak
HIV	1981	Chimpanzees	Sexual contact, blood transfusions, sharing needles	Flu-like illness initially, then a range of opportunistic infections	Varies widely depending on access to treatment; can be high without treatment
SARS	2002	Bats, civet cats	Respiratory droplets, close contact	Fever, cough, shortness of breath	~10%
MERS	2012	Bats, camels	Respiratory droplets, close contact	Fever, cough, shortness of breath	~35%
COVID-19	2019	Likely bats	Respiratory droplets, close contact	Fever, cough, shortness of breath, loss of taste/smell	Varies depending on age and underlying health conditions; generally <2% currently

3. ZOONOTIC TRANSMISSION

Zoonotic transmission where viruses pass from animals to humans is one of the biggest drivers of human virus emergence. Zoonotic pathogens include HIV, Ebola, SARS, and COVID-19, and more than 70 percent of emerging infectious diseases are of this origin. Usually, zoonotic viruses live in animal reservoirs, like bats and rodents, rodents, and primates, and are spread from these animals, either directly or by intermediate hosts, to humans (Weiss & Sankaran, 2022).

- **Animal Reservoirs:** Some animals become natural hosts of viruses. As one example, bats serve as reservoirs for many coronaviruses (including SARS coronavirus) and filoviruses (such as Ebola). The conditions that lead most easily to viral spillover are high population density and close quarters in some animal species, as in agricultural settings (El-Sayed & Kamel, 2021).

- **Intermediate Hosts:** An intermediate host that can carry the virus from a virus animal reservoir to humans may cause zoonotic transmission. An example from the SARS outbreak was that civet cats functioned as an intermediate host between bats and humans. The case of COVID-19 has implicated pangolins as possible intermediaries, although conclusive evidence has not yet been investigated (Ul-Rahman et al., 2023).
- **Human encroachment:** Human encroachment on wildlife habitats is one major factor that creates more contact with zoonotic reservoirs that harbor disease syndromes capable of transmission between animals and humans. Increased closeness of humans to animals from deforestation, hunting, and consumption of wild animals, and to agriculture that spreads into the wild habitats, increases the risk of cross-species transmission. The globalization of the food industry, exotic animal markets, and the wildlife trade have also facilitated other zoonotic transmission pathways (Marrana, 2022; Saleem et al., 2023).

3.1. Viral Mutations and Adaptations:

The emergence of infectious disease is, in part, a product of viral evolution. RNA viruses in particular tend to have a high mutation rate and lack error-correcting means during replication. These mutations allow viruses to more rapidly adapt to new environments, hosts, or conditions to enhance their survival and suitability to infect new species (DeFILIPPIS & VILLARREAL, 2007).

- **Genetic Drift and Genetic Shift:** Genetic drift occurs when a mutation in the viral genome comprised of small, random viral mutation accumulates over time, which may thereby change the virus's ability to infect new hosts or evade immune responses (Sanjuán & Domingo-Calap, 2016). A more abrupt form of genetic change known as the genetic shift occurs when two viruses exchange genetic material, as when different strains of influenza viruses co-infect a host and recombine. Strains can shift from being given a genetic shift containing major changes in viral characteristics that can be more virulent or more transmissible (Holmes, 2009; Stern & Andino, 2016).
- **Recombination and Adaptation:** Indeed, both viral and bacterial strains can also recombine with others of their family during transition what intermediary hosts, producing novel strains with increased transmissibility or (in the case of viruses) pathogenicity. For example, SARS-CoV-2 has under-

gone mutations that allowed it to some extent to avoid immune response and spread globally. Analysis of these adaptations allows the virus to more tightly bind to human cell receptors, including the *ACE2* receptor for SARS-CoV-2 and the *CXCR4* receptor for MERS-CoV (Telenti et al., 2022).

Vaccine and drug development are threatened by rapid viral evolution, whose emergence may render current treatments less effective in the future through resistance. This characteristic necessitates real-time monitoring of viral genomes to track mutations and respond to medical interventions.

3.2. Environmental Changes

The emergence and spread of new viral pathogens are fueled in part by environmental changes, many of which are due to human activity. All disrupt ecosystems and change pathogen transmission dynamics deforestation, climate change, urbanization, and agricultural practices (Nova et al., 2022).

- **Deforestation and Habitat Loss:** The closer humans come to destroying forests and other natural habitats, the more likely they are to transmit zoonotic viruses to their own species. For example, the Ebola virus is known to have deforestation origins in Africa where the loss of habitat sends animals into contact with humans (Tazerji et al., 2022).
- **Climate Change:** Disease vectors, including mosquitoes and ticks, that spread viruses like Zika, dengue, or West Nile, can change with climate changes. This contributes to vector proliferation and to an expansion of their habitats to places that have thus far been 'uninfected' by these viruses (Asad & Carpenter, 2018).
- **Urbanization and Population Density:** The fertile ground for the rapid spread of viruses is quickly becoming dense urbanization, where people live near one another. Outbreaks of smaller louseborne zoonoses can occur, and the ease of transmission makes outbreaks more likely in crowded urban environments, where sanitation as well as healthcare infrastructure are lacking. Urban areas, which to date encompass a higher proportion of humanity, became hot spots for the spread of both COVID-19 and emerging viruses like Zika (Gangopadhayya & Bhukya, 2023; Prothero). Viral diseases and their transmission from animal to human and their effect are shown in Figure 3.

Figure 3. Viral diseases and their transmission from animal to human

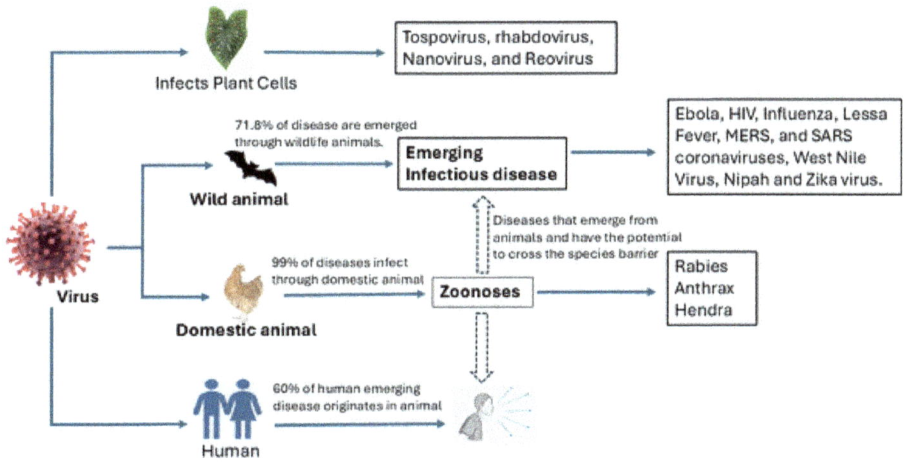

4. CURRENT ANTIVIRAL THERAPEUTICS

Over the last few decades, there have been considerable achievements in the development of antiviral therapeutics, which have turned what were often fatal infections into very controllable diseases. While vaccines continue to be the leading line of defense against viral infections, antiviral drugs are important for treating individuals who become infected and for treating individuals with diseases where no effective vaccine is available or who develop symptoms despite vaccines. Reducing viral replication, easing symptoms, and preventing complications, antiviral drugs are of critical value. Therapeutic interventions for RNA virus infections are shown in Figure 4. While antiviral drug development is hampered by the rapidly evolving multiplication rates of many viruses and their complex biology, the latter pose ongoing challenges (De Clercq & Li, 2016; Loaiza et al., 2023). Table 3. shows antiviral therapies against a few viral diseases with their mechanisms of action.

Figure 4. Therapeutic interventions for RNA virus infections

Antiviral drugs	RNA-based therapeutics	mRNA vaccines
➢ **Influenza Virus** •M2 ion channel blockers e.g damantanes. •RNA-dependent RNA polymerase inhibitors, e.g. favipiravir ➢ **Hepatitits C** •IFNα monotherapy. •Direct-acting antivirals, protease inhibitors e.g. boceprevir and telaprevir. ➢ **SARS-CoV-2** • Direct-acting antivirals, e.g remdesivir, molnupiravir •RNA replication inhibitors, e.g. nirmatrelvir, favipiravir ➢ **HIV** •Azidothymidine (AZT), first approved •Non-nucleoside reverse transcriptase inhibitors e.g doravirine	➢ **RNA interference (RNAi)** •Silencing of RNA expression by using ds RNA •Effective in HIV, Influenza, SARS-C0V treatment ➢ **Antisense oligonucleotides** •Occupancy-mediated degradation target mRNA cleavage •Occupancy-only mechanism: splice switching, blocking of miRNA binding to target RNA ➢ **CRISPR-based genome editing** •Edit genome sequence to irreversible knockout/knock-in of a target gene •Application in detection of SARS-CoV-2, and Zika virus etc. ➢ **Micro RNA (miRNA)** •Have a role in tumor progression and predict radiation necrosis	•mRNA introduced as a vaccine expresses the required functional protein •More efficient, quicker to design and produce, less expensive • mRNA-based vaccination against SARS-CoV-2 • Nanoparticles-based delivery system able to increase stability, immunogenicity, and effectiveness of mRNA vaccines •Three types of mRNA vaccines available are, conventional mRNA, self-amplifying mRNA, and trans-amplifying mRNA •First two mRNA vaccines for SARS-CoV-2, the Pfizer- BioNTech vaccine and the Moderna vaccine

4.1. Mechanisms of Action

The specific type of antiviral drug uses a different point in the viral life cycle to target (prevent entry into the host cell, block viral replication or assembly of new virions) (Kausar et al., 2021). Here are some primary mechanisms by which antiviral drugs combat viral infections:

- **Entry Inhibitors:** Entry inhibitors prevent the virus from attaching to or entering host cells, preventing the virus from starting its process of infection. Entry of viruses into host cells depends on the links between the surface proteins of the virus and receptors on a host cell. Entry inhibitors disrupt interactions that offer these viruses an opportunity to infect and replicate (Kausar et al., 2021).
- **Fusion Inhibitors:** Fusion inhibitors block a step that the virus requires to enter a host cell: fusion of the virus' envelope to the host cell membrane. Testing of such drugs consists of cell culture and animal tests and is hampered somewhat by poor correlations with the actual clinical situation. For example, enfuvirtide, which is used in human HIV medicine to prevent the virus from fusing virally encoded genetic material with a target cell membrane, is an example of a fusion inhibitor (Fallarino, 2018).

- **Receptor Blockers:** Two of the sorts of entry inhibitors do so by binding to cell receptors on a host cell and preventing viral attachment. One HIV entry inhibitor is maraviroc, an HIV entry inhibitor that binds to viral host cells via their CCR five receptor and stops HIV from attaching and entering the cell (Ralli, 2022).

These inhibitors prevent viral entry and thereby reduce the initial viral load to lower the chance of the virus settling into the host and becoming entrenched. The mechanism has an attractive application when multiple stages of the viral life cycle are targeted in combination therapies.

4.2. Replication Blockers

Replication blockers inhibit viral genome replication, an obligatory step for viral propagation on the host. Replication blockers are most commonly nucleoside or nucleotide analogs that mimic the natural building blocks of viral DNA or RNA, and so incorporate themselves into the viral genome and stop replication (Hermann, 2016).

- **Nucleoside and Nucleotide Analogs:** These compounds are like natural nucleotides in structure. Replication determines that they are incorporated into the viral genome and, during this process, they cause premature termination of the genetic chain, i.e., replication is halted. For example, acyclovir is a nucleoside analog used against herpesviruses. Viral enzymes phosphorylate it, incorporate it into the viral DNA, and cause chain termination, so the virus can't replicate any further (Ch'ien et al., 2018).
- **Polymerase Inhibitors:** Polymerase inhibitors are specifically meant to block a viral polymerase enzyme that is used to make viral RNA and viral DNA. Used for COVID, it is a polymerase inhibitor and it interferes with the RNA-dependent RNA polymerase in SARS CoV-2 so it cannot replicate so well (Tian et al., 2021).

4.3. Protease Inhibitors

Inhibitors of protease inhibit the processing of viral proteins limiting the formation of the virus mature virions. Most viruses, including HIV and hepatitis C, make lots of long polyprotein chains that have to be cut to pieces by viral proteases to make individual proteins (Sutherland, 2014).

- **HIV Protease Inhibitors:** Protease inhibitors included in HIV treatment like lopinavir and ritonavir are drugs. These drugs block the cells' ability to produce mature virions, which contain proteins and other viral material that can infect new cells if assembled (Voshavar, 2019).

- **Hepatitis C Protease Inhibitors:** Protease inhibitors targeting hepatitis C causing infection include boceprevir and telaprevir which inhibit the NS3/4A protease enzyme, an enzyme that is required for viral replication and for processing the viral polyprotein. Blocking this protease stops the viral life cycle and suppresses the virus (Nourbakhsh, 2013).

In combination therapy with protease inhibitors, they are extremely effective, as they act against essential steps for viral assembly with a resulting decrease of viral load and there is no spread of infection.

5. EXISTING TREATMENT LIMITATIONS

To date, however, treatments for the virus still have several problems. The challenges are things like drug resistance, side effects, and the fact that you can't always reach the latent viral reservoirs.

5.1. Resistance

Antiviral drugs are one of the principal limitations of resistance for viruses that mutate rapidly, such as HIV and influenza. Because viruses replicate errors are introduced, which may include resistance to a specific drug. If resistance occurs, it's usually a result of reduced efficacy of the drug and we need to develop new therapies or drug combinations (Irwin et al., 2016).

- **HIV Resistance:** Because HIV has such a high mutation rate, single-agent therapies tend to quickly evolve resistance to the treatment (which is why combination therapies, or HAART, are the norm in HIV treatment) (Bock & Lengauer, 2012).
- **Influenza Resistance:** Over recent years, certain antiviral drugs, including those against influenza viruses such as the H1N1 and H3N2 subtypes, became resistant. Worldwide resistance to amantadine and rimantadine, the antiviral drugs that block viral uncoating, is common and has been shown to impair drug efficacy. Resistance problems have also arisen in neuraminidase inhibitors, such as oseltamivir, especially in immunocompromised patients receiving prolonged treatment (Lampejo, 2020).

Viruses can also make the therapy resistant, the therapist needs to keep vigilance, adhere to the guidelines, as well as develop or seek new therapies or if available combinations of drugs to continue to serve the treatment options.

5.2. Side Effects

Like many pharmacological agents, antiviral drugs can cause adverse effects in patients and interfere with adherence to treatment regimens and ultimately with health outcomes. The side effects vary from drug mechanism, virus synergized with, and individual patient factors.

- **HIV Treatment Side Effects:** Long-term use of antiretroviral therapies such as protease inhibitors and NRTIs can result in gastrointestinal consequences, fatigue, and more serious effects including lipodystrophy, liver toxicity, and cardiovascular risk. Patients on long-term antiretroviral therapy are prone to cumulative side effects; so they require regular monitoring and sometimes changes in drug regimens (Gagnon & Holmes, 2016; Montessori et al., 2004).
- **Hepatitis C Drug Side Effects:** These drugs are often well-tolerated (but) may cause headache, fatigue, and nausea. Worst case, side effects are mild, but patients can still develop liver or kidney toxicity. Caution needs to be taken in people with existing health conditions (Cani et al., 2019).
- **COVID-19 Antiviral Side Effects:** For example, remdesivir can cause nausea, elevated liver enzymes, as well as in rare cases kidney failure. Infusion-related reactions may include fever, chills, anaphylaxis, and other rare events in the setting of monoclonal antibody therapeutics. While often manageable, however, these side effects restrict antivirals to some extent particularly so of regarding the that they can be used in vulnerable patients or people with multiple comorbidities (Malin et al., 2020).

Table 3. Antiviral therapies against a few viral diseases with their mechanisms of action.

Virus	Current Antiviral Therapies	Mechanism of Action	Limitations
Herpes Simplex Virus	Acyclovir, Valacyclovir, Famciclovir	Nucleoside analogues	Resistance can develop, especially in immunocompromised individuals
COVID-19	Remdesivir, Paxlovid, Molnupiravir	RNA polymerase inhibitor, Protease inhibitor, RNA analog	Varying efficacy, potential for resistance development, side effects
HIV	Combination antiretroviral therapy (cART)	Various, including reverse transcriptase inhibitors, protease inhibitors, integrase inhibitors	Lifelong treatment required, drug interactions, resistance development
Influenza	Oseltamivir, Zanamivir, Peramivir, Baloxavir	Neuraminidase inhibitors, Endonuclease inhibitor	Resistance development, limited efficacy, side effects
Hepatitis C	Direct-acting antivirals	Various, targeting specific viral proteins	High cost, potential drug interactions
Hepatitis B Virus	Tenofovir, Entecavir, Pegylated interferon	Nucleotide/nucleoside analogs, Immunomodulator	Lifelong treatment is often required (for nucleos(t)ide analogs), interferon has significant side effects, and functional cure is the goal but rarely achieved
Cytomegalovirus	Ganciclovir, Valganciclovir, Foscarnet, Cidofovir	Inhibit viral DNA polymerase	Significant side effects and resistance can develop

6. ANTIVIRAL DRUG DEVELOPMENT:

There are many reasons why antiviral drugs are inherently so hard to develop: viruses manipulate the host's cellular machinery for replication, so there are limited opportunities to target the viruses in host reaction, among other reasons. The chance of side effects is increased by this overlap. Taken together, high mutation rates of viruses, particularly RNA viruses, render these diseases to be highly resistant to drugs, and these new therapies continually have to be developed. To address these challenges we need innovation in drug design; targeting less mutable, conserved viral structures and exploiting combination therapies to lessen the possibility of resistance (Dunne, 2007).

6.1. Therapeutic Strategies

Traditional drug development processes are generally time-consuming (taking years to go from discovery to approval). As needs related to urgent health have been arising, researchers are looking into alternative therapeutic strategies to develop better treatments more rapidly. Several of these, including drug repurposing, immunotherapy, and RNA-based therapies, have received a great deal of attention as possible means to address today's and future viral threats (Rao et al., 2023).

6.2. Drug Repurposing

Drug repurposing or drug repositioning is defined as using drugs invented for one disease on another. There are many advantages of this approach including shorter development timelines and an abundance of previous safety data, making the regulatory approval timeline shortened. When we're approaching viral outbreaks, drugs that are already approved for other conditions can provide valuable, immediate treatment (Parvathaneni et al., 2019).

6.2.1. Repurposed Drugs for Viral Infections:

Many candidates against emerging viral infections have emerged from drug repurposing. Here are some notable examples: A few repurposed drugs with their antiviral activity are shown in Table 4.

- **Remdesivir:** Remdesivir was first created to treat the Ebola virus, and was repurposed as one of the first COVID-19 treatments. It is an RNA polymerase inhibitor, that is, it blocks the enzyme that allows the transfer of viral replication to occur in RNA viruses. Remdesivir was effective in Ebola trials, but had never been successful in SARS-CoV-2; promise against the virus led to its emergency use authorization for COVID, but it has yet to be approved by the FDA for humans. Repurposed based on remdesivir, broad-spectrum antivirals provided the potential for repurposing of these drugs for emerging pathogens and highlighted the value of having drugs that can be rapidly mobilized during outbreaks (Malin et al., 2020).
- **Hydroxychloroquine and Chloroquine:** In addition, these antimalarial drugs were considered a potential treatment for COVID-19. Earlier studies indicated that they might block viral entry into cells, but more recent studies were contradictory, and safety and efficacy were cause for concern, making their use ultimately limited. The repurposing of hydroxychloroquine during the pandemic showed the power and limits of repurposed drugs, the

case that keen clinical trials to confirm efficacy in new contexts are essential (Hernandez et al., 2020).

- **Favipiravir:** Favipiravir is a drug developed in Japan for influenza and re-purposed to treat diverse instances of viral infections including Ebola and COVID-19. In some countries, the drug was authorized for emergency use during the COVID-19 pandemic for mild to moderate cases. Favipiravir has broad-spectrum activity because of drugs using common viral targets could act against numerous pathogens is beneficial (Manabe et al., 2021).

Drug repurposing is still a promising strategy for the rapid response to viral outbreaks. Knowledge of the pharmacological properties of drugs that are already approved, enables researchers to shorten drug development time and costs making a critical therapeutic bridge until disease-specific drugs are developed.

Table 4. Repurposed drugs with their antiviral activity.

Drug	Original Indication	Repurposed
Paxlovid (Nirmatrelvir/ Ritonavir)	HIV	COVID-19
Cidofovir	Cytomegalovirus retinitis	Other herpesviruses, smallpox
Remdesivir	Ebola virus disease	COVID-19
Interferon-alpha	Various cancers, viral hepatitis	Some viral infections
Favipiravir	Influenza	COVID-19
Acyclovir	Herpes simplex virus	Varicella-zoster virus, Epstein-Barr virus
Valganciclovir	Cytomegalovirus	Prevention of CMV in transplant recipients
Molnupiravir	Influenza	COVID-19

6.3. Immunotherapy

Immunotherapy uses the immune system to fight against viral infections either by boosting the immune response or by using immune parts to go straight for the virus. Immunotherapy against viral infection has been tried out in several ways, including monoclonal antibodies, convalescent plasma therapy, and immune modulation.

6.3.1. Monoclonal Antibodies

Lab-engineered antibodies called Monoclonal antibodies (mAbs), bind to specific virus proteins and neutralize the virus before the virus can infect host cells. They are especially useful to treat viral infections as they are quickly made and can be tuned to attack particular viral structures (Mokhtary et al., 2022).

- **COVID-19 Monoclonal Antibodies:** During the COVID-19 pandemic monoclonal antibodies that target the spike protein of SARS-CoV2, and thereby prevent the virus from binding to human cells, were developed as bamlanivimab, casirivimab, and indevimab. In high-risk patients, these antibodies proved effective at reducing viral load and reducing the severity of symptoms. Nevertheless, the detection of SARS-CoV-2 mutants with spike protein mutative events necessitated some flexibility in monoclonal antibody development (Pinna et al., 2021).
- **Ebola Treatment with Monoclonal Antibodies:** During Ebola virus outbreaks monoclonal antibodies were developed and approved for use such as REGN-EB3 and mAb114. These antibodies are directed to the viral glycoprotein, neutralizing the virus and improving dramatically survival rates in patients with Ebola virus disease (Tshiani Mbaya et al., 2021).

Significant promise for the treatment of several viral infections has been shown by monoclonal antibodies. Because their specificity and virus neutralization capacity can provide powerful tools against severe infections, these antibodies are particularly valuable in those instances in which vaccines are unavailable or ineffective.

6.3.2. Convalescent Plasma Therapy

Practicing convalescent plasma therapy involves transfusing someone who recovers from the virus with plasma containing antibodies to the virus into someone who is still actively infected with the virus. For over a century, this approach has been used to treat infectious diseases and is in reemergence for emerging viral infections.

- **COVID-19 and Convalescent Plasma:** As an emergency treatment of severe cases during the early stages of the COVID-19 pandemic, convalescent plasma therapy was used. Initial studies showed that it could offer some benefits but larger studies came up empty, and evidence of effectiveness in COVID-19 treatment is unsettled. Yet, at the beginning of the pandemic, convalescent plasma therapy was an interim treatment method before any other treatment options existed (Chen et al., 2020; Roback & Guarner, 2020).

- **Ebola Virus Disease:** During Ebola outbreaks, convalescent plasma therapy was also used, with only limited success. In resource-limited settings, convalescent plasma was a viable option until logistical challenges of antibody variability and blood type matching were presented (van Griensven et al., 2016).

Although it's no longer a new treatment, convalescent plasma therapy remains a useful option now particularly in the event of an outbreak for treating emerging infections in the first stage. Despite this, the variability of antibody concentrations and the logistics of plasma collection and administration are two more reasons why more standardized immunotherapies are needed.

6.3.3. Immune Modulation

This differs from immune potentiating, which reduces the antiviral function of the body, and immune suppression, in which inflammation of the body is increased. An overactive immune response can create excessive inflammation and tissue damage, a common feature of virally induced disease.

- **Immune Modulation in COVID-19:** For example, to help tamp down the immune system's inflammatory response, corticosteroids such as dexamethasone have been shown to work — in moderate to severe COVID-19 cases to improve survival rates. Dexamethasone reduces the immune response and prevents the 'cytokine storm' of severe COVID-19. Other drugs aimed at immune modulation, like tocilizumab, which stops interleukin-6, have also been used to try and reduce inflammation even further (Parthasarathy et al., 2022; Rabaan et al., 2021).
- **Autoimmune and Viral Infections:** Given that immune response may contribute to the pathology of an infection immune-modulating drugs are particularly beneficial for such infections. For example, immune modulation for viral hepatitis can reduce liver inflammation and fibrosis and prevent long-term liver damage (Alijotas-Reig et al., 2020).

6.4. RNA-Based Therapies

RNA-based therapeutics have evolved as a frontier of antiviral treatment, exploiting the RNA molecules' flexibility and ability to adapt and target viral genes and stimulate the immune response. Rapid design and adaptation of these therapies hold a great advantage in being able to rapidly respond to new viruses.

6.4.1. RNA Interference (RNAi), si RNA

Two techniques that silence specific viral genes to stop replication are small interfering RNA (siRNA) and RNA interference (RNAi). We can trick RNAi (a naturally occurring cellular mechanism that rips viral RNA out of a cell) into destroying the virus by designing siRNA molecules that are complementary to viral RNA (Kim & Rossi, 2007).

- **Hepatitis and HIV:** siRNA has been tested for viral infections, such as hepatitis B and HIV, where it has been used to target essential viral genes to lower viral load. The research of siRNA is very early but it offers a very specific way of antiviral therapy, allowing for targeting the viral genes without affecting the host genetics material (Flisiak et al., 2018).
- **COVID-19 Research:** SiRNA approaches were also tested for SARS-CoV-2 targeting of specific genes to inhibit replication during the COVID-19 pandemic. Delivery of siRNA molecules to infected cells continues to be problematic, and continued efforts to optimize delivery systems to maximize treatment efficacy are ongoing (Sajid et al., 2021).

6.5. mRNA Therapies

mRNA-based therapies are therapeutic agents consisting of synthetic mRNA that directs cells to make proteins that spur an immune response or interfere with viral replication. The demonstrated success of mRNA vaccines for COVID-19 showed that mRNA technology could be invaluable for rapid response vaccine development.

- **COVID-19 mRNA Vaccines:** Due to mRNA technology, Pfizer BioNTech, and Moderna vaccines provided a breakthrough solution during the era of the COVID pandemic. These vaccines employ the use of mRNA encoding the SARS-CoV-2 spike protein to elicit the immune response to the virus, as a typical use of mRNA technology for preventive and therapeutic applications, suggesting a way to enable rapid vaccine deployment to future pandemics (Hogan & Pardi, 2022).
- **Potential for Other Viral Infections:** mRNA technology could be used to treat other viral infections including influenza and Zika. The ability to encode specific viral proteins into mRNA therapies for stimulation of targeted immune responses furnishes a flexible platform for application in a range of antiviral situations (Chaudhary et al., 2021; Pardi & Krammer, 2024).

RNA-based therapies are highly adaptable and translatable, powerful tools for response to novel viral threats. Despite these, significant challenges of stability, delivery, and scalability must be overcome to realize the potential of these therapies.

7. ANTIVIRAL THERAPEUTICS: THE FUTURE

The future of antiviral therapeutics is in the multitiered framework that combines targeted and broad-spectrum antiviral therapeutics to overcome the complexity and plasticity of viruses. Recent technological advancements have made transforming the treatment landscape for antivirals possible.

7.1. Broad spectrum Antiviral Agents

Broad spectrum antivirals are one of the lessons learned from past outbreaks: they are those antivirals that cover as many viruses or entire families of viruses at once, "especially good when a new virus appears since they are the first line of defense until better-targeted treatments or vaccines are found." Therefore, broad-spectrum antivirals target common enzymes or structures of the replication of viruses like a virus polymerase or proteases. The idea is to come up with a therapeutic response that works against existing as well as future unknown viruses to enhance preparedness against any future outbreak (Tompa et al., 2021).

Whether in this case or for example, remdesivir was initially developed to fight against Ebola, this was repurposed during the time of the COVID pandemic. While very little effective, it demonstrated the potential value of broad-spectrum antivirals as quick responders to new viruses. Indeed, future development of broad-spectrum agents with increased efficacy and decreased side effects will be aimed at a rapidly deployable strategy against newly emerging viral pathogens (Lo et al., 2020).

7.2. Antiviral Nanotechnology

New possibilities for increasing the efficacy and delivery of antiviral drugs have been made available by nanotechnology. For instance, most nanoparticles can be engineered to directly deliver antiviral agents to infected cells with increased drug bioavailability and decreased side effects. Furthermore, nanoparticles can be made to the home of affected parts of the virus or infected cells to focus on therapeutic effects while minimizing the destruction of the healthy cells (Chakravarty & Vora, 2021).

Liposomal and polymeric nanoparticles are two examples of applications of nanotechnology in antiviral therapy: these encapsulate drugs to protect them from degradation until they reach their target. The other promising application is to gold nanoparticles which have poted italics as a way to bind to viral surfaces and inhibit italics entry italics into host cells. With future work, nanotechnology could develop into a powerful means of creating highly efficient and targeted antiviral therapies that overcome all viral threats (Maus et al., 2021).

7.3. Artificial Intelligence (AI) and Machine Learning in Antiviral Development

Artificial intelligence (AI) and machine learning (ML) in antiviral research accelerate drug target identification, predict drug efficacy, and guide drug design into drug discovery and development. AI was also incredibly useful in developing COVID-19 therapeutics quickly: researchers used AI models first to predict which of the existing drugs might be able to work against SARS-CoV-2 (Keshavarzi Arshadi et al., 2020; Sarkar et al., 2023). In the future, AI and ML may become more and more integral to predictively identify early viral threats and develop targeted antiviral responses to viral threats before outbreaks even occur. Researchers can then simulate how new viruses might interact with host cells, and use their computational power to get a head start on preemptive drug development (Khanna et al., 2024).

7.4. Prophylactic Antivirals and Pre-Exposure Prophylaxis:

Preventive treatments for people at risk of infection are given, such as healthcare workers or patients who are close to an infected patient, or immunocompromised individuals, which are prophylactic antivirals or PreExposure Prophylaxis (PrEP). The effectiveness of PrEP in the prevention of HIV transmission has shown that prophylactic antivirals have the potential to be effective in high-risk populations (Soriano et al., 2022). A promising area of research is to develop long-acting antivirals that protect for a long period (up to several months). Although these antivirals could be administered in regions with ongoing outbreaks or to travelers to high-risk areas, their comprehensive approach will make them more appealing on a global scale. Prophylactic antivirals could also be rolled out early during an outbreak to give an initial buffer until vaccines are a reality. The approach would not only limit the spread of the virus but also relieve the burden on health systems by reducing the number of cases needing medical treatment (Soriano et al., 2023).

7.5. Preparedness and Collaborative Global Efforts

The outbreak has taught us a critical lesson: emerging viruses are an international threat that can only be managed internationally. It's only through effective combating of viral threats that collaboration between governments, international organizations, and private sector stakeholders is required. The Coalition for Epidemic Preparedness Innovations (CEPI) and the World Health Organization's Health Emergencies Program are funding and coordinating research to accelerate the availability of antiviral therapeutics and vaccines during outbreaks, globally (Jacobsen et al., 2016; Peeri et al., 2020).

Table 5. Future directions with their potential benefits

Future Directions	Description	Potential Benefits	Challenges
Immunotherapies	Enhancing the immune system's ability to clear viral infections	Long-lasting protection, potential for personalized treatment	Complex to develop, potential for immune-related adverse events
Novel drug delivery systems	Improving drug delivery to target tissues, enhancing efficacy, and reducing side effects	Increased drug concentration at the site of infection, reduced systemic toxicity.	Development of effective and safe delivery systems
Broad-spectrum antivirals	Drugs that target common viral mechanisms, effective against multiple viruses	Reduced need for specific diagnostics, potential for rapid response to emerging viruses	Difficulty in finding conserved targets, potential for toxicity
Host-directed therapies	Targeting host factors essential for viral replication	Less likely to lead to resistance development, potential for synergistic effects with existing antivirals	The potential for disrupting normal cellular functions requires a detailed understanding of host-virus interactions
CRISPR-based therapies	Gene editing technologies to target viral genomes or host factors	Potential for a cure, ability to target viruses that are difficult to treat with traditional antivirals	Safety concerns, ethical considerations, delivery challenges

CONCLUSION

Emerging viruses represent a danger to world health because they cause widespread sickness, death, and economic upheaval. These zoonotic infections, which are frequently spread from animals to people, put a strain on healthcare systems despite being vital to both medical research and social stability. Antiviral medications are vital instruments in the fight against these infections, but because of their resistance, creating effective therapies is difficult. Antiviral medications target different stages of the viral life cycle, however, they have drawbacks such as side effects, treatment resistance, and difficulty accessing dormant viral reservoirs. To combat these risks, novel approaches to medication development, focusing on less modifiable viral structures, and employing combination treatments are required. Other strategies to provide more effective treatments for viral infections, drug repurposing, immunotherapy, and RNA-based therapeutics are being investigated. Using medications developed for one condition for another is known as drug repurposing, which reduces development times and provides a wealth of prior safety information. The development of effective antiviral therapeutics is not only a medical imperative but also a global necessity to safeguard public health and economic stability. Future research must focus on identifying and targeting conserved viral structures that are less prone to mutations, thereby minimizing the risk of resistance. Global collaborations and partnerships are essential to accelerate the pace of antiviral development. Coordinated efforts among governments, academia, and the pharmaceutical industry are crucial to ensure equitable access to antiviral therapies, especially in low- and middle-income countries that are disproportionately affected by viral outbreaks.

REFERENCES

Abbink, P., Stephenson, K. E., & Barouch, D. H. (2018). Zika virus vaccines. *Nature Reviews. Microbiology, 16*(10), 594–600. DOI: 10.1038/s41579-018-0039-7 PMID: 29921914

Abiha, U., Patel, N., Goel, N., & Sharma, M. (2023). Innovative Therapeutic Efforts for Tackling Future Viral Pandemics. In *RNA Viruses and Neurological Disorders* (pp. 206–239). CRC Press. DOI: 10.1201/9781003285823-16

Afrough, B., Dowall, S., & Hewson, R. (2019). Emerging viruses and current strategies for vaccine intervention. *Clinical and Experimental Immunology, 196*(2), 157–166. DOI: 10.1111/cei.13295 PMID: 30993690

Alijotas-Reig, J., Esteve-Valverde, E., Belizna, C., Selva-O'Callaghan, A., Pardos-Gea, J., Quintana, A., Mekinian, A., Anunciacion-Llunell, A., & Miró-Mur, F. (2020). Immunomodulatory therapy for the management of severe COVID-19. Beyond the anti-viral therapy: A comprehensive review. *Autoimmunity Reviews, 19*(7), 102569. DOI: 10.1016/j.autrev.2020.102569 PMID: 32376394

Antonelli, G., Falasca, F., & Turriziani, O. (2021). Mechanisms of Action of Antiviral Agents. In *Practical Handbook of Microbiology* (pp. 789–800). CRC Press. DOI: 10.1201/9781003099277-56

Asad, H., & Carpenter, D. O. (2018). Effects of climate change on the spread of zika virus: A public health threat. *Reviews on Environmental Health, 33*(1), 31–42. DOI: 10.1515/reveh-2017-0042 PMID: 29500926

Barth, R. F., Buja, L. M., Barth, A. L., Carpenter, D. E., & Parwani, A. V. (2021). A comparison of the clinical, viral, pathologic, and immunologic features of severe acute respiratory syndrome (SARS), middle east respiratory syndrome (MERS), and coronavirus 2019 (COVID-19) diseases. *Archives of Pathology & Laboratory Medicine, 145*(10), 1194–1211. DOI: 10.5858/arpa.2020-0820-SA PMID: 34232978

Bock, C., & Lengauer, T. (2012). Managing drug resistance in cancer: Lessons from HIV therapy. *Nature Reviews. Cancer, 12*(7), 494–501. DOI: 10.1038/nrc3297 PMID: 22673150

Cani, E., Park, T. E., & Kavanagh, R. (2019). Antiviral drugs. [). Elsevier.]. *Side Effects of Drugs Annual, 41*, 301–319. DOI: 10.1016/bs.seda.2019.10.005

Caron, A., Jori, F., de Nys, H., Herbinger, I., Liégeois, F., Mouinga-Ondémé, A., & Ratiarison, S. (2022). Emergence/Re-emergence of Infectious Agents and Epidemic Risks in Central African Forests. *The Forests of the Congo Basin: State of the Forests, 2021*, 265.

Ch'ien, L. T., Schabel, F. M., & Alford, C. A. (2018). Arabinosyl nucleosides and nucleotides. In *Selective inhibitors of viral functions* (pp. 227–256). CRC Press. DOI: 10.1201/9781351076562-11

Chakravarty, M., & Vora, A. (2021). Nanotechnology-based antiviral therapeutics. *Drug Delivery and Translational Research, 11*(3), 748–787. DOI: 10.1007/s13346-020-00818-0 PMID: 32748035

Chan, J. F.-W., To, K. K.-W., Chen, H., & Yuen, K.-Y. (2015). Cross-species transmission and emergence of novel viruses from birds. *Current Opinion in Virology, 10*, 63–69. DOI: 10.1016/j.coviro.2015.01.006 PMID: 25644327

Chaudhary, N., Weissman, D., & Whitehead, K. A. (2021). mRNA vaccines for infectious diseases: Principles, delivery and clinical translation. *Nature Reviews. Drug Discovery, 20*(11), 817–838. DOI: 10.1038/s41573-021-00283-5 PMID: 34433919

Chen, L., Xiong, J., Bao, L., & Shi, Y. (2020). Convalescent plasma as a potential therapy for COVID-19. *The Lancet. Infectious Diseases, 20*(4), 398–400. DOI: 10.1016/S1473-3099(20)30141-9 PMID: 32113510

Crawford, D. H. (2021). *Viruses: The invisible enemy*. Oxford University Press. DOI: 10.1093/oso/9780192845030.001.0001

De Clercq, E., & Li, G. (2016). Approved antiviral drugs over the past 50 years. *Clinical Microbiology Reviews, 29*(3), 695–747. DOI: 10.1128/CMR.00102-15 PMID: 27281742

DeFilippis, V. R., & Villarreal, L. P. (2007). An introduction to the evolutionary ecology of viruses. Viral ecology, 125.

Di Santo, R. (2014). Inhibiting the HIV integration process: Past, present, and the future. *Journal of Medicinal Chemistry, 57*(3), 539–566. DOI: 10.1021/jm400674a PMID: 24025027

Dunne, M. (2007). Antiretroviral drug development: The challenge of cost and access. *AIDS (London, England), 21*(Suppl 4), S73–S79. DOI: 10.1097/01.aids.0000279709.47298.8a PMID: 17620756

El-Sayed, A., & Kamel, M. (2021). Coronaviruses in humans and animals: The role of bats in viral evolution. *Environmental Science and Pollution Research International, 28*(16), 19589–19600. DOI: 10.1007/s11356-021-12553-1 PMID: 33655480

Ellwanger, J. H., Fearnside, P. M., Ziliotto, M., Valverde-Villegas, J. M., Veiga, A. B. G., Vieira, G. F., Bach, E., Cardoso, J. C., Müller, N. F. D., Lopes, G., Caesar, L., Kulmann-Leal, B., Kaminski, V. L., Silveira, E. S., Spilki, F. R., Weber, M. N., Almeida, S. D. M., Hora, V. P. D., & Chies, J. A. B. (2022). Synthesizing the connections between environmental disturbances and zoonotic spillover. *Anais da Academia Brasileira de Ciências, 94*(suppl 3), e20211530. DOI: 10.1590/0001-3765202220211530 PMID: 36169531

Fallarino, L. (2018). Anti HIV-1 gene therapy approach combining multiple siRNAs with the membrane-anchored fusion inhibitor C-peptide maC46.

Flisiak, R., Jaroszewicz, J., & Łucejko, M. (2018). siRNA drug development against hepatitis B virus infection. *Expert Opinion on Biological Therapy, 18*(6), 609–617. DOI: 10.1080/14712598.2018.1472231 PMID: 29718723

Gagnon, M., & Holmes, D. (2016). Body–drug assemblages: Theorizing the experience of side effects in the context of HIV treatment. *Nursing Philosophy, 17*(4), 250–261. DOI: 10.1111/nup.12136 PMID: 27435229

Gangopadhayya, A., & Bhukya, P. L. (2023). Factors Contributing to the Emergence of Viral Diseases. In *Emerging Human Viral Diseases, Volume I: Respiratory and Haemorrhagic Fever* (pp. 3-69). Springer. DOI: 10.1007/978-981-99-2820-0_1

Graham, B. S., & Sullivan, N. J. (2018). Emerging viral diseases from a vaccinology perspective: Preparing for the next pandemic. *Nature Immunology, 19*(1), 20–28. DOI: 10.1038/s41590-017-0007-9 PMID: 29199281

Herder, M., Graham, J. E., & Gold, R. (2020). From discovery to delivery: Public sector development of the r VSV-ZEBOV Ebola vaccine. *Journal of Law and the Biosciences, 7*(1), lsz019. DOI: 10.1093/jlb/lsz019 PMID: 34221434

Hermann, T. (2016). Small molecules targeting viral RNA. *Wiley Interdisciplinary Reviews. RNA, 7*(6), 726–743. DOI: 10.1002/wrna.1373 PMID: 27307213

Hernandez, A. V., Roman, Y. M., Pasupuleti, V., Barboza, J. J., & White, C. M. (2020). Hydroxychloroquine or chloroquine for treatment or prophylaxis of COVID-19: A living systematic review. *Annals of Internal Medicine, 173*(4), 287–296. DOI: 10.7326/M20-2496 PMID: 32459529

Hogan, M. J., & Pardi, N. (2022). mRNA vaccines in the COVID-19 pandemic and beyond. *Annual Review of Medicine, 73*(1), 17–39. DOI: 10.1146/annurev-med-042420-112725 PMID: 34669432

Holmes, E. C. (2009). The evolutionary genetics of emerging viruses. *Annual Review of Ecology, Evolution, and Systematics, 40*(1), 353–372. DOI: 10.1146/annurev.ecolsys.110308.120248

Irwin, K. K., Renzette, N., Kowalik, T. F., & Jensen, J. D. (2016). Antiviral drug resistance as an adaptive process. *Virus Evolution, 2*(1), vew014. DOI: 10.1093/ve/vew014 PMID: 28694997

Ison, M. G. (2016). Antiviral treatments. *Clinics in Chest Medicine, 38*(1), 139–153. DOI: 10.1016/j.ccm.2016.11.008 PMID: 28159156

Jacobsen, K. H., Aguirre, A. A., Bailey, C. L., Baranova, A. V., Crooks, A. T., Croitoru, A., Delamater, P. L., Gupta, J., Kehn-Hall, K., Narayanan, A., Pierobon, M., Rowan, K. E., Schwebach, J. R., Seshaiyer, P., Sklarew, D. M., Stefanidis, A., & Agouris, P. (2016). Lessons from the Ebola outbreak: Action items for emerging infectious disease preparedness and response. *EcoHealth, 13*(1), 200–212. DOI: 10.1007/s10393-016-1100-5 PMID: 26915507

Karban, Š. (2023). The evolution of SARS-CoV-2 virus during the pandemic of COVID-19.

Kausar, S., Said Khan, F., Ishaq Mujeeb Ur Rehman, M., Akram, M., Riaz, M., Rasool, G., Hamid Khan, A., Saleem, I., Shamim, S., & Malik, A. (2021). A review: Mechanism of action of antiviral drugs. *International Journal of Immunopathology and Pharmacology, 35*, 20587384211002621. DOI: 10.1177/20587384211002621 PMID: 33726557

Keshavarzi Arshadi, A., Webb, J., Salem, M., Cruz, E., Calad-Thomson, S., Ghadirian, N., Collins, J., Diez-Cecilia, E., Kelly, B., Goodarzi, H., & Yuan, J. S. (2020). Artificial intelligence for COVID-19 drug discovery and vaccine development. *Frontiers in Artificial Intelligence, 3*, 65. DOI: 10.3389/frai.2020.00065 PMID: 33733182

Khanna, A., El Barachi, M., Jain, S., Kumar, M., & Nayyar, A. (2024). *Artificial Intelligence and machine learning in drug design and development*. John Wiley & Sons. DOI: 10.1002/9781394234196

Kim, D. H., & Rossi, J. J. (2007). Strategies for silencing human disease using RNA interference. *Nature Reviews. Genetics, 8*(3), 173–184. DOI: 10.1038/nrg2006 PMID: 17304245

Klatt, E. C. (2017). Pathology of HIV/AIDS. *The Mercer University of Medicine Savannah*, 6-50.

Kost, G. J. (2020). Geospatial hotspots need point-of-care strategies to stop highly infectious outbreaks: Ebola and coronavirus. *Archives of Pathology & Laboratory Medicine, 144*(10), 1166–1190. DOI: 10.5858/arpa.2020-0172-RA PMID: 32298139

Lampejo, T. (2020). Influenza and antiviral resistance: An overview. *European Journal of Clinical Microbiology & Infectious Diseases, 39*(7), 1201–1208. DOI: 10.1007/s10096-020-03840-9 PMID: 32056049

Lauring, A. S., Frydman, J., & Andino, R. (2013). The role of mutational robustness in RNA virus evolution. *Nature Reviews. Microbiology, 11*(5), 327–336. DOI: 10.1038/nrmicro3003 PMID: 23524517

Lo, M. K., Albariño, C. G., Perry, J. K., Chang, S., Tchesnokov, E. P., Guerrero, L., Chakrabarti, A., Shrivastava-Ranjan, P., Chatterjee, P., McMullan, L. K., Martin, R., Jordan, R., Götte, M., Montgomery, J. M., Nichol, S. T., Flint, M., Porter, D., & Spiropoulou, C. F. (2020). Remdesivir targets a structurally analogous region of the Ebola virus and SARS-CoV-2 polymerases. *Proceedings of the National Academy of Sciences of the United States of America, 117*(43), 26946–26954. DOI: 10.1073/pnas.2012294117 PMID: 33028676

Loaiza, J. D., Chvatal-Medina, M., Hernandez, J. C., & Rugeles, M. T. (2023). Integrase inhibitors: Current protagonists in antiretroviral therapy. *Immunotherapy, 15*(17), 1477–1495. DOI: 10.2217/imt-2023-0011 PMID: 37822251

Longdon, B., Brockhurst, M. A., Russell, C. A., Welch, J. J., & Jiggins, F. M. (2014). The evolution and genetics of virus host shifts. *PLoS Pathogens, 10*(11), e1004395. DOI: 10.1371/journal.ppat.1004395 PMID: 25375777

Lourenço, J., Maia de Lima, M., Faria, N. R., Walker, A., Kraemer, M. U., Villabona-Arenas, C. J., Lambert, B., Marques de Cerqueira, E., Pybus, O. G., Alcantara, L. C., & Recker, M. (2017). Epidemiological and ecological determinants of Zika virus transmission in an urban setting. *eLife, 6*, e29820. DOI: 10.7554/eLife.29820 PMID: 28887877

Mackay, I. M., & Arden, K. E. (2015). MERS coronavirus: Diagnostics, epidemiology and transmission. *Virology Journal, 12*(1), 1–21. DOI: 10.1186/s12985-015-0439-5 PMID: 26695637

Malin, J. J., Suárez, I., Priesner, V., Fätkenheuer, G., & Rybniker, J. (2020). Remdesivir against COVID-19 and other viral diseases. *Clinical Microbiology Reviews, 34*(1), 00162–00120. DOI: 10.1128/CMR.00162-20 PMID: 33055231

Manabe, T., Kambayashi, D., Akatsu, H., & Kudo, K. (2021). Favipiravir for the treatment of patients with COVID-19: A systematic review and meta-analysis. *BMC Infectious Diseases*, *21*(1), 489. DOI: 10.1186/s12879-021-06164-x PMID: 34044777

Marrana, M. (2022). Epidemiology of disease through the interactions between humans, domestic animals, and wildlife. In *One Health* (pp. 73–111). Elsevier.

Maus, A., Strait, L., & Zhu, D. (2021). Nanoparticles as delivery vehicles for antiviral therapeutic drugs. *Engineered Regeneration*, *2*, 31–46. DOI: 10.1016/j.engreg.2021.03.001 PMID: 38620592

Meganck, R. M., & Baric, R. S. (2021). Developing therapeutic approaches for twenty-first-century emerging infectious viral diseases. *Nature Medicine*, *27*(3), 401–410. DOI: 10.1038/s41591-021-01282-0 PMID: 33723456

Mokhtary, P., Pourhashem, Z., Mehrizi, A. A., Sala, C., & Rappuoli, R. (2022). Recent progress in the discovery and development of monoclonal antibodies against viral infections. *Biomedicines*, *10*(8), 1861. DOI: 10.3390/biomedicines10081861 PMID: 36009408

Montessori, V., Press, N., Harris, M., Akagi, L., & Montaner, J. S. (2004). Adverse effects of antiretroviral therapy for HIV infection. *Canadian Medical Association Journal*, *170*(2), 229–238. PMID: 14734438

Naseer, S., Khalid, S., Parveen, S., Abbass, K., Song, H., & Achim, M. V. (2023). COVID-19 outbreak: Impact on global economy. *Frontiers in Public Health*, *10*, 1009393. DOI: 10.3389/fpubh.2022.1009393 PMID: 36793360

Nourbakhsh, M. (2013). Role of TG Lipases, Arylacetamide Deacetylase and Triacylglycerol Hydrolase, in Hepatitis C Virus Life Cycle.

Nova, N., Athni, T. S., Childs, M. L., Mandle, L., & Mordecai, E. A. (2022). Global change and emerging infectious diseases. *Annual Review of Resource Economics*, *14*(1), 333–354. DOI: 10.1146/annurev-resource-111820-024214 PMID: 38371741

Pardi, N., & Krammer, F. (2024). mRNA vaccines for infectious diseases—Advances, challenges and opportunities. *Nature Reviews. Drug Discovery*, *23*(11), 1–24. DOI: 10.1038/s41573-024-01042-y PMID: 39367276

Parthasarathy, U., Martinelli, R., Vollmann, E. H., Best, K., & Therien, A. G. (2022). The impact of DAMP-mediated inflammation in severe COVID-19 and related disorders. *Biochemical Pharmacology*, *195*, 114847. DOI: 10.1016/j.bcp.2021.114847 PMID: 34801526

Parvathaneni, V., Kulkarni, N. S., Muth, A., & Gupta, V. (2019). Drug repurposing: A promising tool to accelerate the drug discovery process. *Drug Discovery Today*, *24*(10), 2076–2085. DOI: 10.1016/j.drudis.2019.06.014 PMID: 31238113

Peeri, N. C., Shrestha, N., Rahman, M. S., Zaki, R., Tan, Z., Bibi, S., Baghbanzadeh, M., Aghamohammadi, N., Zhang, W., & Haque, U. (2020). The SARS, MERS and novel coronavirus (COVID-19) epidemics, the newest and biggest global health threats: What lessons have we learned? *International Journal of Epidemiology*, *49*(3), 717–726. DOI: 10.1093/ije/dyaa033 PMID: 32086938

Pinna, S. M., Lupia, T., Scabini, S., Vita, D., De Benedetto, I., Gaviraghi, A., Colasanto, I., Varese, A., Cattel, F., & De Rosa, F. G. (2021). Monoclonal antibodies for the treatment of COVID-19 patients: An umbrella to overcome the storm? *International Immunopharmacology*, *101*, 108200. DOI: 10.1016/j.intimp.2021.108200 PMID: 34607231

Prothero, R. M. Chapter V: Urbanization: A Hot-Bed Of Vector-Borne Diseases. In *Demography and Vector-Borne Diseases* (pp. 59-84). CRC Press.

Rabaan, A. A., Al-Ahmed, S. H., Muhammad, J., Khan, A., Sule, A. A., Tirupathi, R., Mutair, A. A., Alhumaid, S., Al-Omari, A., Dhawan, M., Tiwari, R., Sharun, K., Mohapatra, R. K., Mitra, S., Bilal, M., Alyami, S. A., Emran, T. B., Moni, M. A., & Dhama, K. (2021). Role of inflammatory cytokines in COVID-19 patients: A review on molecular mechanisms, immune functions, immunopathology and immunomodulatory drugs to counter cytokine storm. *Vaccines*, *9*(5), 436. DOI: 10.3390/vaccines9050436 PMID: 33946736

Ralli, G. (2022). Blocking the entry of HIV into host cells through co-receptor inhibition.

Rao, M. R., Ghadge, I. S., Kulkarni, S. P., & Asthana, T. (2023). Computational techniques for drug repurposing: A paradigm shift in drug discovery. *Current Drug Therapy*, *18*(4), 271–284. DOI: 10.2174/1574885518666230207143523

Roback, J. D., & Guarner, J. (2020). Convalescent plasma to treat COVID-19: Possibilities and challenges. *Journal of the American Medical Association*, *323*(16), 1561–1562. DOI: 10.1001/jama.2020.4940 PMID: 32219429

Saag, M. S., Gandhi, R. T., Hoy, J. F., Landovitz, R. J., Thompson, M. A., Sax, P. E., Smith, D. M., Benson, C. A., Buchbinder, S. P., Del Rio, C., Eron, J. J.Jr, Fätkenheuer, G., Günthard, H. F., Molina, J.-M., Jacobsen, D. M., & Volberding, P. A. (2020). Antiretroviral drugs for treatment and prevention of HIV infection in adults: 2020 recommendations of the International Antiviral Society–USA Panel. *Journal of the American Medical Association, 324*(16), 1651–1669. DOI: 10.1001/jama.2020.17025 PMID: 33052386

Sajid, M. I., Moazzam, M., Cho, Y., Kato, S., Xu, A., Way, J., Lohan, S., & Tiwari, R. K. (2021). siRNA Therapeutics for the Therapy of COVID-19 and Other Coronaviruses. *Molecular Pharmaceutics, 18*(6), 2105–2121. DOI: 10.1021/acs.molpharmaceut.0c01239 PMID: 33945284

Saleem, M., Ramzan, F., Nazar, M., Khan, M., Mahfooz, A., Zaka, F., Sultan, T., Raza, A., Ahmad, A., & Shakir, A. (2023). Mitigate the Risk of Zoonosis. *Zoonosis, Unique Scientific Publishers, Faisalabad, Pakistan, 1*, 643-658.

Sanjuán, R., & Domingo-Calap, P. (2016). Mechanisms of viral mutation. *Cellular and Molecular Life Sciences : CMLS, 73*(23), 4433–4448. DOI: 10.1007/s00018-016-2299-6 PMID: 27392606

Sarkar, C., Das, B., Rawat, V. S., Wahlang, J. B., Nongpiur, A., Tiewsoh, I., Lyngdoh, N. M., Das, D., Bidarolli, M., & Sony, H. T. (2023). Artificial intelligence and machine learning technology driven modern drug discovery and development. *International Journal of Molecular Sciences, 24*(3), 2026. DOI: 10.3390/ijms24032026 PMID: 36768346

Schenk, E., García, P., & Capron, A. National Academies of Sciences, E., & Medicine. (2021). Therapeutics. In *Public Health Lessons for Non-Vaccine Influenza Interventions: Looking Past COVID-19*. National Academies Press (US).

Schwartz, D. A., & Graham, A. L. (2020). Potential maternal and infant outcomes from coronavirus 2019-nCoV (SARS-CoV-2) infecting pregnant women: Lessons from SARS, MERS, and other human coronavirus infections. *Viruses, 12*(2), 194. DOI: 10.3390/v12020194 PMID: 32050635

Shenge, J. A., & Opayele, A. V. (2020). The impact and control of emerging and re-emerging viral diseases in the environment: an African perspective. *Current Microbiological Research in Africa: Selected Applications for Sustainable Environmental Management*, 185-202.

Soriano, V., Moreno-Torres, V., de Mendoza, C., Fernández-Montero, J. V., Treviño, A., Corral, O., de Jesús, F., & Barreiro, P. (2023). Pre-exposure prophylaxis of non-HIV viral infections and the role of long-acting antivirals. *AIDS Reviews*, *25*(4). Advance online publication. DOI: 10.24875/AIDSRev.M23000066 PMID: 38092029

Soriano, V., Treviño, A., de Mendoza, C., Moreno-Torres, V., Pintos, I., Barreiro, P., & Corral, O. (2022). Pre-Exposure Prophylaxis for viral infections other than HIV. *Le Infezioni in Medicina*, *30*(3), 362. PMID: 36148176

Stern, A., & Andino, R. (2016). Viral evolution: it is all about mutations. In *Viral pathogenesis* (pp. 233–240). Elsevier. DOI: 10.1016/B978-0-12-800964-2.00017-3

Strauss, J. H., & Strauss, E. G. (2012). Emerging and reemerging viral diseases. Viruses and Human Disease, 325.

Sutherland, K. (2014). *Contribution of Gag and protease to variation in susceptibility to protease inhibitors between different strains of HIV-1* UCL (University College London)].

Tazerji, S. S., Nardini, R., Safdar, M., Shehata, A. A., & Duarte, P. M. (2022). An overview of anthropogenic actions as drivers for emerging and re-emerging zoonotic diseases. *Pathogens (Basel, Switzerland)*, *11*(11), 1376. DOI: 10.3390/pathogens11111376 PMID: 36422627

Telenti, A., Hodcroft, E. B., & Robertson, D. L. (2022). The evolution and biology of SARS-CoV-2 variants. *Cold Spring Harbor Perspectives in Medicine*, *12*(5), a041390. DOI: 10.1101/cshperspect.a041390 PMID: 35444005

Tian, L., Qiang, T., Liang, C., Ren, X., Jia, M., Zhang, J., Li, J., Wan, M., YuWen, X., Li, H., Cao, W., & Liu, H. (2021). RNA-dependent RNA polymerase (RdRp) inhibitors: The current landscape and repurposing for the COVID-19 pandemic. *European Journal of Medicinal Chemistry*, *213*, 113201. DOI: 10.1016/j.ejmech.2021.113201 PMID: 33524687

Tompa, D. R., Immanuel, A., Srikanth, S., & Kadhirvel, S. (2021). Trends and strategies to combat viral infections: A review on FDA approved antiviral drugs. *International Journal of Biological Macromolecules*, *172*, 524–541. DOI: 10.1016/j.ijbiomac.2021.01.076 PMID: 33454328

Tshiani Mbaya, O., Mukumbayi, P., & Mulangu, S. (2021). insights on current FDA-approved monoclonal antibodies against Ebola virus infection. *Frontiers in Immunology*, *12*, 721328. DOI: 10.3389/fimmu.2021.721328 PMID: 34526994

Ul-Rahman, A., Sbabbir, M., Rasheed, M., Bano, N., Shahid, M., Habib, M., Mehmood, R., Shafi, N., Khera, H., & Kabeer, S. (2023). Zoonotic and reverse zoonotic transmission of SARS-CoV-2 virus: A perspective on human-animal interface. *Zoonosis, Unique Scientific Publishers, Faisalabad, Pakistan, 3*, 269-282.

Ullah, A. A., & Ferdous, J. (2022). *The post-pandemic world and global politics*. Springer. DOI: 10.1007/978-981-19-1910-7

van Griensven, J., De Weiggheleire, A., Delamou, A., Smith, P. G., Edwards, T., Vandekerckhove, P., Bah, E. I., Colebunders, R., Herve, I., Lazaygues, C., Haba, N., & Lynen, L. (2016). The use of Ebola convalescent plasma to treat Ebola virus disease in resource-constrained settings: A perspective from the field. *Clinical Infectious Diseases*, *62*(1), 69–74. DOI: 10.1093/cid/civ680 PMID: 26261205

Vecchio, R., Gentile, L., Tafuri, S., Costantino, C., & Odone, A. (2024). Exploring future perspectives and pipeline progression in vaccine research and development. *Annali di igiene : medicina preventiva e di comunità*, *36*(4), 446–461. PMID: 38436081

Voshavar, C. (2019). Protease inhibitors for the treatment of HIV/AIDS: Recent advances and future challenges. *Current Topics in Medicinal Chemistry*, *19*(18), 1571–1598. DOI: 10.2174/1568026619666190619115243 PMID: 31237209

Weiss, R. A., & Sankaran, N. (2022). Emergence of epidemic diseases: Zoonoses and other origins. *Faculty Reviews*, *11*, 11. DOI: 10.12703/r/11-2 PMID: 35156099

KEY TERMS AND DEFINITIONS

Cross-Species Transmission: Cross-Species Transmission (also known as spillover) refers to the process by which a pathogen (such as a virus, bacterium, or parasite) is transmitted from one species to another, potentially establishing infection in the new host.

Prophylaxis: Prophylaxis refers to measures taken to prevent the onset or spread of a disease or condition. It can involve medical, behavioral, or environmental interventions designed to protect individuals or populations from potential health risks.

Zoonotic Transmission: Zoonotic Transmission refers to the process by which infectious diseases are transmitted from animals to humans. Diseases that are spread through zoonotic transmission are called **zoonoses** (singular: zoonosis).

Human encroachment: Human Encroachment refers to the expansion of human activities, settlements, and infrastructure into natural ecosystems, wildlife habitats, or undeveloped areas.

Genetic Drift: Genetic drift is a random process that leads to changes in the frequency of alleles (gene variants) in a population over time.

Genetic Shift: Genetic shift, also referred to as antigenic shift, is a specific term often used in virology to describe a major and sudden change in the genetic composition of a virus.

Pre Exposure Prophylaxis (PrEP): Pre-Exposure Prophylaxis is a preventive medical strategy in which individuals who are at high risk of acquiring an infectious disease take medication to reduce their chances of becoming infected.

CRISPR: stands for Clustered Regularly Interspaced Short Palindromic Repeats. It is a natural system found in bacteria and archaea that functions as an adaptive immune defense against invading viruses.

Compilation of References

Abbink, P., Stephenson, K. E., & Barouch, D. H. (2018). Zika virus vaccines. *Nature Reviews. Microbiology, 16*(10), 594–600. DOI: 10.1038/s41579-018-0039-7 PMID: 29921914

Abbott, T. R., Dhamdhere, G., Liu, Y., Lin, X., Goudy, L., Zeng, L., Chemparathy, A., Chmura, S., Heaton, N. S., Debs, R., Pande, T., Endy, D., La Russa, M. F., Lewis, D. B., & Qi, L. S. (2020). Development of CRISPR as an Antiviral Strategy to Combat SARS-CoV-2 and Influenza. *Cell, 181*(4), 865–876.e12. DOI: 10.1016/j.cell.2020.04.020 PMID: 32353252

Abdullah, Y., & Hossain, D. (2021). Telemedicine solutions for viral disease management during COVID-19 pandemic. *Frontiers in Public Health, 9*, 557419.

Abiha, U., Patel, N., Goel, N., & Sharma, M. (2023). Innovative Therapeutic Efforts for Tackling Future Viral Pandemics. In *RNA Viruses and Neurological Disorders* (pp. 206–239). CRC Press. DOI: 10.1201/9781003285823-16

Adams, D., Gonzalez-Duarte, A., O'Riordan, W. D., Yang, C.-C., Ueda, M., Kristen, A. V., Tournev, I., Schmidt, H. H., Coelho, T., Berk, J. L., Lin, K.-P., Vita, G., Attarian, S., Planté-Bordeneuve, V., Mezei, M. M., Campistol, J. M., Buades, J., Brannagan, T. H.III, Kim, B. J., & Suhr, O. B. (2018). Patisiran, an RNAi therapeutic, for hereditary transthyretin amyloidosis. *The New England Journal of Medicine, 379*(1), 11–21. DOI: 10.1056/NEJMoa1716153 PMID: 29972753

Adamson, C. S., Chibale, K., Goss, R. J. M., Jaspars, M., Newman, D. J., & Dorrington, R. A. (2021, March 21). Antiviral drug discovery: Preparing for the next pandemic. *Chemical Society Reviews, 50*(6), 3647–3655. DOI: 10.1039/D0CS01118E PMID: 33524090

Adams, R., & Kotlyar, S. (2021). AI-driven diagnostics and antiviral care in the COVID-19 era. *Journal of Medical Internet Research, 23*(3), e23964.

Adegboye, O., Field, M. A., Kupz, A., Pai, S., Sharma, D., Smout, M., Wangchuk, P., Wong, Y., & Loiseauc, C. (2021). Natural-product-based solutions for tropical infectious diseases. *Clinical Microbiology Reviews*, *34*(4), e00348–e00320. DOI: 10.1128/CMR.00348-20 PMID: 34494873

Adewuyi, O. S., Orevba, J., Yaria, J., Bamgboye, E., & Dairo, D. (2024). Factors that determine the level of non-adherence to HAART among HIV-positive pregnant women in Kaduna State. *World Journal of Biology Pharmacy and Health Sciences*, *19*(03), 458–471. DOI: 10.30574/wjbphs.2024.19.3.0653

Adnan, K., & Suri, H. (2020). Using mobile health technologies for improving adherence to antiretroviral therapy. *Journal of AIDS & Clinical Research*, *11*(3), 1–9.

Afrough, B., Dowall, S., & Hewson, R. (2019). Emerging viruses and current strategies for vaccine intervention. *Clinical and Experimental Immunology*, *196*(2), 157–166. DOI: 10.1111/cei.13295 PMID: 30993690

Ahmad, I., Ahmed, T., Rizvi, A., Kumar, N., & Khan, M. (2023). Advances in CRISPR/Cas9 for antiviral therapy: Challenges and future perspectives. *Trends in Biotechnology*, *41*(1), 56–70.

Ahmed, A. E. (2019). Diagnostic delays in Middle East respiratory syndrome coronavirus patients and health systems. *Journal of Infection and Public Health*, *12*(6), 767–771. DOI: 10.1016/j.jiph.2019.04.002 PMID: 31006635

Ahmed, R., & Hu, Z. (2021). The emerging role of mRNA vaccines: Lessons learned from the COVID-19 pandemic. *Nature Reviews. Immunology*, *21*(9), 569–570.

Ahn, D. G., Lee, W., Choi, J. K., Kim, S. J., Plant, E. P., Almazán, F., Taylor, D. R., Enjuanes, L., & Oh, J. W. (2011). Interference of ribosomal frameshifting by antisense peptide nucleic acids suppresses SARS coronavirus replication. *Antiviral Research*, *91*(1), 1–10. DOI: 10.1016/j.antiviral.2011.04.009 PMID: 21549154

Akbarzadeh, A., Rezaei-Sadabady, R., Davaran, S., Joo, S. W., Zarghami, N., Hanifehpour, Y., Samiei, M., Kouhi, M., & Nejati-Koshki, K. (2013). Liposome: Classification, preparation, and applications. *Nanoscale Research Letters*, *8*(1), 1–9. DOI: 10.1186/1556-276X-8-102 PMID: 23432972

Akhter, M. H., Beg, S., Tarique, M., Malik, A., Afaq, S., Choudhry, H., & Hosawi, S. (2021). Receptor-based targeting of engineered nanocarrier against solid tumors: Recent progress and challenges ahead. *Biochimica et Biophysica Acta (BBA)-. Biochimica et Biophysica Acta. G, General Subjects*, *1865*(2), 129777. DOI: 10.1016/j.bbagen.2020.129777 PMID: 33130062

Akram, N., Saeed, F., Afzaal, M., Shah, Y. A., Qamar, A., Faisal, Z., Ghani, S., Ateeq, H., Akhtar, M. N., Tufail, T., Hussain, M., Asghar, A., Rasheed, A., & Jbawi, E. A. (2023). Gut microbiota and synbiotic foods: Unveiling the relationship in COVID-19 perspective. *Food Science & Nutrition*, *11*(3), 1166–1177. DOI: 10.1002/fsn3.3162 PMID: 36911846

Alam, S., Bhuiyan, F. R., Emon, T. H., & Hasan, M. (2021). Prospects of nutritional interventions in the care of COVID-19 patients. *Heliyon*, *7*(2), e06285. DOI: 10.1016/j.heliyon.2021.e06285 PMID: 33615017

Al-Antary, M., & Seida, M. (2020). Digital interventions in the management of hepatitis C: A global perspective. *International Journal of Medical Informatics*, *136*, 104082.

Albuquerque, P. C., Zicker, F., & Fonseca, B. P. (2022, December). Advancing drug repurposing research: Trends, collaborative networks, innovation and knowledge leaders. *Drug Discovery Today*, *27*(12), 103396. DOI: 10.1016/j.drudis.2022.103396 PMID: 36241041

Alcántara, J. M., & Ugarte, G. (2017). Telemedicine as a tool for the management of hepatitis C in rural areas. *Telemedicine Journal and e-Health*, *23*(6), 503–507. PMID: 27911652

Alhammad, N., Alajlani, M., Abd-Alrazaq, A., Epiphaniou, G., & Arvanitis, T. (2024, May 31). Patients' Perspectives on the Data Confidentiality, Privacy, and Security of mHealth Apps: Systematic Review. *Journal of Medical Internet Research*, *26*, e50715. DOI: 10.2196/50715 PMID: 38820572

Alijotas-Reig, J., Esteve-Valverde, E., Belizna, C., Selva-O'Callaghan, A., Pardos-Gea, J., Quintana, A., Mekinian, A., Anunciacion-Llunell, A., & Miró-Mur, F. (2020). Immunomodulatory therapy for the management of severe COVID-19. Beyond the anti-viral therapy: A comprehensive review. *Autoimmunity Reviews*, *19*(7), 102569. DOI: 10.1016/j.autrev.2020.102569 PMID: 32376394

Ali, R., Awadalla, E., Amin, Y., Fouad, S. S., Ahmed, M. A.-E. B., Hassan, M. H., Abdel-Kahaar, E., & Abdel-Aziz, R. H. (2024). The deleterious effects of sofosbuvir and ribavirin (antiviral drugs against hepatitis C virus) on different body systems in male albino rats regarding reproductive, hematological, biochemical, hepatic, and renal profiles and histopathological changes. *Scientific Reports*, *14*(1), 5682. DOI: 10.1038/s41598-024-55950-5 PMID: 38453980

Al-Omairi, O., Elgalib, A., & Al Kindi, H. (2019). HIV Drug Resistance among Patients Failing Therapy at a Tertiary Center in Oman: A Case Record Review. *Oman Medical Journal*, *34*(6), 490–495. DOI: 10.5001/omj.2019.91 PMID: 31745412

Amrouche, T., Lammi, S., & Drider, D. (2024). Probiotics and prebiotics intervention in respiratory and digestive infections linked to COVID-19. *Probiotics and Antimicrobial Proteins*. Advance online publication. DOI: 10.1007/s12602-024-10404-2 PMID: 39614066

Anagnostis, P., Livadas, S., Goulis, D. G., Rees, M., & Lambrinoudaki, I. (2023). Vitamin D, Menopausal Health and COVID-19: Critical Appraisal of Current Data. *Journal of Clinical Medicine*, *12*(3), 916. DOI: 10.3390/jcm12030916 PMID: 36769563

Anand, K., Ziebuhr, J., Wadhwani, P., Mesters, J. R., & Hilgenfeld, R. (2003). Coronavirus main proteinase (3CLpro) structure: Basis for design of anti-SARS drugs. *Science*, *300*(5626), 1763–1767. DOI: 10.1126/science.1085658 PMID: 12746549

Anderson, M., Forman, R., & Mossialos, E. (2021, October). Navigating the role of the EU Health Emergency Preparedness and Response Authority (HERA) in Europe and beyond. *The Lancet Regional Health. Europe*, *9*, 100203. DOI: 10.1016/j.lanepe.2021.100203 PMID: 34661186

Andreakos, E., & Tsiodras, S. (2020). COVID-19: Lambda interferon against viral load and hyperinflammation. *EMBO Molecular Medicine*, *12*(6), e12465. DOI: 10.15252/emmm.202012465 PMID: 32333818

Antonelli, G., Falasca, F., & Turriziani, O. (2021). Mechanisms of Action of Antiviral Agents. In *Practical Handbook of Microbiology* (pp. 789–800). CRC Press. DOI: 10.1201/9781003099277-56

Appaji, R., Thomas, L., & Baral, T., N, S., Chaithra, Saravu, K., ... & Rao, M. (2024). Characteristics of isoniazid-induced psychosis: A systematic review of case reports and case series. *European Journal of Clinical Pharmacology*, *80*(11), 1725–1740.

Arankalle, V. A., Shrivastava, S., Cherian, S., Gunjikar, R. S., Walimbe, A. M., Jadhav, S. M., Sudeep, A. B., & Mishra, A. C. (2007). Genetic divergence of Chikungunya viruses in India (1963-2006) with special reference to the 2005-2006 explosive epidemic. *The Journal of General Virology*, *88*(Pt 7), 1967–1976. DOI: 10.1099/vir.0.82714-0 PMID: 17554030

Asad, H., & Carpenter, D. O. (2018). Effects of climate change on the spread of zika virus: A public health threat. *Reviews on Environmental Health*, *33*(1), 31–42. DOI: 10.1515/reveh-2017-0042 PMID: 29500926

Ashique, S., Mishra, N., Mohanto, S., Garg, A., Taghizadeh-Hesary, F., Gowda, B. H. J., & Chellappan, D. K. (2024). Application of artificial intelligence (AI) to control COVID-19 pandemic: Current status and future prospects. *Heliyon*, *10*(4), e25754. DOI: 10.1016/j.heliyon.2024.e25754 PMID: 38370192

Asselah, T., Marcellin, P., & Schinazi, R. F. (2018). Treatment of hepatitis C virus infection with direct-acting antiviral agents: 100% cure?. *Liver international: official journal of the International Association for the Study of the Liver, 38 Suppl 1*(Suppl 1), 7–13. https://doi.org/DOI: 10.1111/liv.13673

Attia, M. F., Anton, N., Wallyn, J., Omran, Z., & Vandamme, T. F. (2019). An overview of active and passive targeting strategies to improve the nanocarriers efficiency to tumour sites. *The Journal of Pharmacy and Pharmacology*, *71*(8), 1185–1198. DOI: 10.1111/jphp.13098 PMID: 31049986

Ávila-Ríos, S., Parkin, N., Swanstrom, R., Paredes, R., Shafer, R., Ji, H., & Kantor, R. (2020). Next-Generation Sequencing for HIV Drug Resistance Testing: Laboratory, Clinical, and Implementation Considerations. *Viruses*, *12*(6), 617. DOI: 10.3390/v12060617 PMID: 32516949

Baddeley, H. J. E., & Isalan, M. (2021). The Application of CRISPR/Cas Systems for Antiviral Therapy. *Frontiers in Genome Editing*, *3*, 745559. DOI: 10.3389/fgeed.2021.745559 PMID: 34723245

Baden, L. R., El Sahly, H. M., Essink, B., Kotloff, K., Frey, S., Novak, R., Diemert, D., Spector, S. A., Rouphael, N., Creech, C. B., McGettigan, J., Khetan, S., Segall, N., Solis, J., Brosz, A., Fierro, C., Schwartz, H., Neuzil, K., Corey, L., & Zaks, T. (2021). Efficacy and safety of the mRNA-1273 SARS-CoV-2 vaccine. *The New England Journal of Medicine*, *384*(5), 403–416. DOI: 10.1056/NEJMoa2035389 PMID: 33378609

Badi, S., & Hussain, K. (2020). Real-time monitoring systems for antiviral treatment adherence: Role of wearables. *Journal of Medical Systems*, *44*(11), 204.

Bahrulolum, H., Tarrahimofrad, H., Rouzbahani, F. N., Nooraei, S., Sameh, M. M., Hajizade, A., & Ahmadian, G. (2023). Potential of CRISPR/Cas system as emerging tools in the detection of viral hepatitis infection. *Virology Journal*, *20*(1), 91. DOI: 10.1186/s12985-023-02048-5 PMID: 37158910

Baker, J. R., Dubacheva, G. V., & Gentry, S. (2021). Nanoparticle vaccines: A review of the latest advancements. *Advanced Drug Delivery Reviews*, *170*, 107–123. DOI: 10.1016/j.addr.2020.10.009

Baker, J. R., & Wang, F. (2021). Nanoparticle vaccines for the prevention of infectious diseases. *Frontiers in Immunology*, *12*, 648–658.

Bakkers, M. J. G., Ritschel, T., Tiemessen, M., Dijkman, J., Zuffianò, A. A., Yu, X., van Overveld, D., Le, L., Voorzaat, R., van Haaren, M. M., de Man, M., Tamara, S., van der Fits, L., Zahn, R., Juraszek, J., & Langedijk, J. P. M. (2024). Efficacious human metapneumovirus vaccine based on AI-guided engineering of a closed prefusion trimer. *Nature Communications*, *15*(1), 6270. DOI: 10.1038/s41467-024-50659-5 PMID: 39054318

Balwani, M., Sardh, E., Ventura, P., Peiró, P. A., Rees, D. C., Stölzel, U., Bissell, D. M., Bonkovsky, H. L., Windyga, J., Anderson, K. E., Parker, C., Silver, S. M., Keel, S. B., Wang, J. D., Stein, P. E., Harper, P., Vassiliou, D., Wang, B., Phillips, J., & Gouya, L. (2020). Phase 3 Trial of RNAi Therapeutic Givosiran for Acute Intermittent Porphyria. *The New England Journal of Medicine*, *382*(24), 2289–2301. DOI: 10.1056/NEJMoa1913147 PMID: 32521132

Banakar, U. V. (2021). *Pharmaceutical dissolution testing, bioavailability, and bioequivalence: Science, applications, and beyond.* John Wiley & Sons.

Barbosu, S. (2024). *Harnessing AI to Accelerate Innovation in the Biopharmaceutical Industry*. Information Technology & Innovation Foundation.

Barghash, R. F., Gemmati, D., Awad, A. M., Elbakry, M. M. M., Tisato, V., Awad, K., & Singh, A. V. (2024). Navigating the COVID-19 Therapeutic Landscape: Unveiling Novel Perspectives on FDA-Approved Medications, Vaccination Targets, and Emerging Novel Strategies. *Molecules (Basel, Switzerland)*, *29*(23), 5564. Advance online publication. DOI: 10.3390/molecules29235564 PMID: 39683724

Barth, R. F., Buja, L. M., Barth, A. L., Carpenter, D. E., & Parwani, A. V. (2021). A comparison of the clinical, viral, pathologic, and immunologic features of severe acute respiratory syndrome (SARS), middle east respiratory syndrome (MERS), and coronavirus 2019 (COVID-19) diseases. *Archives of Pathology & Laboratory Medicine*, *145*(10), 1194–1211. DOI: 10.5858/arpa.2020-0820-SA PMID: 34232978

Bashshur, R. L., Shannon, G. W., Krupinski, E. A., & Grigsby, J. (2013). The role of telemedicine in the future of healthcare. *Telemedicine Journal and e-Health*, *19*(4), 282–285.

Bauer, A. N., Williams, J. F., Pokhrel, L. R., Garcia, S., Majumdar, N., Eells, J. B., Cook, P. P., & Akula, S. M. (2024). Evaluating Molecular Mechanism of Viral Inhibition of Aerosolized Smart Nano-Enabled Antiviral Therapeutic (SNAT) on SARS-CoV-2-Infected Hamsters. *Toxics*, *12*(7), 495. DOI: 10.3390/toxics12070495 PMID: 39058147

Beans, C. (2022). Researchers getting closer to a "universal" flu vaccine. *Proceedings of the National Academy of Sciences of the United States of America, 119*(5), e2123477119. DOI: 10.1073/pnas.2123477119 PMID: 35082157

Begines, B., Ortiz, T., Pérez-Aranda, M., Martínez, G., Merinero, M., Argüelles-Arias, F., & Alcudia, A. (2020). Polymeric nanoparticles for drug delivery: Recent developments and future prospects. *Nanomaterials (Basel, Switzerland), 10*(7), 1403. DOI: 10.3390/nano10071403 PMID: 32707641

Bekbolatova, M., Mayer, J., Ong, C. W., & Toma, M. (2024). Transformative Potential of AI in Healthcare: Definitions, Applications, and Navigating the Ethical Landscape and Public Perspectives. *Healthcare (Basel), 12*(2), 125. DOI: 10.3390/healthcare12020125 PMID: 38255014

Ben-Shabat, S., Yarmolinsky, L., Porat, D., & Dahan, A. (2020). Antiviral effect of phytochemicals from medicinal plants: Applications and drug delivery strategies. *Drug Delivery and Translational Research, 10*(2), 354–367. DOI: 10.1007/s13346-019-00691-6 PMID: 31788762

Benson, M. D., Waddington-Cruz, M., Berk, J. L., Polydefkis, M., Dyck, P. J., Wang, A. K., Planté-Bordeneuve, V., Barroso, F. A., Merlini, G., Obici, L., Scheinberg, M., Brannagan, T. H.III, Litchy, W. J., Whelan, C., Drachman, B. M., Adams, D., Heitner, S. B., Conceição, I., Schmidt, H. H., & Coelho, T. (2018). Inotersen Treatment for Patients with Hereditary Transthyretin Amyloidosis. *The New England Journal of Medicine, 379*(1), 22–31. DOI: 10.1056/NEJMoa1716793 PMID: 29972757

Bente, D. A., Forrester, N. L., Watts, D. M., McAuley, A. J., Whitehouse, C. A., & Bray, M. (2013). Crimean-Congo hemorrhagic fever: History, epidemiology, pathogenesis, clinical syndrome and genetic diversity. *Antiviral Research, 100*(1), 159–189. DOI: 10.1016/j.antiviral.2013.07.006 PMID: 23906741

Berger, C. T., & Kim, A. Y. (2012). IL28B polymorphisms as a pretreatment predictor of response to HCV treatment. *Infectious Disease Clinics of North America, 26*(4), 863–877. DOI: 10.1016/j.idc.2012.08.010 PMID: 23083820

Bezbaruah, R., Chavda, V. P., Nongrang, L., Alom, S., Deka, K., Kalita, T., Ali, F., Bhattacharjee, B., & Vora, L. (2022). Nanoparticle-Based Delivery Systems for Vaccines. *Vaccines, 10*(11), 1946. DOI: 10.3390/vaccines10111946 PMID: 36423041

Bhargavi, G., & Subbian, S. (2024). The causes and consequences of trained immunity in myeloid cells. *Frontiers in Immunology, 15*, 1365127. DOI: 10.3389/fimmu.2024.1365127 PMID: 38665915

Bhatia, S., & Bhatia, S. (2016). Nanoparticles types, classification, characterization, fabrication methods and drug delivery applications. In *Natural polymer drug delivery systems: Nanoparticles, plants, and algae* (pp. 33-93). DOI: 10.1007/978-3-319-41129-3_2

Biswas, M., Sawajan, N., Rungrotmongkol, T., Sanachai, K., Ershadian, M., & Sukasem, C. (2022). Pharmacogenetics and Precision Medicine Approaches for the Improvement of COVID-19 Therapies. *Frontiers in Pharmacology*, *13*, 835136. DOI: 10.3389/fphar.2022.835136 PMID: 35250581

Bock, C., & Lengauer, T. (2012). Managing drug resistance in cancer: Lessons from HIV therapy. *Nature Reviews. Cancer*, *12*(7), 494–501. DOI: 10.1038/nrc3297 PMID: 22673150

Bodsworth, N. J., Crooks, R. J., Borelli, S., & Vejlsgaard, G. (2018). Valacyclovir for the suppression of recurrent genital herpes simplex virus infection. *Journal of Medical Virology*, *90*(5), 831–838. DOI: 10.1002/jmv.25034

Boersma, S., Rabouw, H. H., Bruurs, L. J. M., Pavlovič, T., van Vliet, A. L. W., Beumer, J., Clevers, H., van Kuppeveld, F. J. M., & Tanenbaum, M. E. (2020). Translation and Replication Dynamics of Single RNA Viruses. *Cell*, *183*(7), 1930–1945.e23. DOI: 10.1016/j.cell.2020.10.019 PMID: 33188777

Boilesen, D. R., Nielsen, K. N., & Holst, P. J. (2021). Novel Antigenic Targets of HPV Therapeutic Vaccines. *Vaccines*, *9*(11), 1262. DOI: 10.3390/vaccines9111262 PMID: 34835193

Bokolo, A. J. (2020). Use of telemedicine and virtual care for remote treatment in response to COVID-19 pandemic. *Journal of Medical Systems*, *44*(7), 132. DOI: 10.1007/s10916-020-01596-5 PMID: 32542571

Bokolo, A. J. (2021). Application of telemedicine and eHealth technology for clinical services in response to COVID-19 pandemic. *Health and Technology*, *11*(4), 359–366. DOI: 10.1007/s12553-020-00516-4 PMID: 33469474

Bolarin, J. A., Oluwatoyosi, M. A., Orege, J. I., Ayeni, E. A., Ibrahim, Y. A., Adeyemi, S. B., Tiamiyu, B. B., Gbadegesin, L. A., Akinyemi, T. O., Odoh, C. K., Umeobi, H. I., & Adeoye, A. B. (2021). Therapeutic drugs for SARS-CoV-2 treatment: Current state and perspective. *International Immunopharmacology*, *90*, 107228. DOI: 10.1016/j.intimp.2020.107228 PMID: 33302035

Borden, K. L., & Culjkovic-Kraljacic, B. (2010). Ribavirin as an anticancer therapy: Acute myeloid leukemia and beyond? *Leukemia & Lymphoma*, *51*(10), 1805–1815. DOI: 10.3109/10428194.2010.496506 PMID: 20629523

Bottino, P., Pastrone, L., Curtoni, A., Bondi, A., Sidoti, F., Zanotto, E., Cavallo, R., Solidoro, P., & Costa, C. (2023). Antiviral Approach to Cytomegalovirus Infection: An Overview of Conventional and Novel Strategies. *Microorganisms*, *11*(10), 2372. DOI: 10.3390/microorganisms11102372 PMID: 37894030

Bower, J. E., & Lim, S. P. (2018). The role of mobile health (mHealth) in the management of viral diseases: HIV and Hepatitis C. *Journal of Viral Therapy*, *45*(8), 78–86.

Braun, E., & Sauter, D. (2019). Furin-mediated protein processing in infectious diseases and cancer. *Clinical & Translational Immunology*, *8*(8), e1073. DOI: 10.1002/cti2.1073 PMID: 31406574

Braveman, P., & Gottlieb, L. (2014). The social determinants of health: it is time to consider the causes of the causes. *Public health reports (Washington, D.C.: 1974)*, *129 Suppl 2*(Suppl 2), 19–31. DOI: 10.1177/00333549141291S206

Broniatowski, D. A., Jamison, A. M., Qi, S. H., AlKulaib, L., Chen, T., Benton, A., Quinn, S. C., & Dredze, M. (2018). Weaponized health communication: Twitter bots and Russian trolls amplify the vaccine debate. *American Journal of Public Health*, *108*(10), 1378–1384. DOI: 10.2105/AJPH.2018.304567 PMID: 30138075

Browne, M., Thomson, P., & Mijan, J.. (2021). Vaccine hesitancy: A systematic review of quantitative studies. *International Journal of Public Health*, *66*, 584–592. DOI: 10.3389/ijph.2021.00059

Bui, L. V., Ha, S. T., Nguyen, H. N., Nguyen, T. T., Nguyen, T. P., Tran, K., Tran, T. V., Nguyen, T. H., Tran, T. H., Pham, N. D., & Bui, H. M. (2021). The contribution of digital health in the response to COVID-19 in Vietnam. *Frontiers in Public Health*, *9*, 672732. DOI: 10.3389/fpubh.2021.672732 PMID: 34540779

Burbank, A. J. (2023). Risk Factors for Respiratory Viral Infections: A Spotlight on Climate Change and Air Pollution. *Journal of Asthma and Allergy*, *16*, 183–194. DOI: 10.2147/JAA.S364845 PMID: 36721739

Burt, F. J., Chen, W., Miner, J. J., Lenschow, D. J., Merits, A., Schnettler, E., Kohl, A., Rudd, P. A., Taylor, A., Herrero, L. J., Zaid, A., Ng, L. F. P., & Mahalingam, S. (2017). Chikungunya virus: An update on the biology and pathogenesis of this emerging pathogen. *The Lancet. Infectious Diseases*, *17*(4), e107–e117. DOI: 10.1016/S1473-3099(16)30385-1 PMID: 28159534

Buti, M., Tsai, N., Petersen, J., Flisiak, R., Gurel, S., Krastev, Z., & Gane, E. (2020). Seven-year efficacy and safety of tenofovir disoproxil fumarate treatment for chronic hepatitis B. *Journal of Hepatology*, *73*(3), 532–541. DOI: 10.1016/j.jhep.2020.04.014

Cai, B., Chang, S., Tian, Y., & Zhen, S. (2023, May). CRISPR/Cas9 for hepatitis B virus infection treatment. *Immunity, Inflammation and Disease, 11*(5), e866. DOI: 10.1002/iid3.866 PMID: 37249290

Cani, E., Park, T. E., & Kavanagh, R. (2019). Antiviral drugs. [). Elsevier.]. *Side Effects of Drugs Annual, 41*, 301–319. DOI: 10.1016/bs.seda.2019.10.005

Cao, J., Huang, D., & Peppas, N. A. (2020). Advanced engineered nanoparticulate platforms to address key biological barriers for delivering chemotherapeutic agents to target sites. *Advanced Drug Delivery Reviews, 167*, 170–188. DOI: 10.1016/j.addr.2020.06.030 PMID: 32622022

Cao, L., Zhao, S., Lou, J., Zheng, H., Chan, R. W. Y., Chong, M. K. C., Chen, Z., Chan, P. K. S., Zee, B. C. Y., & Wang, M. H. (2021). Differential Influence of Age on the Relationship between Genetic Mismatch and A(H1N1)pdm09 Vaccine Effectiveness. *Viruses, 13*(4), 619. DOI: 10.3390/v13040619 PMID: 33916601

Carlsson, L., Clarke, J. C., Yen, C., Gregoire, F., Albery, T., Billger, M., Egnell, A. C., Gan, L. M., Jennbacken, K., Johansson, E., Linhardt, G., Martinsson, S., Sadiq, M. W., Witman, N., Wang, Q. D., Chen, C. H., Wang, Y. P., Lin, S., Ticho, B., & Fritsche-Danielson, R. (2018). Biocompatible, Purified *VEGF-A* mRNA Improves Cardiac Function after Intracardiac Injection 1 Week Post-myocardial Infarction in Swine. *Molecular Therapy. Methods & Clinical Development, 9*, 330–346. DOI: 10.1016/j.omtm.2018.04.003 PMID: 30038937

Caron, A., Jori, F., de Nys, H., Herbinger, I., Liégeois, F., Mouinga-Ondémé, A., & Ratiarison, S. (2022). Emergence/Re-emergence of Infectious Agents and Epidemic Risks in Central African Forests. *The Forests of the Congo Basin: State of the Forests, 2021*, 265.

CDC. (2023). Vaccine side effects. Centers for Disease Control and Prevention. Retrieved from CDC website.

Čelková, A., Búcsi, A., Klacsová, M., Fazekaš, T., Martínez, J. C., & Uhríková, D. (2024). Oseltamivir phosphate interaction with model membranes. *Biochimica et Biophysica Acta (BBA)-. Biomembranes, 1866*(3), 184273. DOI: 10.1016/j.bbamem.2024.184273 PMID: 38211646

Centers for Disease Control and Prevention (CDC). (2023). Possible Side Effects After Vaccines. Retrieved from https://www.cdc.gov

Centers for Disease Control and Prevention (CDC). (2023). Varicella and Shingles Vaccines. Retrieved from https://www.cdc.gov

Centers for Disease Control and Prevention. (2020). Measles Cases and Outbreaks. Retrieved March 28, 2025, from https://www.cdc.gov/measles/data-research/index .html

Centers for Disease Control and Prevention. (2022). *HIV/AIDS: Basics*. https:// www.cdc.gov/hiv/basics

Centers for Disease Control and Prevention. (2023). History of measles. Retrieved March 28, 2025, from https://www.cdc.gov/measles/about/history.html

Centers for Disease Control and Prevention. (2023). *Influenza (Flu)*. https://www .cdc.gov/flu

Centers for Disease Control and Prevention. (2023). *Possible side effects after getting a COVID-19 vaccine*. https://www.cdc.gov/coronavirus/2019-ncov/vaccines/ expect/after.html

Centers for Disease Control and Prevention. (2023). Underutilization of Influenza Antiviral Treatment Among Children and Adolescents—United States, 2023–24. *Morbidity and Mortality Weekly Report*, *73*(45), 1012–1016. Retrieved March 28, 2025, from https://www.cdc.gov/mmwr/volumes/73/wr/mm7345a2.htm

Ch'ien, L. T., Schabel, F. M., & Alford, C. A. (2018). Arabinosyl nucleosides and nucleotides. In *Selective inhibitors of viral functions* (pp. 227–256). CRC Press. DOI: 10.1201/9781351076562-11

Chakrabarti, S., Roy, P. K., Chattopadhyay, P., & Mazumder, B. (2019). Prospects of nanotechnology in brain targeting drug delivery. In *Nanotechnology* (pp. 273–314). CRC Press. DOI: 10.1201/9781351111874-10

Chakraborty, S. S., Ray Dutta, J., Ganesan, R., & Minary, P. (2025). The Evolution of Nucleic Acid-Based Diagnosis Methods from the (pre)CRISPR to CRISPR era and the Associated Machine/Deep Learning Approaches in Relevant RNA Design. *Methods in Molecular Biology (Clifton, N.J.)*, *2847*, 241–300. DOI: 10.1007/978-1-0716-4079-1_17 PMID: 39312149

Chakravarty, M., & Vora, A. (2021). Nanotechnology-based antiviral therapeutics. *Drug Delivery and Translational Research*, *11*(3), 748–787. DOI: 10.1007/s13346-020-00818-0 PMID: 32748035

Chan, J. F.-W., To, K. K.-W., Chen, H., & Yuen, K.-Y. (2015). Cross-species transmission and emergence of novel viruses from birds. *Current Opinion in Virology*, *10*, 63–69. DOI: 10.1016/j.coviro.2015.01.006 PMID: 25644327

Chaudhary, N., Weissman, D., & Whitehead, K. A. (2021). mRNA vaccines for infectious diseases: Principles, delivery and clinical translation. *Nature Reviews. Drug Discovery*, *20*(11), 817–838. DOI: 10.1038/s41573-021-00283-5 PMID: 34433919

Chaurasia, M., & Ahmad, B. (2021). Artificial intelligence in the prediction of viral outbreaks and drug efficacy. *Computational Biology and Chemistry*, *91*, 107442.

Chehelgerdi, M., Chehelgerdi, M., Khorramian-Ghahfarokhi, M., Shafieizadeh, M., Mahmoudi, E., Eskandari, F., Rashidi, M., Arshi, A., & Mokhtari-Farsani, A. (2024). Comprehensive review of CRISPR-based gene editing: Mechanisms, challenges, and applications in cancer therapy. *Molecular Cancer*, *23*(9), 9. Advance online publication. DOI: 10.1186/s12943-023-01925-5 PMID: 38195537

Chen, C. C., Yu, X., Kuo, C. J., Min, J., Chen, S., Ma, L., Liu, K., & Guo, R. T. (2021). Overview of antiviral drug candidates targeting coronaviral 3C-like main proteases. *The FEBS Journal*, *288*(17), 5089–5121. DOI: 10.1111/febs.15696 PMID: 33400393

Chen, D., & Zhang, T. (2021). Digital health strategies for influenza prevention: A systematic review. *Influenza and Other Respiratory Viruses*, *15*(3), 331–340. PMID: 33665959

Cheng, H., Deng, H., Ma, D., Gao, M., Zhou, Z., Li, H., Liu, S., & Teng, T. (2024). Insight into the natural regulatory mechanisms and clinical applications of the CRISPR–Cas system. *Heliyon*, *10*(20), e39538. DOI: 10.1016/j.heliyon.2024.e39538 PMID: 39502233

Chen, K. K., Tsung-Ning Huang, D., & Huang, L. M. (2022). SARS-CoV-2 variants - Evolution, spike protein, and vaccines. *Biomedical Journal*, *45*(4), 573–579. DOI: 10.1016/j.bj.2022.04.006 PMID: 35526825

Chen, L., & Liang, J. (2020). An overview of functional nanoparticles as novel emerging antiviral therapeutic agents. *Materials Science and Engineering C*, *112*, 110924. DOI: 10.1016/j.msec.2020.110924 PMID: 32409074

Chen, L., Xiong, J., Bao, L., & Shi, Y. (2020). Convalescent plasma as a potential therapy for COVID-19. *The Lancet. Infectious Diseases*, *20*(4), 398–400. DOI: 10.1016/S1473-3099(20)30141-9 PMID: 32113510

Chen, R., Wang, T., Song, J., Pu, D., He, D., Li, J., Yang, J., Li, K., Zhong, C., & Zhang, J. (2021). Antiviral drug delivery system for enhanced bioactivity, better metabolism and pharmacokinetic characteristics. *International Journal of Nanomedicine*, *16*, 4959–4984. DOI: 10.2147/IJN.S315705 PMID: 34326637

Chen, T., Li, S., Deng, D., Zhang, W., Zhang, J., & Shen, Z. (2024). Key role of interferon regulatory factor 1 (IRF-1) in regulating liver disease: progress and outlook. IRF-1 : . *Journal of Zhejiang University. Science. B.*, *25*(6), 451–470. DOI: 10.1631/jzus.B2300159 PMID: 38910492

Chen, X., Zheng, M., Lin, S., Huang, M., Chen, S., & Chen, S. (2025). The application of CRISPR/Cas9–based genome-wide screening to disease research. *Molecular and Cellular Probes*, *79*, 102004. DOI: 10.1016/j.mcp.2024.102004 PMID: 39709065

Chen, Y., & Zhang, Y. (2021). mRNA vaccines: A new era in vaccinology. *Frontiers in Immunology*, *12*, 654.

Chen, Y., & Zhang, Y. (2021). Nanotechnology in antiviral drug delivery: A review. *Nanomedicine (London)*, *16*(1), 1–15.

Cherry, J. D. (2004). The chronology of the 2002-2003 SARS mini pandemic. *Paediatric Respiratory Reviews*, *5*(4), 262–269. DOI: 10.1016/j.prrv.2004.07.009 PMID: 15531249

Chilot, D., Woldeamanuel, Y., & Manyazewal, T. (2021). Real-time impact of COVID-19 on clinical care and treatment of patients with tuberculosis: A multicenter cross-sectional study in Addis Ababa, Ethiopia. *Annals of Global Health*, *87*(1), 109. DOI: 10.5334/aogh.3481 PMID: 34824990

Chilot, D., Woldeamanuel, Y., & Manyazewal, T. (2022). COVID-19 burden on HIV patients attending antiretroviral therapy in Addis Ababa, Ethiopia: A multicenter cross-sectional study. *Frontiers in Medicine*, *9*, 741862. DOI: 10.3389/fmed.2022.741862 PMID: 35308528

Chirico, F., & Teixeira da Silva, J. A. (2022). COVID-19 health policies: The need for transparent data sharing between scientists, governments, and policymakers. *Oman Medical Journal*, *37*(1), e418. PMID: 36188880

Chis, A. A., Dobrea, C., Morgovan, C., Arseniu, A. M., Rus, L. L., Butuca, A., Juncan, A. M., Totan, M., Vonica-Tincu, A. L., Cormos, G., Muntean, A. C., Muresan, M. L., Gligor, F. G., & Frum, A. (2020). Applications and limitations of dendrimers in biomedicine. *Molecules (Basel, Switzerland)*, *25*(17), 3982. DOI: 10.3390/molecules25173982 PMID: 32882920

Chitalia, V. C., & Munawar, A. H. (2020, October 15). A painful lesson from the COVID-19 pandemic: The need for broad-spectrum, host-directed antivirals. *Journal of Translational Medicine*, *18*(1), 390. DOI: 10.1186/s12967-020-02476-9 PMID: 33059719

Choi, W., Cha, S., & Kim, K. (2024). Navigating the CRISPR/Cas Landscape for Enhanced Diagnosis and Treatment of Wilson's Disease. *Cells, 13*(14), 1214. DOI: 10.3390/cells13141214 PMID: 39056796

Chokwassanasakulkit, T., Oti, V. B., Idris, A., & McMillan, N. A. (2024, October 23). SiRNAs as antiviral drugs - Current status, therapeutic potential and challenges. *Antiviral Research, 232*, 106024. DOI: 10.1016/j.antiviral.2024.106024 PMID: 39454759

Chou, T. C., Maggirwar, N. S., & Marsden, M. D. (2024). HIV Persistence, Latency, and Cure Approaches: Where Are We Now? *Viruses, 16*(7), 1163. DOI: 10.3390/v16071163 PMID: 39066325

Chowdhury, M. E., & Sultana, M. (2020). Mobile health technologies for the management of HIV and viral hepatitis. *The Lancet. Infectious Diseases, 20*(7), e180–e189.

Chung, W., Lee, C., & Lee, S. (2021). The future of digital health technologies: Trends, opportunities, and challenges. *Healthcare Informatics Research, 27*(3), 189–201. PMID: 34384201

CICC Research. CICC Global Institute xinran. liu@ cicc. com. cn. (2024). Pharmaceuticals: Chinese Companies to Increase Efforts in Innovation and Expand to Higher-Added-Value Processes. In The Reshaping of China's Industry Chains (pp. 291-311). Singapore: Springer Nature Singapore.

Clinical Trials Arena. (2023). Self-amplifying RNA and vaccine innovation: Getting from concept to clinic. *Clinical Trials Arena*. Retrieved from https://www.clinicaltrialsarena.com

Clipman, S. J., Solomon, S. S., Srikrishnan, A. K., McFall, A. M., Gomathi, S., Saravanan, S., Anand, S., Vasudevan, C. K., Kumar, M. S., Celentano, D. D., Mehta, S. H., & Lucas, G. M. (2022). Antiretroviral Drug Resistance in HIV Sequences From People Who Inject Drugs and Men Who Have Sex With Men Across 21 Cities in India. *Open Forum Infectious Diseases, 9*(10), ofac481. Advance online publication. DOI: 10.1093/ofid/ofac481 PMID: 36225747

Cochran, C., & Sorensen, D. (2016). Wearable health technologies and their potential to transform the healthcare industry. *Journal of Medical Systems, 40*(4), 75.

Coelho, T., Adams, D., Silva, A., Lozeron, P., Hawkins, P. N., Mant, T., Perez, J., Chiesa, J., Warrington, S., Tranter, E., Munisamy, M., Falzone, R., Harrop, J., Cehelsky, J., Bettencourt, B. R., Geissler, M., Butler, J. S., Sehgal, A., Meyers, R. E., & Suhr, O. B. (2013). Safety and efficacy of RNAi therapy for transthyretin amyloidosis. *The New England Journal of Medicine*, *369*(9), 819–829. DOI: 10.1056/NEJMoa1208760 PMID: 23984729

Cohen, M. S., Chen, Y. Q., McCauley, M., Gamble, T., Hosseinipour, M. C., Kumarasamy, N., Hakim, J. G., Kumwenda, J., Grinsztejn, B., Pilotto, J. H., Godbole, S. V., Mehendale, S., Chariyalertsak, S., Santos, B. R., Mayer, K. H., Hoffman, I. F., Eshleman, S. H., Piwowar-Manning, E., Wang, L., & Fleming, T. R.HPTN 052 Study Team. (2011). Prevention of HIV-1 infection with early antiretroviral therapy. *The New England Journal of Medicine*, *365*(6), 493–505. DOI: 10.1056/NEJMoa1105243 PMID: 21767103

Cojocaru, F. D., Botezat, D., Gardikiotis, I., Uritu, C. M., Dodi, G., Trandafir, L., Rezus, C., Rezus, E., Tamba, B. I., & Mihai, C. T. (2020). Nanomaterials designed for antiviral drug delivery transport across biological barriers. *Pharmaceutics*, *12*(2), 171. DOI: 10.3390/pharmaceutics12020171 PMID: 32085535

Collaborators, G. B. D. A. R. (2022, December 17). Global mortality associated with 33 bacterial pathogens in 2019: A systematic analysis for the Global Burden of Disease Study 2019. *Lancet*, *400*(10369), 2221–2248. DOI: 10.1016/S0140-6736(22)02185-7 PMID: 36423648

Cook, N. L., & Lauer, M. S. (2021). Biomedical research COVID-19 impact assessment: Lessons learned and compelling needs. *NAM Perspectives*. Advance online publication. DOI: 10.31478/202107e PMID: 34611607

Costa, B., & Vale, N. (2023, January 3). Modulating Immune Response in Viral Infection for Quantitative Forecasts of Drug Efficacy. *Pharmaceutics*, *15*(1), 167. DOI: 10.3390/pharmaceutics15010167 PMID: 36678799

Crawford, D. H. (2021). *Viruses: The invisible enemy*. Oxford University Press. DOI: 10.1093/oso/9780192845030.001.0001

Crux, N. B., & Elahi, S. (2017). Human Leukocyte Antigen (HLA) and Immune Regulation: How Do Classical and Non-Classical HLA Alleles Modulate Immune Response to Human Immunodeficiency Virus and Hepatitis C Virus Infections? *Frontiers in Immunology*, *8*, 832. DOI: 10.3389/fimmu.2017.00832 PMID: 28769934

Cui, L., Watanabe, S., Miyanaga, K., Kiga, K., Sasahara, T., Aiba, Y., Tan, X. E., Veeranarayanan, S., Thitiananpakorn, K., Nguyen, H. M., & Wannigama, D. L. (2024). A Comprehensive Review on Phage Therapy and Phage-Based Drug Development. *Antibiotics (Basel, Switzerland)*, *13*(9), 870. DOI: 10.3390/antibiotics13090870 PMID: 39335043

Cummins, N., & Schuller, B. W. (2020). Five crucial challenges in digital health. *Frontiers in Digital Health*, *2*, 536203. DOI: 10.3389/fdgth.2020.536203 PMID: 34713029

da Costa Castilho, M., de Filippis, A., Machado, L., de Lima Calvanti, T., Lima, M., Fonseca, V., Giovanetti, M., Docena, C., Neto, A. M., Bôtto-Menezes, C. H. A., Kara, E. O., de La Barrera, R., Modjarrad, K., Giozza, S. P., Pereira, G. F., Alcantara, L. C. J., Broutet, N. J. N., Calvet, G. A., Wallau, G. L., & Franca, R. (2024). Evidence of Zika Virus Reinfection by Genome Diversity and Antibody Response Analysis, Brazil. *Emerging Infectious Diseases*, *30*(2), 310–320. DOI: 10.3201/eid3002.230122 PMID: 38270216

Dasgupta, S., Roy, A., & Bhowmick, S. (2021). Telemedicine for HIV care: Current status and future prospects. *Current HIV/AIDS Reports*, *18*(1), 45–52.

Das, K., & Arnold, E. (2013). HIV-1 reverse transcriptase and antiviral drug resistance. Part 1. *Current Opinion in Virology*, *3*(2), 111–118. DOI: 10.1016/j.coviro.2013.03.012 PMID: 23602471

Davis, A. M., Engkvist, O., Fairclough, R. J., Feierberg, I., Freeman, A., & Iyer, P. (2021, June). Public–Private Partnerships: Compound and Data Sharing in Drug Discovery and Development. *SLAS Discovery*, *26*(5), 604–619. DOI: 10.1177/2472555220982268 PMID: 33586501

Dawood, F. S., Iuliano, A. D., Reed, C., Meltzer, M. I., Shay, D. K., Cheng, P. Y., Bandaranayake, D., Breiman, R. F., Brooks, W. A., Buchy, P., Feikin, D. R., Fowler, K. B., Gordon, A., Hien, N. T., Horby, P., Huang, Q. S., Katz, M. A., Krishnan, A., Lal, R., & Widdowson, M. A. (2012). Estimated global mortality associated with the first 12 months of 2009 pandemic influenza A H1N1 virus circulation: A modelling study. *The Lancet. Infectious Diseases*, *12*(9), 687–695. DOI: 10.1016/S1473-3099(12)70121-4 PMID: 22738893

De Clercq, E., & Li, G. (2016). Approved Antiviral Drugs over the Past 50 Years. *Clinical Microbiology Reviews*, *29*(3), 695–747. DOI: 10.1128/CMR.00102-15 PMID: 27281742

Deehan, E. C., Al Antwan, S., Witwer, R. S., Guerra, P., John, T., & Monheit, L. (2024). Revisiting the concepts of prebiotic and prebiotic effect in light of scientific and regulatory progress—A consensus paper from the Global Prebiotic Association. *Advances in Nutrition*, *15*(12), 100329. DOI: 10.1016/j.advnut.2024.100329 PMID: 39481540

DeFilippis, V. R., & Villarreal, L. P. (2007). An introduction to the evolutionary ecology of viruses. Viral ecology, 125.

Delshadi, R., Bahrami, A., McClements, D. J., Moore, M. D., & Williams, L. (2021). Development of nanoparticle-delivery systems for antiviral agents: A review. *Journal of Controlled Release*, *331*, 30–44. DOI: 10.1016/j.jconrel.2021.01.017 PMID: 33450319

Dey, N., Kamatchi, C., Vickram, A. S., Anbarasu, K., Thanigaivel, S., Palanivelu, J., Pugazhendhi, A., & Ponnusamy, V. K. (2022). Role of nanomaterials in deactivating multiple drug resistance efflux pumps – A review. *Environmental Research*, *204*, 111968. DOI: 10.1016/j.envres.2021.111968 PMID: 34453898

Di Santo, R. (2014). Inhibiting the HIV integration process: Past, present, and the future. *Journal of Medicinal Chemistry*, *57*(3), 539–566. DOI: 10.1021/jm400674a PMID: 24025027

Di Stefano, M., Ismail, M. H., Leitner, T., Faleo, G., Alwazzeh, M. J., Mbisa, J. L., Fiore, J. R., & Santantonio, T. A. (2023). A novel candidate hepatitis C virus genotype 4 subtype identified by next generation sequencing full-genome characterization in a patient from Saudi Arabia. *Frontiers in Microbiology*, *14*, 1285367. DOI: 10.3389/fmicb.2023.1285367 PMID: 38029191

Dijkman, R., Jebbink, M. F., Koekkoek, S. M., Deijs, M., Jónsdóttir, H. R., Molenkamp, R., Ieven, M., Goossens, H., Thiel, V., & van der Hoek, L. (2013). Isolation and characterization of current human coronavirus strains in primary human epithelial cell cultures reveal differences in target cell tropism. *Journal of Virology*, *87*(11), 6081–6090. DOI: 10.1128/JVI.03368-12 PMID: 23427150

Dimitrievska, M., Bansal, D., Vitale, M., Strouboulis, J., Miccio, A., Nicolaides, K. H., El Hoss, S., Shangaris, P., & Jacków-Malinowska, J. (2024). Revolutionizing healing: Gene editing's breakthrough against sickle cell disease. *Blood Reviews*, *65*, 101185. DOI: 10.1016/j.blre.2024.101185 PMID: 38493007

Dinesh, P., & Nair, S. R. (2020). Digital health interventions in the management of infectious diseases: A review. *Infection, Disease & Health*, *25*(1), 1–8. PMID: 31916947

Domachowske, J. B. (2024, July 12). Understanding the Regulatory Pathways Used to Develop, Evaluate, Authorize, and Approve New Drugs and Vaccines in the United States. *Journal of the Pediatric Infectious Diseases Society*, *13*(Supplement_2), S93–S102. DOI: 10.1093/jpids/piae036 PMID: 38995086

Domingues, W., Folgosi, V. Â., Sanabani, S. S., Leite, P. D.Junior, Assone, T., & Casseb, J. (2024). Novel approaches for HTLV-1 therapy: Innovative applications of CRISPR-Cas9. *Revista do Instituto de Medicina Tropical de São Paulo*, *66*, e48. DOI: 10.1590/s1678-9946202466048 PMID: 39194140

Dopazo, J., Maya-Miles, D., García, F., Lorusso, N., Calleja, M. Á., Pareja, M. J., López-Miranda, J., Rodríguez-Baño, J., Padillo, J., Túnez, I., & Romero-Gómez, M. (2021, May 26). Implementing Personalized Medicine in COVID-19 in Andalusia: An Opportunity to Transform the Healthcare System. *Journal of Personalized Medicine*, *11*(6), 475. DOI: 10.3390/jpm11060475 PMID: 34073493

Dou, D., Revol, R., Östbye, H., Wang, H., & Daniels, R. (2018). Influenza A Virus Cell Entry, Replication, Virion Assembly and Movement. *Frontiers in Immunology*, *9*, 1581. DOI: 10.3389/fimmu.2018.01581 PMID: 30079062

Drost, M., Diamanti, E., Fuhrmann, K., Goes, A., Shams, A., Haupenthal, J., Koch, M., Hirsch, A. K., & Fuhrmann, G. (2021). Bacteriomimetic liposomes improve antibiotic activity of a novel energy-coupling factor transporter inhibitor. *Pharmaceutics*, *14*(1), 4. DOI: 10.3390/pharmaceutics14010004 PMID: 35056900

Duan, X. P., Qin, B. D., Jiao, X. D., Liu, K., Wang, Z., & Zang, Y. S. (2024). New clinical trial design in precision medicine: Discovery, development and direction. *Signal Transduction and Targeted Therapy*, *9*(1), 57. DOI: 10.1038/s41392-024-01760-0 PMID: 38438349

Duan, Y., Luo, J., Zhao, L., Zhang, X., Miao, J., Moher, D., & Bian, Z. (2022). Reporting and data sharing level for COVID-19 vaccine trials: A cross-sectional study. *EBioMedicine*, *78*, 103962. DOI: 10.1016/j.ebiom.2022.103962 PMID: 35339894

Dube, E.. (2021). Vaccine hesitancy: Causes, consequences, and the way forward. *The Lancet. Infectious Diseases*, *21*(8), e268–e276.

Dube, T., Ghosh, A., Mishra, J., Kompella, U. B., & Panda, J. J. (2021). Repurposed drugs, molecular vaccines, immune-modulators, and nanotherapeutics to treat and prevent COVID-19 associated with SARS-CoV-2, a deadly nanovector. *Advanced Therapeutics*, *4*(2), 2000172. DOI: 10.1002/adtp.202000172 PMID: 33173808

Dubey, A. K., & Mostafavi, E. (2023). Biomaterials-mediated CRISPR/Cas9 delivery: Recent challenges and opportunities in gene therapy. *Frontiers in Chemistry, 11*, 1259435. DOI: 10.3389/fchem.2023.1259435 PMID: 37841202

Dunne, M. (2007). Antiretroviral drug development: The challenge of cost and access. *AIDS (London, England), 21*(Suppl 4), S73–S79. DOI: 10.1097/01.aids.0000279709.47298.8a PMID: 17620756

Durmuş, S., & Ülgen, K. Ö. (2017). Comparative interactomics for virus-human protein-protein interactions: DNA viruses versus RNA viruses. *FEBS Open Bio, 7*(1), 96–107. DOI: 10.1002/2211-5463.12167 PMID: 28097092

Du, S., Hu, X., Menendez-Arias, L., Zhan, P., & Liu, X. (2024, March). Target-based drug design strategies to overcome resistance to antiviral agents: Opportunities and challenges. *Drug Resistance Updates, 73*, 101053. DOI: 10.1016/j.drup.2024.101053 PMID: 38301487

Dust, K., Carpenter, M., Chen, J. C., Grant, C., McCorrister, S., Westmacott, G. R., & Severini, A. (2022). Human Papillomavirus 16 E6 and E7 Oncoproteins Alter the Abundance of Proteins Associated with DNA Damage Response, Immune Signaling and Epidermal Differentiation. *Viruses, 14*(8), 1764. DOI: 10.3390/v14081764 PMID: 36016386

Düzgüneş, N., Fernandez-Fuentes, N., & Konopka, K. (2021). Inhibition of Viral Membrane Fusion by Peptides and Approaches to Peptide Design. *Pathogens (Basel, Switzerland), 10*(12), 1599. DOI: 10.3390/pathogens10121599 PMID: 34959554

Eckstein, F. (2023). Advances in RNA chemistry for antisense oligonucleotide therapeutics. *Nucleic Acids Research, 51*(6), 2529–2536. DOI: 10.1093/nar/gkad013 PMID: 36881759

Edwards, C. E., Yount, B. L., Graham, R. L., Leist, S. R., Hou, Y. J., Dinnon, K. H.III, Sims, A. C., Swanstrom, J., Gully, K., Scobey, T. D., Cooley, M. R., Currie, C. G., Randell, S. H., & Baric, R. S. (2020). Swine acute diarrhea syndrome coronavirus replication in primary human cells reveals potential susceptibility to infection. *Proceedings of the National Academy of Sciences of the United States of America, 117*(43), 26915–26925. DOI: 10.1073/pnas.2001046117 PMID: 33046644

Eini, P., Hatami, M., Eini, P., & Poorolajal, J. (2023). Evaluation and Comparison of Serum Vitamin D Levels between Injection Drug Users with and without Chronic Hepatitis C Infection in Hamadan-Iran. *Maedica, 18*(3), 437–441. DOI: 10.26574/maedica.2023.18.3.437 PMID: 38023763

El Khoury, C. J. (2024, July 30). Application of Geographic Information Systems (GIS) in the Study of Prostate Cancer Disparities: A Systematic Review. *Cancers (Basel), 16*(15), 2715. Advance online publication. DOI: 10.3390/cancers16152715 PMID: 39123443

Elliott, J. L., & Kutluay, S. B. (2020). Going beyond Integration: The Emerging Role of HIV-1 Integrase in Virion Morphogenesis. *Viruses, 12*(9), 1005. DOI: 10.3390/v12091005 PMID: 32916894

Ellwanger, J. H., & Chies, J. A. B. (2021). Zoonotic spillover: Understanding basic aspects for better prevention. *Genetics and Molecular Biology, 44*(1, Suppl 1), e20200355. DOI: 10.1590/1678-4685-gmb-2020-0355 PMID: 34096963

Ellwanger, J. H., Fearnside, P. M., Ziliotto, M., Valverde-Villegas, J. M., Veiga, A. B. G., Vieira, G. F., Bach, E., Cardoso, J. C., Müller, N. F. D., Lopes, G., Caesar, L., Kulmann-Leal, B., Kaminski, V. L., Silveira, E. S., Spilki, F. R., Weber, M. N., Almeida, S. D. M., Hora, V. P. D., & Chies, J. A. B. (2022). Synthesizing the connections between environmental disturbances and zoonotic spillover. *Anais da Academia Brasileira de Ciências, 94*(suppl 3), e20211530. DOI: 10.1590/0001-3765202220211530 PMID: 36169531

El-Sayed, A., & Kamel, M. (2021). Coronaviruses in humans and animals: The role of bats in viral evolution. *Environmental Science and Pollution Research International, 28*(16), 19589–19600. DOI: 10.1007/s11356-021-12553-1 PMID: 33655480

Elste, J., Saini, A., Mejia-Alvarez, R., Mejía, A., Millán-Pacheco, C., Swanson-Mungerson, M., & Tiwari, V. (2024). Significance of Artificial Intelligence in the Study of Virus–Host Cell Interactions. *Biomolecules, 14*(8), 911. DOI: 10.3390/biom14080911 PMID: 39199298

Epstein, J. H., & Anthony, S. J. (2017). Viral discovery as a fool for pandemic preparedness. *Revue Scientifique et Technique (International Office of Epizootics), 36*(2), 499–512. DOI: 10.20506/rst.36.2.2669 PMID: 30152468

Eryildiz, B., Yavuzturk Gul, B., & Koyuncu, I. (2022, October). A sustainable approach for the removal methods and analytical determination methods of antiviral drugs from water/wastewater: A review. *Journal of Water Process Engineering, 49*, 103036. DOI: 10.1016/j.jwpe.2022.103036 PMID: 35966450

Essa, S., Siddique, I., Saad, M., & Raghupathy, R. (2021). Modulation of Production of Th1/Th2 Cytokines in Peripheral Blood Mononuclear Cells and Neutrophils by Hepatitis C Virus Infection in Chronically Infected Patients. *Pathogens (Basel, Switzerland), 10*(11), 1519. DOI: 10.3390/pathogens10111519 PMID: 34832674

Estévez-Herrera, J., Pérez-Yanes, S., Cabrera-Rodríguez, R., Márquez-Arce, D., Trujillo-González, R., Machado, J. D., Madrid, R., & Valenzuela-Fernández, A. (2021). Zika Virus Pathogenesis: A Battle for Immune Evasion. *Vaccines*, *9*(3), 294. DOI: 10.3390/vaccines9030294 PMID: 33810028

European Centre for Disease Prevention and Control. (2023). *Dengue cases January–December 2023.*

Fader, K. A., Zhang, J., Menetski, J. P., Thadhani, R. I., Antman, E. M., Friedman, G. S., Ramaiah, S. K., & Vaidya, V. S. (2021, January). A Biomarker-Centric Approach to Drug Discovery and Development: Lessons Learned from the Coronavirus Disease 2019 Pandemic. *The Journal of Pharmacology and Experimental Therapeutics*, *376*(1), 12–20. DOI: 10.1124/jpet.120.000204 PMID: 33115823

Faivre, N., Verollet, C., & Dumas, F. (2024). The chemokine receptor CCR5: Multifaceted hook for HIV-1. *Retrovirology*, *21*(2), 2. Advance online publication. DOI: 10.1186/s12977-024-00634-1 PMID: 38263120

Fallarino, L. (2018). Anti HIV-1 gene therapy approach combining multiple siRNAs with the membrane-anchored fusion inhibitor C-peptide maC46.

Farah, F. H., & Farah, F. H. (2019). Nanocarriers as delivery systems for therapeutics agents. *International Journal of Pharmaceutical Sciences and Research*, *10*, 3487–3507.

Faria, M. J., Lopes, C. M., das Neves, J., & Lúcio, M. (2021). Lipid nanocarriers for anti-HIV therapeutics: A focus on physicochemical properties and biotechnological advances. *Pharmaceutics*, *13*(8), 1294. DOI: 10.3390/pharmaceutics13081294 PMID: 34452255

Fasano, M., Ieva, F., Ciarallo, M., Caccianotti, B., & Santantonio, T. A. (2024). Acute Hepatitis C: Current Status and Future Perspectives. *Viruses*, *16*(11), 1739. DOI: 10.3390/v16111739 PMID: 39599853

Fatima, H., Singh, D., Muhammad, H., Acharya, S., & Aziz, M. A. (2025). Improving the use of CRISPR/Cas9 gene editing machinery as a cancer therapeutic tool with the help of nanomedicine. 3 Biotech, 15(1), 17. https://doi.org/DOI: 10.1007/s13205-024-04186-1

Feld, J. J., & Jacobson, I. M. (2020). The evolution of hepatitis C treatment. *Nature Reviews. Gastroenterology & Hepatology*, *17*(10), 533–548. DOI: 10.1038/s41575-020-0313-6

Feng, J. Y., Du Pont, V., Babusis, D., Gordon, C. J., Tchesnokov, E. P., Perry, J. K., Duong, V., Vijjapurapu, A., Zhao, X., Chan, J., Cohen, C., Juneja, K., Cihlar, T., Götte, M., & Bilello, J. P. (2022). The Nucleoside/Nucleotide Analogs Tenofovir and Emtricitabine Are Inactive against SARS-CoV-2. *Molecules (Basel, Switzerland), 27*(13), 4212. DOI: 10.3390/molecules27134212 PMID: 35807457

Figueroa, S. M., Fleischmann, D., & Goepferich, A. (2021). Biomedical nanoparticle design: What we can learn from viruses. *Journal of Controlled Release, 329*, 552–569. DOI: 10.1016/j.jconrel.2020.09.045 PMID: 33007365

Financial Times. (2024). The Covid-era tech that could reinvent cancer care. Retrieved from https://www.ft.com/content/efff95f4-e7c4-4a3c-bccd-f449fecac27e

Finkel, R. S., Mercuri, E., Darras, B. T., Connolly, A. M., Kuntz, N. L., Kirschner, J., Chiriboga, C. A., Saito, K., Servais, L., Tizzano, E., Topaloglu, H., Tulinius, M., Montes, J., Glanzman, A. M., Bishop, K., Zhong, Z. J., Gheuens, S., Bennett, C. F., Schneider, E., & De Vivo, D. C. (2017). Nusinersen versus sham control in infantile-onset spinal muscular atrophy. *The New England Journal of Medicine, 377*(18), 1723–1732. DOI: 10.1056/NEJMoa1702752 PMID: 29091570

Fisher, C. A., & Zorn, M. (2018). The potential of mobile health technology in HIV care. *American Journal of Public Health, 108*(5), 609–611. PMID: 29617612

Flies, E. J., Toi, C., Weinstein, P., Doggett, S. L., & Williams, C. R. (2015). Converting Mosquito Surveillance to Arbovirus Surveillance with Honey-Baited Nucleic Acid Preservation Cards. *Vector Borne and Zoonotic Diseases (Larchmont, N.Y.), 15*(7), 397–403. DOI: 10.1089/vbz.2014.1759 PMID: 26186511

Flisiak, R., Jaroszewicz, J., & Łucejko, M. (2018). siRNA drug development against hepatitis B virus infection. *Expert Opinion on Biological Therapy, 18*(6), 609–617. DOI: 10.1080/14712598.2018.1472231 PMID: 29718723

Florio, M., & Gamba, S. (2021). Biomed Europa: After the coronavirus, a public infrastructure to overcome the pharmaceutical oligopoly. *Annals of Public and Cooperative Economics, 92*(3), 387–409. DOI: 10.1111/apce.12341

Foo, C. S., Abdelnabi, R., Kaptein, S. J. F., Zhang, X., Ter Horst, S., Mols, R., Delang, L., Rocha-Pereira, J., Coelmont, L., Leyssen, P., Dallmeier, K., Vergote, V., Heylen, E., Vangeel, L., Chatterjee, A. K., Annaert, P. P., Augustijns, P. F., De Jonghe, S., Jochmans, D., & Neyts, J. (2022). HIV protease inhibitors Nelfinavir and Lopinavir/Ritonavir markedly improve lung pathology in SARS-CoV-2-infected Syrian hamsters despite lack of an antiviral effect. *Antiviral Research, 202*, 105311. DOI: 10.1016/j.antiviral.2022.105311 PMID: 35390430

Forrester, N. L., Wertheim, J. O., Dugan, V. G., Auguste, A. J., Lin, D., Adams, A. P., Chen, R., Gorchakov, R., Leal, G., Estrada-Franco, J. G., Pandya, J., Halpin, R. A., Hari, K., Jain, R., Stockwell, T. B., Das, S. R., Wentworth, D. E., Smith, M. D., Kosakovsky Pond, S. L., & Weaver, S. C. (2017). Evolution and spread of Venezuelan equine encephalitis complex alphavirus in the Americas. *PLoS Neglected Tropical Diseases*, *11*(8), e0005693. DOI: 10.1371/journal.pntd.0005693 PMID: 28771475

Frericks, N., Klöhn, M., Lange, F., Pottkämper, L., Carpentier, A., & Steinmann, E. (2025). Host-targeting antivirals for chronic viral infections of the liver. *Antiviral Research*, *234*, 106062. DOI: 10.1016/j.antiviral.2024.106062 PMID: 39716667

Gagnon, M., & Holmes, D. (2016). Body–drug assemblages: Theorizing the experience of side effects in the context of HIV treatment. *Nursing Philosophy*, *17*(4), 250–261. DOI: 10.1111/nup.12136 PMID: 27435229

Gangopadhayya, A., & Bhukya, P. L. (2023). Factors Contributing to the Emergence of Viral Diseases. In *Emerging Human Viral Diseases, Volume I: Respiratory and Haemorrhagic Fever* (pp. 3-69). Springer. DOI: 10.1007/978-981-99-2820-0_1

Gangwal, A., Ansari, A., Ahmad, I., Azad, A. K., & Wan Sulaiman, W. M. A. (2024, September). Current strategies to address data scarcity in artificial intelligence-based drug discovery: A comprehensive review. *Computers in Biology and Medicine*, *179*, 108734. DOI: 10.1016/j.compbiomed.2024.108734 PMID: 38964243

García-Cruz, J. C., Huelgas-Méndez, D., Jiménez-Zúñiga, J. S., Rebollar-Juárez, X., Hernández-Garnica, M., Fernández-Presas, A. M., Husain, F. M., Alenazy, R., Alqasmi, M., Albalawi, T., Alam, P., & García-Contreras, R. (2023). Myriad applications of bacteriophages beyond phage therapy. *PeerJ*, *11*, e15272. DOI: 10.7717/peerj.15272 PMID: 37101788

Garnett, L. E., & Strong, J. E. (2019). Lassa fever: With 50 years of study, hundreds of thousands of patients and an extremely high disease burden, what have we learned? *Current Opinion in Virology*, *37*, 123–131. DOI: 10.1016/j.coviro.2019.07.009 PMID: 31479990

Gastañaduy, P. A.. (2019). Measles — United States, January 1–April 26, 2019. *Morbidity and Mortality Weekly Report*, *68*(17), 396–400. DOI: 10.15585/mmwr.mm6817e2 PMID: 31048675

Gavi, The Vaccine Alliance. (2020, September 17). *India draws lessons from polio eradication initiative*. Retrieved March 28, 2025, from https://www.gavi.org/vaccineswork/india-draws-lessons-polio-eradication-initiative

Gavi, the Vaccine Alliance. (2023). *Annual Progress Report 2023*. Retrieved March 28, 2025, from https://www.gavi.org/progress-report

Gawande, M. S., Zade, N., Kumar, P., Gundewar, S., Weerarathna, I. N., & Verma, P. (2025). The role of artificial intelligence in pandemic responses: From epidemiological modeling to vaccine development. *Molecular Biomedicine*, *6*(1), 1. DOI: 10.1186/s43556-024-00238-3 PMID: 39747786

Gazal, S., Sharma, N., Gazal, S., Tikoo, M., Shikha, D., Badroo, G. A., Rashid, M., & Lee, S. J. (2022). Nipah and Hendra viruses: Deadly zoonotic paramyxoviruses with the potential to cause the next pandemic. *Pathogens (Basel, Switzerland)*, *11*(12), 1419. DOI: 10.3390/pathogens11121419 PMID: 36558753

Geng, N., Ma, L., Jin, Y., Lu, J., Zheng, Y., Wang, J., Wang, X., & Chen, X. (2024). Prediction Model for the Clearance of Hepatitis B Surface Antigen in Patients with Chronic Hepatitis B before Interferon Therapy: A Prospective Case–Control Study. *Diagnostics (Basel)*, *14*(1), 118. DOI: 10.3390/diagnostics14010118 PMID: 38201427

Getachew, E., Woldeamanuel, Y., & Manyazewal, T. (2022). Capacity and readiness assessment of healthcare facilities for digital health interventions against tuberculosis and HIV in Addis Ababa, Ethiopia. *Frontiers in Digital Health*, *4*, 821390. DOI: 10.3389/fdgth.2022.821390 PMID: 35295619

Ghamdi, E., MashariAayadhAlotaibi, K. N. A., MeshalBijadAlotaibi, M., Mufadhi-Alanazi, F., Asiree, A. H., Malki, S. M., & Mutairi, A. M. (2023). A comprehensive analysis of global health challenges: Public health interventions, vaccine distribution hurdles, and the socio-economic roots of health disparities. *Chelonian Research Foundation*, *18*(1), 272–282.

Ghandour, G., & Fawzy, M. (2021). COVID-19 and mobile health applications: A step forward in digital health innovation. *Journal of Medical Internet Research*, *23*(6), e23732.

Ghandour, M., & Bernard, A. (2020). Mobile health interventions for managing HIV and viral hepatitis: A systematic review. *BMC Infectious Diseases*, *20*, 516.

Ghasemi, M., Amini-Rarani, M., Shaarbafchi Zadeh, N., & Karimi, S. (2022). Role of Public–Private Partnerships in Primary Healthcare Services Worldwide: A Scoping Review. *Health Scope*, *11*(3). Advance online publication. DOI: 10.5812/jhealthscope-129176

Ghosh, U., Sayef Ahammed, K., Mishra, S., & Bhaumik, A. (2022). The emerging roles of silver nanoparticles to target viral life cycle and detect viral pathogens. *Chemistry, an Asian Journal, 17*(5), e202101149. DOI: 10.1002/asia.202101149 PMID: 35020270

Giladi, H., Ketzinel-Gilad, M., Rivkin, L., Felig, Y., Nussbaum, O., & Galun, E. (2003). Small interfering RNA inhibits hepatitis B virus replication in mice. *Molecular Therapy : the Journal of the American Society of Gene Therapy, 8*(5), 769–776. DOI: 10.1016/S1525-0016(03)00244-2 PMID: 14599810

Giovanetti, M., Pannella, G., Altomare, A., Rocchi, G., Guarino, M., Ciccozzi, M., Riva, E., & Gherardi, G. (2024). Exploring the Interplay between COVID-19 and Gut Health: The Potential Role of Prebiotics and Probiotics in Immune Support. *Viruses, 16*(3), 370. DOI: 10.3390/v16030370 PMID: 38543736

Glasgow, A., Glasgow, J., Limonta, D., Solomon, P., Lui, I., Zhang, Y., Nix, M. A., Rettko, N. J., Zha, S., Yamin, R., Kao, K., Rosenberg, O. S., Ravetch, J. V., Wiita, A. P., Leung, K. K., Lim, S. A., Zhou, X. X., Hobman, T. C., Kortemme, T., & Wells, J. A. (2020). Engineered ACE2 receptor traps potently neutralize SARS-CoV-2. *Proceedings of the National Academy of Sciences of the United States of America, 117*(45), 28046–28055. DOI: 10.1073/pnas.2016093117 PMID: 33093202

Global Polio Eradication Initiative. (2022). *Polio eradication: Status report.* https://polioeradication.org/

Global Polio Eradication Initiative. (2023). *Global Polio Eradication Initiative: Annual Report 2022.* World Health Organization. https://polioeradication.org/resources/annual-report-2022-pre-publication-version/contentReference[oaicite:1]index=1

Godwin, P. O., Polsonetti, B., Caron, M. F., & Oppelt, T. F. (2024). Remdesivir for the Treatment of COVID-19: A Narrative Review. *Infectious Diseases and Therapy, 13*(1), 1–19. DOI: 10.1007/s40121-023-00900-3 PMID: 38193988

Gorbea, C., Mosbruger, T., & Cazalla, D. (2017). A viral Sm-class RNA base-pairs with mRNAs and recruits microRNAs to inhibit apoptosis. *Nature, 550*(7675), 275–279. DOI: 10.1038/nature24034 PMID: 28976967

Gorib, S., & Ayaz, M. (2020). Digital health platforms in the management of infectious diseases: A review. *Telemedicine Journal and e-Health, 26*(3), 250–256.

Gour, N., & Sharma, S. (2020). Artificial intelligence and its role in advancing antiviral therapies. *Future Microbiology, 15*(12), 981–993.

Graham, B. S., & Sullivan, N. J. (2018). Emerging viral diseases from a vaccinology perspective: Preparing for the next pandemic. *Nature Immunology*, *19*(1), 20–28. DOI: 10.1038/s41590-017-0007-9 PMID: 29199281

Greenwood, B. (2014). The contribution of vaccination to global health: Past, present and future. *Philosophical Transactions of the Royal Society of London. Series B, Biological Sciences*, *369*(1645), 20130433. DOI: 10.1098/rstb.2013.0433 PMID: 24821919

Gubler, D. J. (2002). Epidemic dengue/dengue hemorrhagic fever as a public health, social and economic problem in the 21st century. *Trends in Microbiology*, *10*(2), 100–103. DOI: 10.1016/S0966-842X(01)02288-0 PMID: 11827812

Gul, S., Batool, Z., Bano, S., Aziz, A., & Tanveer, S. H. (2024). Comprehensive Details on Human Papillomavirus (HPV) Infections and Associated Malignancies: Awareness about Prophylaxis & Treatment Options Available for HPV Infections. *Bangladesh Journal of Medical Science*, *23*(4), 967–983. DOI: 10.3329/bjms.v23i4.76504

Gunter, H. M.. (2021). mRNA vaccines for COVID-19: A new era of vaccinology. *Current Opinion in Immunology*, *71*, 63–69. DOI: 10.1016/j.coi.2021.07.001

Gupta, M., Singh, R., & Desai, N. (2020). Mobile health applications for managing antiviral therapy: A review of the literature. *Health Informatics Journal*, *26*(3), 311–326.

Gupta, P., & Sharma, A. (2021). The future of antiviral drugs and mHealth applications in managing viral diseases. *Health and Technology*, *11*(3), 281–289.

Gurunathan, S., Qasim, M., Choi, Y., Do, J. T., Park, C., Hong, K., Kim, J. H., & Song, H. (2020). Antiviral potential of nanoparticles—Can nanoparticles fight against coronaviruses? *Nanomaterials (Basel, Switzerland)*, *10*(9), 1645. DOI: 10.3390/nano10091645 PMID: 32825737

Gu, W., Wu, C., Chen, J., & Xiao, Y. (2013). Nanotechnology in the targeted drug delivery for bone diseases and bone regeneration. *International Journal of Nanomedicine*, •••, 2305–2317. DOI: 10.2147/IJN.S44393 PMID: 23836972

Haas, P., Muralidharan, M., Krogan, N. J., Kaake, R. M., & Hüttenhain, R. (2021). Proteomic Approaches to Study SARS-CoV-2 Biology and COVID-19 Pathology. *Journal of Proteome Research*, *20*(2), 1133–1152. DOI: 10.1021/acs.jproteome.0c00764 PMID: 33464917

Hadi, R., Poddar, A., Sonnaila, S., Bhavaraju, V. S. M., & Agrawal, S. (2024). Advancing CRISPR-Based Solutions for COVID-19 Diagnosis and Therapeutics. *Cells*, *13*(21), 1794. DOI: 10.3390/cells13211794 PMID: 39513901

Haghani, M., Bliemer, M. C. J., Goerlandt, F., & Li, J. (2020, September). The scientific literature on Coronaviruses, COVID-19 and its associated safety-related research dimensions: A scientometric analysis and scoping review. *Safety Science*, *129*, 104806. DOI: 10.1016/j.ssci.2020.104806 PMID: 32382213

Hagiwara, S., Nishida, N., & Kudo, M. (2015). Antiviral therapy for chronic hepatitis B: Combination of nucleoside analogs and interferon. *World Journal of Hepatology*, *7*(23), 2427–2431. DOI: 10.4254/wjh.v7.i23.2427 PMID: 26483864

Haidar, G.. (2021). Booster Doses and COVID-19 Vaccine Effectiveness in High-Risk Populations. *Journal of the American Medical Association*, *326*(11), 1–5.

Haidar, G., Dorritie, K., Fusco, H., Bogdanovich, T., & Samanta, P. (2021). Invasive mold infections after chimeric antigen receptor-modified T-cell therapy: A case series, review of the literature, and implications for prophylaxis. *Clinical Infectious Diseases*, *73*(11), e1369–e1374. DOI: 10.1093/cid/ciaa1328 PMID: 32780792

Hameed, M., Najafi, M., Cheeti, S., Sheokand, A., Mago, A., & Desai, S. (2022, November). Factors influencing international collaboration on the prevention of COVID-19. *Public Health*, *212*, 95–101. DOI: 10.1016/j.puhe.2022.08.017 PMID: 36272205

Hannan, A., & Rahman, M. (2021). mHealth interventions for managing chronic viral diseases: A systematic review. *Journal of Medical Internet Research*, *23*(7), e24653.

Hansen, J., Baum, A., Pascal, K. E., Russo, V., Giordano, S., Wloga, E., Fulton, B. O., Yan, Y., Koon, K., Patel, K., Chung, K. M., Hermann, A., Ullman, E., Cruz, J., Rafique, A., Huang, T., Fairhurst, J., Libertiny, C., Malbec, M., & Kyratsous, C. A. (2020). Studies in humanized mice and convalescent humans yield a SARS-CoV-2 antibody cocktail. *Science*, *369*(6506), 1010–1014. DOI: 10.1126/science.abd0827 PMID: 32540901

Harding, F. A., Stickler, M. M., Razo, J., & DuBridge, R. B. (2010). The immunogenicity of humanized and fully human antibodies: Residual immunogenicity resides in the CDR regions. *mAbs*, *2*(3), 256–265. DOI: 10.4161/mabs.2.3.11641 PMID: 20400861

Hasanzadeh, A., Hamblin, M. R., Kiani, J., Noori, H., Hardie, J. M., Karimi, M., & Shafiee, H. (2022). Could artificial intelligence revolutionize the development of nanovectors for gene therapy and mRNA vaccines? *Nano Today*, *47*, 101665. DOI: 10.1016/j.nantod.2022.101665 PMID: 37034382

Hashem, H., Abufaraj, M., Tbakhi, A., & Sultan, I. (2020). Obstacles and Considerations Related to Clinical Trial Research During the COVID-19 Pandemic. *Frontiers in Medicine*, *7*, 598038. DOI: 10.3389/fmed.2020.598038 PMID: 33425945

Hawley, C., & Bonner, S. (2019). Telemedicine for hepatitis C treatment in underserved populations. *Journal of Viral Hepatitis*, *26*(6), 756–763.

Healy, C. M.. (2021). Safety and immunogenicity of the Novavax COVID-19 vaccine. *The New England Journal of Medicine*, *384*(24), 2320–2331. DOI: 10.1056/NEJMoa2103061

Hegazy, S. K., Tharwat, S., & Hassan, A. H. (2023). Comparing the efficacy of regen-cov, remdesivir, and favipiravir in reducing invasive mechanical ventilation need in hospitalized COVID-19 patients. *World Journal of Clinical Cases*, *11*(26), 6105–6121. DOI: 10.12998/wjcc.v11.i26.6105 PMID: 37731581

Henao-Restrepo, A. M., Camacho, A., Longini, I. M., Watson, C. H., Edmunds, W. J., Egger, M., Carroll, M. W., Dean, N. E., Diatta, I., Doumbia, M., Draguez, B., Duraffour, S., Enwere, G., Grais, R., Gunther, S., Gsell, P.-S., Hossmann, S., Watle, S. V., Kondé, M. K., & Kieny, M.-P. (2017). Efficacy and effectiveness of an rVSV-vectored vaccine in preventing Ebola virus disease: Final results from the Guinea ring vaccination, open-label, cluster-randomised trial (Ebola Ça Suffit!). *Lancet*, *389*(10068), 505–518. DOI: 10.1016/S0140-6736(16)32621-6 PMID: 28017403

Heneghan, C. J., Onakpoya, I., Jones, M. A., Doshi, P., Del Mar, C. B., Hama, R., Thompson, M. J., Spencer, E. A., Mahtani, K. R., Nunan, D., Howick, J., & Jefferson, T. (2016). Neuraminidase inhibitors for influenza: A systematic review and meta-analysis of regulatory and mortality data. *Health Technology Assessment*, *20*(42), 1–242. DOI: 10.3310/hta20420 PMID: 27246259

Herder, M., Graham, J. E., & Gold, R. (2020). From discovery to delivery: Public sector development of the r VSV-ZEBOV Ebola vaccine. *Journal of Law and the Biosciences*, *7*(1), lsz019. DOI: 10.1093/jlb/lsz019 PMID: 34221434

Hermann, T. (2016). Small molecules targeting viral RNA. *Wiley Interdisciplinary Reviews. RNA*, *7*(6), 726–743. DOI: 10.1002/wrna.1373 PMID: 27307213

Hernandez Carballo, I., Bakola, M., & Stuckler, D. (2022). The impact of air pollution on COVID-19 incidence, severity, and mortality: A systematic review of studies in Europe and North America. *Environmental Research*, *215*(Pt 1), 114155. DOI: 10.1016/j.envres.2022.114155 PMID: 36030916

Hernandez, A. V., Roman, Y. M., Pasupuleti, V., Barboza, J. J., & White, C. M. (2020). Hydroxychloroquine or chloroquine for treatment or prophylaxis of COVID-19: A living systematic review. *Annals of Internal Medicine*, *173*(4), 287–296. DOI: 10.7326/M20-2496 PMID: 32459529

Hernandez, J., & Roberts, L. (2021). Telemedicine and its role in the future of antiviral care: A review. *Telemedicine Journal and e-Health*, *27*(7), 781–788.

He, Y., Shen, M., Wang, X., Yin, A., Liu, B., Zhu, J., & Zhang, Z. (2024). Suppression of Interferon Response and Antiviral Strategies of Bunyaviruses. *Tropical Medicine and Infectious Disease*, *9*(9), 205. DOI: 10.3390/tropicalmed9090205 PMID: 39330894

He, Y., Zhou, J., Gao, H., Liu, C., Zhan, P., & Liu, X. (2024, February 5). Broad-spectrum antiviral strategy: Host-targeting antivirals against emerging and re-emerging viruses. *European Journal of Medicinal Chemistry*, *265*, 116069. DOI: 10.1016/j.ejmech.2023.116069 PMID: 38160620

Himeidan, Y. E., Kweka, E. J., Mahgoub, M. M., El Rayah, E. A., & Ouma, J. O. (2014). Recent outbreaks of rift valley Fever in East Africa and the middle East. *Frontiers in Public Health*, *2*, 169. DOI: 10.3389/fpubh.2014.00169 PMID: 25340047

Hirsch, M. S., & Wensing, A. M. (2019). Management of HIV-1 infection in adults. *Lancet*, *394*(10200), 381–392.

Hoang, P. T., Luong, Q. X. T., Ayun, R. Q., Lee, Y., Oh, K. J., Kim, T., Lee, T. K., & Lee, S. (2024). A synergistic therapy against influenza virus A/H1N1/PR8 by a HA1 specific neutralizing single-domain VL and an RNA hydrolyzing scFv. *Frontiers in Microbiology*, *15*, 1355599. DOI: 10.3389/fmicb.2024.1355599 PMID: 38706966

Ho, D., Quake, S. R., McCabe, E. R. B., Chng, W. J., Chow, E. K., Ding, X., Gelb, B. D., Ginsburg, G. S., Hassenstab, J., Ho, C. M., Mobley, W. C., Nolan, G. P., Rosen, S. T., Tan, P., Yen, Y., & Zarrinpar, A. (2020). Enabling Technologies for Personalized and Precision Medicine. *Trends in Biotechnology*, *38*(5), 497–518. DOI: 10.1016/j.tibtech.2019.12.021 PMID: 31980301

Hoffmann Dahl, E., Mbala, P., Juchet, S., Touré, A., Montoyo, A., Serra, B., Kojan, R., D'Ortenzio, E., Blomberg, B., & Jaspard, M. (2024). Improving Ebola virus disease outbreak control through targeted postexposure prophylaxis. *The Lancet. Global Health, 12*(10), e1730–e1736. DOI: 10.1016/S2214-109X(24)00255-9 PMID: 39270687

Hogan, M. J., & Pardi, N. (2022). mRNA vaccines in the COVID-19 pandemic and beyond. *Annual Review of Medicine, 73*(1), 17–39. DOI: 10.1146/annurev-med-042420-112725 PMID: 34669432

Holmes, E. C. (2009). The evolutionary genetics of emerging viruses. *Annual Review of Ecology, Evolution, and Systematics, 40*(1), 353–372. DOI: 10.1146/annurev.ecolsys.110308.120248

Holmes, E. C., Dudas, G., Rambaut, A., & Andersen, K. G. (2016). The evolution of Ebola virus: Insights from the 2013-2016 epidemic. *Nature, 538*(7624), 193–200. DOI: 10.1038/nature19790 PMID: 27734858

Hossain, M. M., & Alam, M. (2021). Role of bioinformatics in antiviral drug discovery and development. *Current Drug Targets, 22*(2), 168–179.

Hossain, M. S., Hami, I., Sawrav, M. S., Rabbi, M. F., Saha, O., Bahadur, N. M., & Rahaman, M. M. (2020). Drug repurposing for prevention and treatment of COVID-19: A clinical landscape. *Discoveries (Craiova, Romania), 8*(4), e121. DOI: 10.15190/d.2020.18 PMID: 33403227

Hou, Y. J., Chiba, S., Halfmann, P., Ehre, C., Kuroda, M., Dinnon, K. H.III, Leist, S. R., Schäfer, A., Nakajima, N., Takahashi, K., Lee, R. E., Mascenik, T. M., Graham, R., Edwards, C. E., Tse, L. V., Okuda, K., Markmann, A. J., Bartelt, L., de Silva, A., & Baric, R. S. (2020). SARS-CoV-2 D614G variant exhibits efficient replication ex vivo and transmission in vivo. *Science, 370*(6523), 1464–1468. DOI: 10.1126/science.abe8499 PMID: 33184236

Howe, A. Y. M., Rodrigo, C., Cunningham, E. B., Douglas, M. W., Dietz, J., Grebely, J., Popping, S., Sfalcin, J. A., Parczewski, M., Sarrazin, C., de Salazar, A., Fuentes, A., Sayan, M., Quer, J., Kjellin, M., Kileng, H., Mor, O., Lennerstrand, J., Fourati, S., & Collaborators, S. H. A. R. E. D. (2022). Characteristics of hepatitis C virus resistance in an international cohort after a decade of direct-acting antivirals. *JHEP Reports : Innovation in Hepatology, 4*(5), 100462. DOI: 10.1016/j.jhepr.2022.100462 PMID: 35434589

Huang, C., & Zhao, M. (2021). The impact of digital health on the treatment of viral hepatitis. *Liver International, 41*(4), 668–676.

Huang, C., Zhao, M., & Wang, L. (2020). Wearable devices for monitoring adherence to antiviral therapy. *BMC Infectious Diseases*, *20*, 481.

Hua, Y., Sahashi, K., Hung, G., Rigo, F., Passini, M. A., Bennett, C. F., & Krainer, A. R. (2010). Antisense correction of SMN2 splicing in the CNS rescues necrosis in a type III SMA mouse model. *Genes & Development*, *24*(15), 1634–1644. DOI: 10.1101/gad.1941310 PMID: 20624852

Hung, I. F., Lung, K. C., Tso, E. Y., Liu, R., Chung, T. W., Chu, M. Y., Ng, Y. Y., Lo, J., Chan, J., Tam, A. R., Shum, H. P., Chan, V., Wu, A. K., Sin, K. M., Leung, W. S., Law, W. L., Lung, D. C., Sin, S., Yeung, P., & Yuen, K. Y. (2020). Triple combination of interferon beta-1b, lopinavir-ritonavir, and ribavirin in the treatment of patients admitted to hospital with COVID-19: An open-label, randomised, phase 2 trial. *Lancet*, *395*(10238), 1695–1704. DOI: 10.1016/S0140-6736(20)31042-4 PMID: 32401715

Hung, S. K., Wu, C. C., Singh, A., Li, J. H., Lee, C., Chou, E. H., Pekosz, A., Rothman, R., & Chen, K. F. (2023). Developing and validating clinical features-based machine learning algorithms to predict influenza infection in influenza-like illness patients. *Biomedical Journal*, *46*(5), 100561. DOI: 10.1016/j.bj.2022.09.002 PMID: 36150651

Hussein, M., Molina, M. A., Berkhout, B., & Herrera-Carrillo, E. (2023, January 13). A CRISPR–Cas Cure for HIV/AIDS. *International Journal of Molecular Sciences*, *24*(2), 1563. DOI: 10.3390/ijms24021563 PMID: 36675077

Hussen, B. M., Najmadden, Z. B., Abdullah, S. R., Rasul, M. F., Mustafa, S. A., Ghafouri-Fard, S., & Taheri, M. (2024). CRISPR/Cas9 gene editing: A novel strategy for fighting drug resistance in respiratory disorders. *Cell Communication and Signaling : CCS*, *22*(1), 329. DOI: 10.1186/s12964-024-01713-8 PMID: 38877530

Hyun, J., Lee, S. K., Kim, J. H., Cho, E.-J., Kim, H.-S., Kim, J.-S., Song, W., & Kim, H. S. (2023). Variant Analysis of the Thymidine Kinase and DNA Polymerase Genes of Herpes Simplex Virus in Korea: Frequency of Acyclovir Resistance Mutations. *Viruses*, *15*(8), 1709. DOI: 10.3390/v15081709 PMID: 37632051

Ikematsu, H., Baba, T., Saito, M. M., Kinoshita, M., Miyazawa, S., Hata, A., Nakano, S., Kitanishi, Y., & Hayden, F. G. (2024). Comparative Effectiveness of Baloxavir Marboxil and Oseltamivir Treatment in Reducing Household Transmission of Influenza: A Post Hoc Analysis of the BLOCKSTONE Trial. *Influenza and Other Respiratory Viruses*, *18*(5), e13302. DOI: 10.1111/irv.13302 PMID: 38706384

Ilunga Kalenga, O., Moeti, M., Sparrow, A., Nguyen, V. K., Lucey, D., & Ghebreyesus, T. A. (2019). The Ongoing Ebola Epidemic in the Democratic Republic of Congo, 2018-2019. *The New England Journal of Medicine, 381*(4), 373–383. DOI: 10.1056/NEJMsr1904253 PMID: 31141654

Inoue, Y., Tanaka, N., Tanaka, Y., Inoue, S., Morita, K., Zhuang, M., Hattori, T., & Sugamura, K. (2007). Clathrin-dependent entry of severe acute respiratory syndrome coronavirus into target cells expressing ACE2 with the cytoplasmic tail deleted. *Journal of Virology, 81*(16), 8722–8729. DOI: 10.1128/JVI.00253-07 PMID: 17522231

Ioele, G., Chieffallo, M., Occhiuzzi, M. A., De Luca, M., Garofalo, A., Ragno, G., & Grande, F. (2022). Anticancer drugs: Recent strategies to improve stability profile, pharmacokinetic and pharmacodynamic properties. *Molecules (Basel, Switzerland), 27*(17), 5436. DOI: 10.3390/molecules27175436 PMID: 36080203

Irwin, K. K., Renzette, N., Kowalik, T. F., & Jensen, J. D. (2016). Antiviral drug resistance as an adaptive process. *Virus Evolution, 2*(1), vew014. DOI: 10.1093/ve/vew014 PMID: 28694997

Ison, M. G. (2016). Antiviral treatments. *Clinics in Chest Medicine, 38*(1), 139–153. DOI: 10.1016/j.ccm.2016.11.008 PMID: 28159156

Ivanov, A. I. (2008). Pharmacological inhibition of endocytic pathways: Is it specific enough to be useful? *Methods in Molecular Biology (Clifton, N.J.), 440*, 15–33. DOI: 10.1007/978-1-59745-178-9_2 PMID: 18369934

Jacobsen, K. H., Aguirre, A. A., Bailey, C. L., Baranova, A. V., Crooks, A. T., Croitoru, A., Delamater, P. L., Gupta, J., Kehn-Hall, K., Narayanan, A., Pierobon, M., Rowan, K. E., Schwebach, J. R., Seshaiyer, P., Sklarew, D. M., Stefanidis, A., & Agouris, P. (2016). Lessons from the Ebola outbreak: Action items for emerging infectious disease preparedness and response. *EcoHealth, 13*(1), 200–212. DOI: 10.1007/s10393-016-1100-5 PMID: 26915507

Jan, E., Mohr, I., & Walsh, D. (2016). A Cap-to-Tail Guide to mRNA Translation Strategies in Virus-Infected Cells. *Annual Review of Virology, 3*(1), 283–307. DOI: 10.1146/annurev-virology-100114-055014 PMID: 27501262

Jennings, L., Kellermann, T., Spinelli, M., Nkantsu, Z., Cogill, D., van Schalkwyk, M., Decloedt, E., van Zyl, G., Orrell, C., & Gandhi, M. (2022). Drug Resistance, Rather than Low Tenofovir Levels in Blood or Urine, Is Associated with Tenofovir, Emtricitabine, and Efavirenz Failure in Resource-Limited Settings. *AIDS Research and Human Retroviruses, 38*(6), 455–462. DOI: 10.1089/aid.2021.0135 PMID: 34779228

Jha, A., & Gupta, R. (2020). The potential of mRNA vaccines in tackling emerging viral diseases. *BioMed Research International, 8657983*. Advance online publication. DOI: 10.1155/2020/8657983

Jiang, H., Zhu, X., Jiao, J., Yan, C., Liu, K., Chen, W., & Qin, P. (2025). CRISPR/dCas9-based hotspot self-assembling SERS biosensor integrated with a smartphone for simultaneous, ultrasensitive and robust detection of multiple pathogens. *Biosensors & Bioelectronics, 270*, 116974. DOI: 10.1016/j.bios.2024.116974 PMID: 39586143

Ji, J. J., Sun, Q. M., Nie, D. Y., Wang, Q., Zhang, H., Qin, F. F., Wang, Q. S., Lu, S. F., Pang, G. M., & Lu, Z. G. (2021). Probiotics protect against RSV infection by modulating the microbiota-alveolar-macrophage axis. *Acta Pharmacologica Sinica, 42*(10), 1630–1641. DOI: 10.1038/s41401-020-00573-5 PMID: 33495515

Jit, M., Ananthakrishnan, A., McKee, M., Wouters, O. J., Beutels, P., & Teerawattananon, Y. (2021, October). Multicountry collaboration in responding to global infectious disease threats: Lessons for Europe from the COVID-19 pandemic. *The Lancet Regional Health. Europe, 9*, 100221. DOI: 10.1016/j.lanepe.2021.100221 PMID: 34642675

Johns Hopkins University. (2023). *COVID-19 dashboard.* https://coronavirus.jhu.edu/map.html

Johnson, D. B., Nebhan, C. A., Moslehi, J. J., & Balko, J. M. (2022). Immune-checkpoint inhibitors: Long-term implications of toxicity. *Nature Reviews. Clinical Oncology, 19*(4), 254–267. DOI: 10.1038/s41571-022-00600-w PMID: 35082367

Johnson, K. A., Zhang, H., & Wang, Y. (2023). mRNA-based vaccines: Innovations in lipid nanoparticle formulations and delivery mechanisms. *Nature Reviews. Drug Discovery, 22*(5), 345–358.

Johnson, M. L., & Gupta, S. (2021). Digital health interventions for enhancing antiviral adherence in HIV patients: A review. *International Journal of Medical Informatics, 150*, 104431.

Johnson, M. L., & Roberts, K. (2021). Overcoming the digital divide in health: Implications for antiviral treatment. *Public Health Reviews, 42*, 160166.

Jones, J. E., Le Sage, V., & Lakdawala, S. S. (2021). Viral and host heterogeneity and their effects on the viral life cycle. *Nature Reviews. Microbiology, 19*(5), 272–282. DOI: 10.1038/s41579-020-00449-9 PMID: 33024309

Jonsson, C. B., Figueiredo, L. T., & Vapalahti, O. (2010). A global perspective on hantavirus ecology, epidemiology, and disease. *Clinical Microbiology Reviews, 23*(2), 412–441. DOI: 10.1128/CMR.00062-09 PMID: 20375360

Jose-Abrego, A., Roman, S., Rebello Pinho, J. R., Gomes-Gouvêa, M. S., & Panduro, A. (2023). High Frequency of Antiviral Resistance Mutations in HBV Genotypes A2 and H: Multidrug Resistance Strains in Mexico. *Journal of Clinical and Translational Hepatology*, *11*(5), 1023–1034. DOI: 10.14218/JCTH.2022.00135S PMID: 37577226

Kalidasan, V., & Theva Das, K. (2020). Lessons Learned From Failures and Success Stories of HIV Breakthroughs: Are We Getting Closer to an HIV Cure? *Frontiers in Microbiology*, *11*, 46. DOI: 10.3389/fmicb.2020.00046 PMID: 32082282

Karban, Š. (2023). The evolution of SARS-CoV-2 virus during the pandemic of COVID-19.

Katsiki, A. D., Karatzas, P. E., De Lastic, H. X., Georgakilas, A. G., Tsitsilonis, O., & Vorgias, C. E. (2024). DExplore: An Online Tool for Detecting Differentially Expressed Genes from mRNA Microarray. *Experimental Biology*, *13*(5), 351. DOI: 10.3390/biology13050351 PMID: 38785833

Katz, J. M.. (2020). Oral vaccines: A practical and effective immunization strategy. *Nature Reviews. Immunology*, *20*(3), 197–210. DOI: 10.1038/s41577-019-0237-7

Katz, J.. (2019). Understanding vaccine hesitancy: A public health issue. *Health Affairs*, *38*(4), 587–592. DOI: 10.1377/hlthaff.2018.05248

Kaufman, E. B., & Kramer, L. D. (2017). Zika virus mosquito vectors: Competence, biology, and vector control. *The Journal of Infectious Diseases*, *216*(suppl_10), S976–S990. DOI: 10.1093/infdis/jix405 PMID: 29267910

Kaur, M., & Tiwari, S. (2020). Exploring the future of nanotechnology in antiviral drug delivery systems. *Nanomedicine; Nanotechnology, Biology, and Medicine*, *28*, 102178.

Kausar, S., Said Khan, F., Ishaq Mujeeb Ur Rehman, M., Akram, M., Riaz, M., Rasool, G., Hamid Khan, A., Saleem, I., Shamim, S., & Malik, A. (2021). A review: Mechanism of action of antiviral drugs. *International Journal of Immunopathology and Pharmacology*, *35*, 20587384211002621. DOI: 10.1177/20587384211002621 PMID: 33726557

Kefi, M., Cardoso-Jaime, V., Saab, S. A., & Dimopoulos, G. (2024). Curing mosquitoes with genetic approaches for malaria control. *Trends in Parasitology*, *40*(6), 487–499. DOI: 10.1016/j.pt.2024.04.010 PMID: 38760256

Kelso, J. M.. (2021). Anaphylaxis After Vaccination: Incidence, Mechanisms, and Management. *The Journal of Allergy and Clinical Immunology*, *147*(4), 1379–1389.

Keshavarzi Arshadi, A., Webb, J., Salem, M., Cruz, E., Calad-Thomson, S., Ghadirian, N., Collins, J., Diez-Cecilia, E., Kelly, B., Goodarzi, H., & Yuan, J. S. (2020). Artificial intelligence for COVID-19 drug discovery and vaccine development. *Frontiers in Artificial Intelligence*, *3*, 65. DOI: 10.3389/frai.2020.00065 PMID: 33733182

Khan, M. M., Ali, M. J., Hanif, H., Maqsood, M. H., Ahmad, I., Alvarez, J. E. G., Catana, M. A., & Lau, D. T. Y. (2022). The dilemma of cytomegalovirus and hepatitis B virus interaction. *Gastroenterology Report*, *10*, goac018. Advance online publication. DOI: 10.1093/gastro/goac018 PMID: 35663152

Khanna, A., El Barachi, M., Jain, S., Kumar, M., & Nayyar, A. (2024). *Artificial Intelligence and machine learning in drug design and development*. John Wiley & Sons. DOI: 10.1002/9781394234196

Khan, S., & Kamal, A. (2018). Telemedicine for HIV care: Enhancing treatment adherence in remote regions. *Global Health Action*, *11*(1), 1512972.

Khan, Y., Sadia, H., Ali Shah, S. Z., Khan, M. N., Shah, A. A., Ullah, N., Ullah, M. F., Bibi, H., Bafakeeh, O. T., Khedher, N. B., Eldin, S. M., Fadhl, B. M., & Khan, M. I. (2022). Classification, synthetic, and characterization approaches to nanoparticles, and their applications in various fields of nanotechnology: A review. *Catalysts*, *12*(11), 1386. DOI: 10.3390/catal12111386

Khatun, F., & Khalil, S. (2020). The role of mobile health applications in managing chronic diseases in low-income countries. *Health Information Science and Systems*, *8*(1), 6. DOI: 10.1007/s13755-020-00109-3 PMID: 31949893

Khoo, T., Lam, D., & Olynyk, J. K. (2021). Impact of modern antiviral therapy of chronic hepatitis B and C on clinical outcomes of liver disease. *World Journal of Gastroenterology*, *27*(29), 4831–4845. DOI: 10.3748/wjg.v27.i29.4831 PMID: 34447229

Khorram-Manesh, A., Burkle, F. M.Jr, & Goniewicz, K. (2024, August). Pandemics: past, present, and future: multitasking challenges in need of cross-disciplinary, transdisciplinary, and multidisciplinary collaborative solutions. *Osong Public Health and Research Perspectives*, *15*(4), 267–285. DOI: 10.24171/j.phrp.2023.0372 PMID: 39039818

Kim, D. H., & Rossi, J. J. (2007). Strategies for silencing human disease using RNA interference. *Nature Reviews. Genetics*, *8*(3), 173–184. DOI: 10.1038/nrg2006 PMID: 17304245

Kim, E. J., Moretti, M. E., Kimathi, A. M., Chan, S. Y., & Wootton, R. (2022). Use of provider-to-provider telemedicine in Kenya during the COVID-19 pandemic. *Frontiers in Public Health*, *10*, 1028999. DOI: 10.3389/fpubh.2022.1028999 PMID: 36478732

Kim, J. I., Park, S., Lee, I., Lee, S., Shin, S., Won, Y., Hwang, M. W., Bae, J. Y., Heo, J., Hyun, H. E., Jun, H., Lim, S. S., & Park, M. S. (2012). GFP-expressing influenza A virus for evaluation of the efficacy of antiviral agents. *Journal of Microbiology (Seoul, Korea)*, *50*(2), 359–362. DOI: 10.1007/s12275-012-2163-9 PMID: 22538668

Kim, J., Yeom, M., Lee, T., Kim, H. O., Na, W., Kang, A., Lim, J. W., Park, G., Park, C., Song, D., & Haam, S. (2020). Porous gold nanoparticles for attenuating infectivity of influenza A virus. *Journal of Nanobiotechnology*, *18*(1), 1. DOI: 10.1186/s12951-020-00611-8 PMID: 32209114

Kim, Y., Lee, H., & Park, J. (2021). Telemedicine and antiviral therapies: Enhancing access in underserved regions. *Telemedicine Journal and e-Health*, *27*(3), 251–260. PMID: 32857682

Kirchdoerfer, R. N., & Ward, A. B. (2019). Structure of the SARS-CoV nsp12 polymerase bound to nsp7 and nsp8 co-factors. *Nature Communications*, *10*(1), 2342. DOI: 10.1038/s41467-019-10280-3 PMID: 31138817

Kiso, M., Uraki, R., Yamayoshi, S., Imai, M., & Kawaoka, Y. (2024, September 20). Drug susceptibility and the potential for drug-resistant SARS-CoV-2 emergence in immunocompromised animals. *iScience*, *27*(9), 110729. DOI: 10.1016/j.isci.2024.110729 PMID: 39280602

Klatt, E. C. (2017). Pathology of HIV/AIDS. *The Mercer University of Medicine Savannah*, 6-50.

Klemm, T., Ebert, G., Calleja, D. J., Allison, C. C., Richardson, L. W., Bernardini, J. P., Lu, B. G., Kuchel, N. W., Grohmann, C., Shibata, Y., Gan, Z. Y., Cooney, J. P., Doerflinger, M., Au, A. E., Blackmore, T. R., van der Heden van Noort, G. J., Geurink, P. P., Ovaa, H., Newman, J., & Komander, D. (2020). Mechanism and inhibition of the papain-like protease, PLpro, of SARS-CoV-2. *The EMBO Journal*, *39*(18), e106275. DOI: 10.15252/embj.2020106275 PMID: 32845033

Kose, N., Fox, J. M., Sapparapu, G., Bombardi, R., Tennekoon, R. N., de Silva, A. D., Elbashir, S. M., Theisen, M. A., Humphris-Narayanan, E., Ciaramella, G., Himansu, S., Diamond, M. S., & Crowe, J. E.Jr. (2019). A lipid-encapsulated mRNA encoding a potently neutralizing human monoclonal antibody protects against chikungunya infection. *Science Immunology*, *4*(35), eaaw6647. Advance online publication. DOI: 10.1126/sciimmunol.aaw6647 PMID: 31101672

Kost, G. J. (2020). Geospatial hotspots need point-of-care strategies to stop highly infectious outbreaks: Ebola and coronavirus. *Archives of Pathology & Laboratory Medicine, 144*(10), 1166–1190. DOI: 10.5858/arpa.2020-0172-RA PMID: 32298139

Koujah, L., Shukla, D., & Naqvi, A. R. (2019). CRISPR–Cas based targeting of host and viral genes as an antiviral strategy. *Seminars in Cell & Developmental Biology, 96*, 53–64. DOI: 10.1016/j.semcdb.2019.04.004 PMID: 30953741

Kumar, A., Combe, E., Mougené, L., Zoulim, F., & Testoni, B. (2024). Applications of CRISPR/Cas as a Toolbox for Hepatitis B Virus Detection and Therapeutics. *Viruses, 16*(10), 1565. DOI: 10.3390/v16101565 PMID: 39459899

Kumari, M., Lu, R. M., Li, M. C., Huang, J. L., Hsu, F. F., Ko, S. H., Ke, F. Y., Su, S. C., Liang, K. H., Yuan, J. P., Chiang, H. L., Sun, C. P., Lee, I. J., Li, W. S., Hsieh, H. P., Tao, M. H., & Wu, H. C. (2022, September 12). A critical overview of current progress for COVID-19: Development of vaccines, antiviral drugs, and therapeutic antibodies. *Journal of Biomedical Science, 29*(1), 68. DOI: 10.1186/s12929-022-00852-9 PMID: 36096815

Kumar, K., & Rani, S. (2021). Impact of digital technologies in antiviral drug development: Current trends and future perspectives. *Biotechnology Progress, 37*(1), e3139.

Kumar, M., Kuroda, K., Dhangar, K., Mazumder, P., Sonne, C., Rinklebe, J., & Kitajima, M. (2020). Potential Emergence of Antiviral-Resistant Pandemic Viruses via Environmental Drug Exposure of Animal Reservoirs. *Environmental Science & Technology, 54*(14), 8503–8505. DOI: 10.1021/acs.est.0c03105 PMID: 32609508

Kumar, N., Sharma, S., Kumar, R., Tripathi, B. N., Barua, S., Ly, H., & Rouse, B. T. (2020). Host-Directed Antiviral Therapy. *Clinical Microbiology Reviews, 33*(3), e00168–e19. DOI: 10.1128/CMR.00168-19 PMID: 32404434

Kumar, S. (2024, June 27). Innovative Drug Discovery Research by Pharmaceutical Companies in India and China. *Journal of Medicinal Chemistry, 67*(12), 9773–9775. DOI: 10.1021/acs.jmedchem.4c01180 PMID: 38857362

Lakhman, S. S., Ma, Q., & Morse, G. D. (2009). Pharmacogenomics of CYP3A: Considerations for HIV treatment. *Pharmacogenomics, 10*(8), 1323–1339. DOI: 10.2217/pgs.09.53 PMID: 19663676

Lal, A., Ashworth, H. C., Dada, S., Hoemeke, L., & Tambo, E. (2022, February). Optimizing Pandemic Preparedness and Response Through Health Information Systems: Lessons Learned From Ebola to COVID-19. *Disaster Medicine and Public Health Preparedness, 16*(1), 333–340. DOI: 10.1017/dmp.2020.361 PMID: 33004102

Lampejo T. (2020). Influenza and antiviral resistance: an overview. *European journal of clinical microbiology & infectious diseases: official publication of the European Society of Clinical Microbiology, 39*(7), 1201–1208. https://doi.org/DOI: 10.1007/s10096-020-03840-9

Landeras-Bueno, S., Wasserman, H., Oliveira, G., VanAernum, Z. L., Busch, F., Salie, Z. L., Wysocki, V. H., Andersen, K., & Saphire, E. O. (2021). Cellular mRNA triggers structural transformation of Ebola virus matrix protein VP40 to its essential regulatory form. *Cell Reports*, *35*(2), 108986. DOI: 10.1016/j.celrep.2021.108986 PMID: 33852858

Lane, M., Luke, C., Bresee, J., Dugan, V., Post, D., Schafer, J., Roberts, P., Wentworth, D., & Ison, M. (2024). Meeting report: Controlled human influenza virus infection model studies: Current status and future directions for innovation. *Influenza and Other Respiratory Viruses*, *18*(10), e13358. DOI: 10.1111/irv.13358 PMID: 39440405

Lanford, R. E., Hildebrandt-Eriksen, E. S., Petri, A., Persson, R., Lindow, M., Munk, M. E., Kauppinen, S., & Ørum, H. (2010). Therapeutic silencing of microRNA-122 in primates with chronic hepatitis C virus infection. *Science*, *327*(5962), 198–201. DOI: 10.1126/science.1178178 PMID: 19965718

Lange, S., Baehr, C., Cakman-Hinrichs, N. I., Cron, K., Fengels, H., Gregor, C., Matschenz, K., Petersen, J., Steinfurth, R., Stoehr, A., Unger, S., Gil Mir, M., Hernández, C., Schwenken, M., & Buggisch, P. (2024). ReLink strategy in diagnosed-but-untreated hepatitis C-positive patients in Germany: report from a single center. Die ReLink-Strategie bei diagnostizierten aber unbehandelten Hepatitis C-positiven Patient*innen in Deutschland: Bericht aus einem einzelnen Zentrum - Hepatitis C ReLink-Projekt Deutschland. *Zeitschrift für Gastroenterologie*, *62*(10), 1701–1707. DOI: 10.1055/a-2349-2767 PMID: 39013433

Langmia, I. M., Just, K. S., Yamoune, S., Brockmöller, J., Masimirembwa, C., & Stingl, J. C. (2021). CYP2B6 Functional Variability in Drug Metabolism and Exposure Across Populations-Implication for Drug Safety, Dosing, and Individualized Therapy. *Frontiers in Genetics*, *12*, 692234. DOI: 10.3389/fgene.2021.692234 PMID: 34322158

Lashgarian, H. E., Valibeik, A., Marzban, A., Karkhane, M., & Shahzamani, K. (2021). The Relationship Between HCV-NS5A Gene Mutations and Resistance to Combination Therapy in Patients with HCV- Genotype 1-B. *Reports of Biochemistry & Molecular Biology*, *10*(2), 233–242. DOI: 10.52547/rbmb.10.2.233 PMID: 34604413

Lauer, S. A., Grantz, K. H., Bi, Q., Jones, F. K., Zheng, Q., Meredith, H. R., Azman, A. S., Reich, N. G., & Lessler, J. (2020). The incubation period of coronavirus disease 2019 (COVID-19) from publicly reported confirmed cases: Estimation and application. *Annals of Internal Medicine*, *172*(9), 577–582. DOI: 10.7326/M20-0504 PMID: 32150748

Lauring, A. S., Frydman, J., & Andino, R. (2013). The role of mutational robustness in RNA virus evolution. *Nature Reviews. Microbiology*, *11*(5), 327–336. DOI: 10.1038/nrmicro3003 PMID: 23524517

Lawitz, E., Mangia, A., Wyles, D., Rodriguez-Torres, M., Hassanein, T., Gordon, S. C., Schultz, M., Davis, M. N., Kayali, Z., Reddy, K. R., Jacobson, I. M., Kowdley, K. V., Nyberg, L., Subramanian, G. M., Hyland, R. H., Arterburn, S., Jiang, D., McNally, J., Brainard, D., & Gane, E. J. (2013). Sofosbuvir for previously untreated chronic hepatitis C infection. *The New England Journal of Medicine*, *368*(20), 1878–1887. DOI: 10.1056/NEJMoa1214853 PMID: 23607594

Lazarevic, I., Banko, A., Miljanovic, D., & Cupic, M. (2024). Hepatitis B Surface Antigen Isoforms: Their Clinical Implications, Utilization in Diagnosis, Prevention and New Antiviral Strategies. *Pathogens (Basel, Switzerland)*, *13*(1), 46. DOI: 10.3390/pathogens13010046 PMID: 38251353

Lee, D., Kang, J., & Kim, K. (2020). Global Collaboration Research Strategies for Sustainability in the Post COVID-19 Era: Analyzing Virology-Related National-Funded Projects. *Sustainability (Basel)*, *12*(16), 6561. Advance online publication. DOI: 10.3390/su12166561

Lee, H. L., & Jang, J. W. (2024). Core protein inhibitors: Opportunities and challenges at the forefront of hepatitis B cure: Editorial on "Phase 1 trial of the safety, pharmacokinetics, and antiviral activity of EDP-514 in untreated viremic chronic hepatitis B patients". *Clinical and Molecular Hepatology*, *30*(4), 692–694. DOI: 10.3350/cmh.2024.0346 PMID: 38741237

Lee, J. H., Hwang, K., & Kim, W. (2023). Machine learning in nanotechnology-based drug delivery systems for targeted therapy. *Advanced Drug Delivery Reviews*, *196*, 114–123.

Lee, Y., & Kim, Y. (2020). Bioinformatics applications in discovering antiviral drugs: A review. *Computer Methods and Programs in Biomedicine*, *196*, 105599.

Legge, D. G., & Kim, S. (2021). Equitable Access to COVID-19 Vaccines: Cooperation around Research and Production Capacity Is Critical. *Journal for Peace and Nuclear Disarmament, 4*(sup1), 73-134. DOI:DOI: 10.1080/25751654.2021.1906591

Lian, W., Wen, L., Zhou, Q., Zhu, W., Duan, W., Xiao, X., Mhungu, F., Huang, W., Li, C., Cheng, W., & Tian, J. (2020). Digital health technologies respond to the COVID-19 pandemic in a tertiary hospital in China: Development and usability study. *Journal of Medical Internet Research*, *22*(7), e24505. DOI: 10.2196/24505 PMID: 33141679

Li, H., Zhang, N., Ma, X., Wang, Y., Yang, F., Wang, W., Huang, Y., Xie, Y., & Du, Y. (2025). Google Trends applications for COVID-19 pandemic: A bibliometric analysis. *Digital Health*, *11*, 20552076241310055. DOI: 10.1177/20552076241310055 PMID: 39758260

Li, H., & Zhao, Y. (2020). Wearable devices for antiviral treatment adherence: A systematic review. *JMIR mHealth and uHealth*, *8*(4), e18934.

Li, J., & Zhang, Y. (2021). CRISPR-based antiviral strategies: Current status and future directions. *Frontiers in Microbiology*, *12*, 654.

Li, J., & Zhang, Y. (2021). Machine learning in antiviral drug discovery: A comprehensive review. *Computational Biology and Chemistry*, *90*, 107408.

Li, N., Stroud, N. J., & Jamieson, K. H. (2017). Understanding the MMR vaccine controversy: An analysis of the role of journalistic tone. *Health Communication*, *32*(4), 395–402. DOI: 10.1080/10410236.2016.1138475

Lin, A., Giuliano, C. J., Palladino, A., John, K. M., Abramowicz, C., Yuan, M. L., Sausville, E. L., Lukow, D. A., Liu, L., Chait, A. R., Galluzzo, Z. C., Tucker, C., & Sheltzer, J. M. (2019). Off-target toxicity is a common mechanism of action of cancer drugs undergoing clinical trials. *Science Translational Medicine*, *11*(509), eaaw8412. DOI: 10.1126/scitranslmed.aaw8412 PMID: 31511426

Lin, C. L., & Kao, J. H. (2015). Hepatitis B virus genotypes and variants. *Cold Spring Harbor Perspectives in Medicine*, *5*(5), a021436. DOI: 10.1101/cshperspect.a021436 PMID: 25934462

Lin, D. M., Koskella, B., & Lin, H. C. (2017). Phage therapy: An alternative to antibiotics in the age of multidrug resistance. *World Journal of Gastrointestinal Pharmacology and Therapeutics*, *8*(3), 162–173. DOI: 10.4292/wjgpt.v8.i3.162 PMID: 28828194

Lindsey, N. P., Staples, J. E., & Fischer, M. (2018). Eastern Equine Encephalitis Virus in the United States, 2003-2016. *The American Journal of Tropical Medicine and Hygiene*, *98*(5), 1472–1477. DOI: 10.4269/ajtmh.17-0927 PMID: 29557336

Ling, K. M., Stick, S. M., & Kicic, A. (2023). Pulmonary bacteriophage and cystic fibrosis airway mucus: Friends or foes? *Frontiers in Medicine*, *10*, 1088494. DOI: 10.3389/fmed.2023.1088494 PMID: 37265479

Linsky, T. W., Vergara, R., Codina, N., Nelson, J. W., Walker, M. J., Su, W., Barnes, C. O., Hsiang, T. Y., Esser-Nobis, K., Yu, K., Reneer, Z. B., Hou, Y. J., Priya, T., Mitsumoto, M., Pong, A., Lau, U. Y., Mason, M. L., Chen, J., Chen, A., & Silva, D. A. (2020). De novo design of potent and resilient hACE2 decoys to neutralize SARS-CoV-2. *Science*, *370*(6521), 1208–1214. DOI: 10.1126/science.abe0075 PMID: 33154107

Linton, N. M., Kobayashi, T., Yang, Y., Hayashi, K., Akhmetzhanov, A. R., Jung, S. M., Yuan, B., Kinoshita, R., & Nishiura, H. (2020). Incubation Period and Other Epidemiological Characteristics of 2019 Novel Coronavirus Infections with Right Truncation: A Statistical Analysis of Publicly Available Case Data. *Journal of Clinical Medicine*, *9*(2), 538. DOI: 10.3390/jcm9020538 PMID: 32079150

Li, S. (2024). Application of Artificial Intelligence in Early Diagnosis of Influenza A (H1N1) Virus Infection. *Molecular Pathogens*, *15*(1), 1–8. DOI: 10.5376/mp.2024.15.0001 PMID: 39860962

Li, S., Li, H., Lian, R., Xie, J., & Feng, R. (2024, June). New perspective of small-molecule antiviral drugs development for RNA viruses. *Virology*, *594*, 110042. DOI: 10.1016/j.virol.2024.110042 PMID: 38492519

Liu, Z., Chen, X., Carter, W., Moruf, A., Komatsu, T. E., Pahwa, S., Chan-Tack, K., Snyder, K., Petrick, N., Cha, K., Lal-Nag, M., Hatim, Q., Thakkar, S., Lin, Y., Huang, R., Wang, D., Patterson, T. A., & Tong, W. (2022). AI-powered drug repurposing for developing COVID-19 treatments. *Reference Module in Biomedical Sciences*, B978-0-12-824010-6.00005-8. DOI: 10.1016/B978-0-12-824010-6.00005-8

Liu, X., & Zhan, Q. (2020). The use of wearable devices for monitoring the effects of antiviral medications: A review. *Current Pharmaceutical Design*, *26*(25), 2913–2923.

Li, W., Moore, M. J., Vasilieva, N., Sui, J., Wong, S. K., Berne, M. A., Somasundaran, M., Sullivan, J. L., Luzuriaga, K., Greenough, T. C., Choe, H., & Farzan, M. (2003). Angiotensin-converting enzyme 2 is a functional receptor for the SARS coronavirus. *Nature*, *426*(6965), 450–454. DOI: 10.1038/nature02145 PMID: 14647384

Li, W., Shi, Z., Yu, M., Ren, W., Smith, C., Epstein, J. H., Wang, H., Crameri, G., Hu, Z., Zhang, H., Zhang, J., McEachern, J., Field, H., Daszak, P., Eaton, B. T., Zhang, S., & Wang, L. F. (2005). Bats are natural reservoirs of SARS-like coronaviruses. *Science*, *310*(5748), 676–679. DOI: 10.1126/science.1118391 PMID: 16195424

Li, X., & Zhang, Y. (2021). Artificial intelligence in the development of mRNA vaccines. *Frontiers in Immunology*, *12*, 654.

Li, Y., Muffat, J., Omer Javed, A., Keys, H. R., Lungjangwa, T., Bosch, I., Khan, M., Virgilio, M. C., Gehrke, L., Sabatini, D. M., & Jaenisch, R. (2019). Genome-wide CRISPR screen for Zika virus resistance in human neural cells. *Proceedings of the National Academy of Sciences of the United States of America*, *116*(19), 9527–9532. DOI: 10.1073/pnas.1900867116 PMID: 31019072

Li, Y., Wan, Y., Liu, P., Zhao, J., Lu, G., Qi, J., Wang, Q., Lu, X., Wu, Y., Liu, W., Zhang, B., Yuen, K. Y., Perlman, S., Gao, G. F., & Yan, J. (2015). A humanized neutralizing antibody against MERS-CoV targeting the receptor-binding domain of the spike protein. *Cell Research*, *25*(11), 1237–1249. DOI: 10.1038/cr.2015.113 PMID: 26391698

Loaiza, J. D., Chvatal-Medina, M., Hernandez, J. C., & Rugeles, M. T. (2023). Integrase inhibitors: Current protagonists in antiretroviral therapy. *Immunotherapy*, *15*(17), 1477–1495. DOI: 10.2217/imt-2023-0011 PMID: 37822251

Lo, M. K., Albariño, C. G., Perry, J. K., Chang, S., Tchesnokov, E. P., Guerrero, L., Chakrabarti, A., Shrivastava-Ranjan, P., Chatterjee, P., McMullan, L. K., Martin, R., Jordan, R., Götte, M., Montgomery, J. M., Nichol, S. T., Flint, M., Porter, D., & Spiropoulou, C. F. (2020). Remdesivir targets a structurally analogous region of the Ebola virus and SARS-CoV-2 polymerases. *Proceedings of the National Academy of Sciences of the United States of America*, *117*(43), 26946–26954. DOI: 10.1073/pnas.2012294117 PMID: 33028676

Lombardo, D., Kiselev, M. A., & Caccamo, M. T. (2019). Smart nanoparticles for drug delivery application: Development of versatile nanocarrier platforms in biotechnology and nanomedicine. *Journal of Nanomaterials*, *2019*(1), 3702518. DOI: 10.1155/2019/3702518

Longdon, B., Brockhurst, M. A., Russell, C. A., Welch, J. J., & Jiggins, F. M. (2014). The evolution and genetics of virus host shifts. *PLoS Pathogens*, *10*(11), e1004395. DOI: 10.1371/journal.ppat.1004395 PMID: 25375777

López, R., & Wang, Y. (2020). Telehealth for chronic viral hepatitis management: Current trends and future directions. *Journal of Clinical and Translational Hepatology*, *8*(4), 345–356.

López, R., & Wang, Y. (2021). Real-time data analytics in antiviral treatment management. *Health and Technology*, *11*(2), 335–348.

Louie, J. K., Yang, S., Acosta, M., Yen, C., Samuel, M. C., Schechter, R., Guevara, H., & Uyeki, T. M. (2012). Treatment with neuraminidase inhibitors for critically ill patients with influenza A (H1N1)pdm09. *Clinical Infectious Diseases*, *55*(9), 1198–1204. DOI: 10.1093/cid/cis636 PMID: 22843781

Loureiro, J. A., & Pereira, M. C. (2020). PLGA-based drug carrier and pharmaceutical applications: The most recent advances. *Pharmaceutics*, *12*(9), 903. DOI: 10.3390/pharmaceutics12090903 PMID: 32971970

Lourenço, J., Maia de Lima, M., Faria, N. R., Walker, A., Kraemer, M. U., Villabona-Arenas, C. J., Lambert, B., Marques de Cerqueira, E., Pybus, O. G., Alcantara, L. C., & Recker, M. (2017). Epidemiological and ecological determinants of Zika virus transmission in an urban setting. *eLife*, *6*, e29820. DOI: 10.7554/eLife.29820 PMID: 28887877

Lu, L., Su, S., Yang, H., & Jiang, S. (2021). Antivirals with common targets against highly pathogenic viruses. *Cell*, *184*(6), 1604–1620. DOI: 10.1016/j.cell.2021.02.013 PMID: 33740455

Luo, C. M., Wang, N., Yang, X. L., Liu, H. Z., Zhang, W., Li, B., Hu, B., Peng, C., Geng, Q. B., Zhu, G. J., Li, F., & Shi, Z. L. (2018). Discovery of Novel Bat Coronaviruses in South China That Use the Same Receptor as Middle East Respiratory Syndrome Coronavirus. *Journal of Virology*, *92*(13), e00116–e00118. DOI: 10.1128/JVI.00116-18 PMID: 29669833

Luo, Z., Liu, L. F., Jiang, Y. N., Tang, L. P., Li, W., Ouyang, S. H., Tu, L. F., Wu, Y. P., Gong, H. B., Yan, C. Y., Jiang, S., Lu, Y. H., Liu, T., Jiang, Z., Kurihara, H., Yu, Y., Yao, X. S., Li, Y. F., & He, R. R. (2020). Novel insights into stress-induced susceptibility to influenza: Corticosterone impacts interferon-β responses by Mfn2-mediated ubiquitin degradation of MAVS. *Signal Transduction and Targeted Therapy*, *5*(1), 202. DOI: 10.1038/s41392-020-00238-z PMID: 32943610

Luvira, V., Schilling, W. H. K., Jittamala, P., Watson, J. A., Boyd, S., Siripoon, T., Ngamprasertchai, T., Almeida, P. J., Ekkapongpisit, M., Cruz, C., Callery, J. J., Singh, S., Tuntipaiboontana, R., Kruabkontho, V., Ngernseng, T., Tubprasert, J., Abdad, M. Y., Keayarsa, S., Madmanee, W., & Group, P. C. (2024, January 15). Clinical antiviral efficacy of favipiravir in early COVID-19 (PLATCOV): An open-label, randomized, controlled, adaptive platform trial. *BMC Infectious Diseases*, *24*(1), 89. DOI: 10.1186/s12879-023-08835-3 PMID: 38225598

Macarrón Palacios, A., Korus, P., Wilkens, B. G. C., Heshmatpour, N., & Patnaik, S. R. (2024). Revolutionizing in vivo therapy with CRISPR/Cas genome editing: Breakthroughs, opportunities and challenges. *Frontiers in Genome Editing*, *6*, 1342193. DOI: 10.3389/fgeed.2024.1342193 PMID: 38362491

Mackay, I. M., & Arden, K. E. (2015). MERS coronavirus: Diagnostics, epidemiology and transmission. *Virology Journal*, *12*(1), 1–21. DOI: 10.1186/s12985-015-0439-5 PMID: 26695637

Madhu, A., Cherian, I., & Gautam, A. K. (2022). Interdisciplinary approach to biomedical research: a panacea to efficient research output during the global pandemic. In *Coronavirus Drug Discovery* (pp. 331-347). DOI:DOI: 10.1016/B978-0-323-85156-5.00018-3

Mahajan, S., Choudhary, S., Kumar, P., & Tomar, S. (2021). Antiviral strategies targeting host factors and mechanisms obliging +ssRNA viral pathogens. *Bioorganic & Medicinal Chemistry*, *46*, 116356. DOI: 10.1016/j.bmc.2021.116356 PMID: 34416512

Mahalakshmi, S., & Kumar, A. (2021). Mobile health applications for chronic disease management: Opportunities and challenges. *Health Informatics Journal*, *27*(2), 1083–1097.

Majumder, J., Taratula, O., & Minko, T. (2019). Nanocarrier-based systems for targeted and site-specific therapeutic delivery. *Advanced Drug Delivery Reviews*, *144*, 57–77. DOI: 10.1016/j.addr.2019.07.010 PMID: 31400350

Makoul, G., & Clayman, M. L. (2010). An integrative model of patient-centered communication in the healthcare setting. *Journal of Health Communication*, *15*(2), 7–24.

Maleki Varnosfaderani, S., & Forouzanfar, M. (2024). The Role of AI in Hospitals and Clinics: Transforming Healthcare in the 21st Century. *Bioengineering (Basel, Switzerland)*, *11*(4), 337. DOI: 10.3390/bioengineering11040337 PMID: 38671759

Malik, J. A., Ahmed, S., Yaseen, Z., Alanazi, M., Alharby, T. N., Alshammari, H. A., & Anwar, S. (2022). Association of SARS-CoV-2 and polypharmacy with gut–lung axis: From pathogenesis to treatment. *ACS Omega*, *7*(38), 33651–33665. DOI: 10.1021/acsomega.2c02524 PMID: 36164411

Malin, J. J., Suárez, I., Priesner, V., Fätkenheuer, G., & Rybniker, J. (2020). Remdesivir against COVID-19 and other viral diseases. *Clinical Microbiology Reviews*, *34*(1), 00162–00120. DOI: 10.1128/CMR.00162-20 PMID: 33055231

Manabe, T., Kambayashi, D., Akatsu, H., & Kudo, K. (2021). Favipiravir for the treatment of patients with COVID-19: A systematic review and meta-analysis. *BMC Infectious Diseases*, *21*(1), 489. DOI: 10.1186/s12879-021-06164-x PMID: 34044777

Mansour, T., & Jang, Y. (2020). Nanomedicine approaches in drug delivery for antiviral therapies: A review. *Journal of Drug Delivery Science and Technology*, *60*, 101987.

Manubolu, K., Yamini, S., Charishma, R., Kumar, N., Kishori, O., Tanusha, K., & Baba, K. H. (2023). Formulation and evaluation of phytosome and plants exhibiting anti-inflammatory activity. *Future Journal of Pharmaceuticals and Health Sciences*, *3*(1), 90–97.

Manzari, M. T., Shamay, Y., Kiguchi, H., Rosen, N., Scaltriti, M., & Heller, D. A. (2021). Targeted drug delivery strategies for precision medicines. *Nature Reviews. Materials*, *6*(4), 351–370. DOI: 10.1038/s41578-020-00269-6 PMID: 34950512

Maqsood, R., Smith, M. F., Holland, L. A., Sullins, R. A., Holland, S. C., Tan, M., Hernandez Barrera, G. M., Thomas, A. W., Islas, M., Kramer, J. L., Nordstrom, L., Mulrow, M., White, M., Murugan, V., & Lim, E. S. (2024). Influenza Virus Genomic Surveillance, Arizona, USA, 2023-2024. *Viruses*, *16*(5), 692. DOI: 10.3390/v16050692 PMID: 38793574

Marcinkowska-Swojak, M., Rakoczy, M., Podkowiński, J., Handschuh, J., Wojciechowski, P., & Handschuh, L. (2024). Od Sangera do sekwencjonowania genomów – przegląd technologii sekwencjonowania DNA [From Sanger to genome sequencing - an overview of DNA sequencing technologies]. *Postepy Biochemii*, *70*(2), 173–189. DOI: 10.18388/pb.2021_534 PMID: 39083466

Marrana, M. (2022). Epidemiology of disease through the interactions between humans, domestic animals, and wildlife. In *One Health* (pp. 73–111). Elsevier.

Maserat, E. (2022). Integration of Artificial Intelligence and CRISPR/Cas9 System for Vaccine Design. *Cancer Informatics*, *21*, 11769351221140102. DOI: 10.1177/11769351221140102 PMID: 36452267

Mashel, T. V., Tarakanchikova, Y. V., Muslimov, A. R., Zyuzin, M. V., Timin, A. S., Lepik, K. V., & Fehse, B. (2020). Overcoming the delivery problem for therapeutic genome editing: Current status and perspective of non-viral methods. *Biomaterials*, *258*, 120282. DOI: 10.1016/j.biomaterials.2020.120282 PMID: 32798742

Maslennikov, R., Ivashkin, V., Efremova, I., Poluektova, E., & Shirokova, E. (2021). Probiotics in hepatology: An update. *World Journal of Hepatology*, *13*(9), 1154–1166. DOI: 10.4254/wjh.v13.i9.1154 PMID: 34630882

Matsumoto, Y.. (2021). Microneedle technology: A promising alternative for vaccine delivery. *Advanced Drug Delivery Reviews*, *178*, 113942. DOI: 10.1016/j.addr.2021.113942

Matthew, A. N., Leidner, F., Lockbaum, G. J., Henes, M., Zephyr, J., Hou, S., Rao, D. N., Timm, J., Rusere, L. N., Ragland, D. A., Paulsen, J. L., Prachanronarong, K., Soumana, D. I., Nalivaika, E. A., Kurt Yilmaz, N., Ali, A., & Schiffer, C. A. (2021). Drug Design Strategies to Avoid Resistance in Direct-Acting Antivirals and Beyond. *Chemical Reviews*, *121*(6), 3238–3270. DOI: 10.1021/acs.chemrev.0c00648 PMID: 33410674

Maus, A., Strait, L., & Zhu, D. (2021). Nanoparticles as delivery vehicles for antiviral therapeutic drugs. *Engineered Regeneration*, *2*, 31–46. DOI: 10.1016/j.engreg.2021.03.001 PMID: 38620592

Mays, J. W., Powell, N. D., Hunzeker, J. T., Hanke, M. L., Bailey, M. T., & Sheridan, J. F. (2012). Stress and the anti-influenza immune response: Repeated social defeat augments clonal expansion of CD8(+)T cells during primary influenza A viral infection. *Journal of Neuroimmunology*, *243*(1-2), 34–42. DOI: 10.1016/j.jneuroim.2011.12.011 PMID: 22244573

Mazziotta, C., Tognon, M., Martini, F., Torreggiani, E., & Rotondo, J. C. (2023). Probiotics Mechanism of Action on Immune Cells and Beneficial Effects on Human Health. *Cells*, *12*(1), 184. DOI: 10.3390/cells12010184 PMID: 36611977

Mbunge, E., Batani, J., Gaobotse, G., & Muchemwa, B. (2022, June). Virtual healthcare services and digital health technologies deployed during coronavirus disease 2019 (COVID-19) pandemic in South Africa: A systematic review. *Global Health Journal (Amsterdam, Netherlands)*, *6*(2), 102–113. DOI: 10.1016/j.glohj.2022.03.001 PMID: 35282399

McCarthy, T. D., Karellas, P., Henderson, S. A., Giannis, M., O'Keefe, D. F., Heery, G., Paull, J. R., Matthews, B. R., & Holan, G. (2005). Dendrimers as drugs: Discovery and preclinical and clinical development of dendrimer-based microbicides for HIV and STI prevention. *Molecular Pharmaceutics*, *4*(4), 312–318. DOI: 10.1021/mp050023q PMID: 16053334

McCombie, W. R., McPherson, J. D., & Mardis, E. R. (2019). Next-Generation Sequencing Technologies. *Cold Spring Harbor Perspectives in Medicine*, *9*(11), a036798. DOI: 10.1101/cshperspect.a036798 PMID: 30478097

Medina, C., García, A. H., Crespo, F. I., Toro, F. I., Mayora, S. J., & De Sanctis, J. B. (2023). A Synopsis of Hepatitis C Virus Treatments and Future Perspectives. *Current Issues in Molecular Biology*, *45*(10), 8255–8276. DOI: 10.3390/cimb45100521 PMID: 37886964

Medina-Cruz, D., Saleh, B., Vernet-Crua, A., Ajo, A., Roy, A. K., & Webster, T. J. (2020). Drug-delivery nanocarriers for skin wound-healing applications. In *Wound healing, tissue repair, and regeneration in diabetes* (pp. 439–488). Academic Press. DOI: 10.1016/B978-0-12-816413-6.00022-8

Meganck, R. M., & Baric, R. S. (2021, March). Developing therapeutic approaches for twenty-first-century emerging infectious viral diseases. *Nature Medicine*, *27*(3), 401–410. DOI: 10.1038/s41591-021-01282-0 PMID: 33723456

Megantara, S., Rusdin, A., Budiman, A., Shamsuddin, S., Mohtar, N., & Muchtaridi, M. (2024). Revolutionizing antiviral therapeutics: Unveiling innovative approaches for enhanced drug efficacy. *International Journal of Nanomedicine*, *19*, 2889–2915. DOI: 10.2147/IJN.S447721 PMID: 38525012

Mehta, S. K., & Pradhan, R. B. (2024, September). Phytochemicals in antiviral drug development against human respiratory viruses. *Drug Discovery Today*, *29*(9), 104107. DOI: 10.1016/j.drudis.2024.104107 PMID: 39032810

Mei, M., & Tan, X. (2021). Current Strategies of Antiviral Drug Discovery for COVID-19. *Frontiers in Molecular Biosciences*, *8*, 671263. DOI: 10.3389/fmolb.2021.671263 PMID: 34055887

Menachery, V. D., Yount, B. L.Jr, Debbink, K., Agnihothram, S., Gralinski, L. E., Plante, J. A., Graham, R. L., Scobey, T., Ge, X. Y., Donaldson, E. F., Randell, S. H., Lanzavecchia, A., Marasco, W. A., Shi, Z. L., & Baric, R. S. (2015). A SARS-like cluster of circulating bat coronaviruses shows potential for human emergence. *Nature Medicine*, *21*(12), 1508–1513. DOI: 10.1038/nm.3985 PMID: 26552008

Menachery, V. D., Yount, B. L.Jr, Sims, A. C., Debbink, K., Agnihothram, S. S., Gralinski, L. E., Graham, R. L., Scobey, T., Plante, J. A., Royal, S. R., Swanstrom, J., Sheahan, T. P., Pickles, R. J., Corti, D., Randell, S. H., Lanzavecchia, A., Marasco, W. A., & Baric, R. S. (2016). SARS-like WIV1-CoV poised for human emergence. *Proceedings of the National Academy of Sciences of the United States of America*, *113*(11), 3048–3053. DOI: 10.1073/pnas.1517719113 PMID: 26976607

Meng, B., Lui, Y. W., Meng, S., Cao, C., & Hu, Y. (2006). Identification of effective siRNA blocking the expression of SARS viral envelope E and RDRP genes. *Molecular Biotechnology*, *33*(2), 141–148. DOI: 10.1385/MB:33:2:141 PMID: 16757801

Milewska, A., Nowak, P., Owczarek, K., Szczepanski, A., Zarebski, M., Hoang, A., Berniak, K., Wojarski, J., Zeglen, S., Baster, Z., Rajfur, Z., & Pyrc, K. (2018). Entry of Human Coronavirus NL63 into the Cell. *Journal of Virology*, *92*(3), e01933–e17. DOI: 10.1128/JVI.01933-17 PMID: 29142129

Millet, J. K., & Whittaker, G. R. (2018). Physiological and molecular triggers for SARS-CoV membrane fusion and entry into host cells. *Virology*, *517*, 3–8. DOI: 10.1016/j.virol.2017.12.015 PMID: 29275820

Mishra, S. K., & Zhang, J. (2020). Telemedicine in the era of digital health for chronic viral infections: A review. *Journal of Clinical Virology*, *129*, 104539.

Model, A. E., Lim, D., Nguyen, T. M., Sreekanth, V., & Choudhary, A. (2022, February). CRISPR-based therapeutics: Current challenges and future applications. *Trends in Pharmacological Sciences*, *43*(2), 151–161. DOI: 10.1016/j.tips.2021.10.012 PMID: 34952739

Mofijur, M., Fattah, I. M. R., Alam, M. A., Islam, A., Ong, H. C., Rahman, S. M. A., Najafi, G., Ahmed, S. F., Uddin, M. A., & Mahlia, T. M. I. (2021, April). Impact of COVID-19 on the social, economic, environmental and energy domains: Lessons learnt from a global pandemic. *Sustainable Production and Consumption*, *26*, 343–359. DOI: 10.1016/j.spc.2020.10.016 PMID: 33072833

Mohamed Abdelnajid, D., Elmowafy, A. Y., Rostaing, L., & Elrakaiby, M. T. (2023). Prediction of response to sofosbuvir-based therapy using serum interleukin-12 and single-nucleotide polymorphism of the interleukin 28B gene as predictive factors in HCV positive genotype-4 patients. *Medicine*, *102*(28), e34125. DOI: 10.1097/MD.0000000000034125 PMID: 37443472

Mohamed, H., Gurrola, T., Berman, R., Collins, M., Sariyer, I. K., Nonnemacher, M. R., & Wigdahl, B. (2022). Targeting CCR5 as a Component of an HIV-1 Therapeutic Strategy. *Frontiers in Immunology*, *12*, 816515. DOI: 10.3389/fimmu.2021.816515 PMID: 35126374

Mohamed, R. M., Gamal, S. M., & Fawzy, M. (2021). Mobile health technology for improving health outcomes: A review of existing applications. *Journal of Healthcare Engineering*, *6625178*. Advance online publication. DOI: 10.1155/2021/6625178

Mohammadi, M., & Rasekh, A. (2021). Big data and AI applications in antiviral care: Challenges and perspectives. *Frontiers in Digital Health*, *3*, 634210.

Mohammed, H., Oljira, L., Roba, K. T., Yimer, G., Fekadu, A., & Manyazewal, T. (2020). Containment of COVID-19 in Ethiopia and implications for tuberculosis care and research. *Infectious Diseases of Poverty*, *9*(1), 131. DOI: 10.1186/s40249-020-00753-9 PMID: 32938497

Mojahed, N., Mohammadkhani, M. A., & Mohamadkhani, A. (2022). Climate Crises and Developing Vector-Borne Diseases: A Narrative Review. *Iranian Journal of Public Health*, *51*(12), 2664–2673. DOI: 10.18502/ijph.v51i12.11457 PMID: 36742229

Mokhtary, P., Pourhashem, Z., Mehrizi, A. A., Sala, C., & Rappuoli, R. (2022). Recent progress in the discovery and development of monoclonal antibodies against viral infections. *Biomedicines*, *10*(8), 1861. DOI: 10.3390/biomedicines10081861 PMID: 36009408

Monoson, A., Schott, E., Ard, K., Kilburg-Basnyat, B., Tighe, R. M., Pannu, S., & Gowdy, K. M. (2023). Air pollution and respiratory infections: the past, present, and future. *Toxicological sciences: an official journal of the Society of Toxicology, 192*(1), 3–14. https://doi.org/DOI: 10.1093/toxsci/kfad003

Montessori, V., Press, N., Harris, M., Akagi, L., & Montaner, J. S. (2004). Adverse effects of antiretroviral therapy for HIV infection. *Canadian Medical Association Journal*, *170*(2), 229–238. PMID: 14734438

Moon, S.. (2023). Overcoming vaccine inequity: Lessons from COVID-19. *Nature Medicine*, *29*(1), 15–19.

Moon, S., Fogaça Vieira, M. C., Ruiz, A. A., & Navarro, D. M. (2022). *New business models for pharmaceutical research and development as a global public good: Considerations for the WHO European Region*. W. R. O. f. *Europe*.

Morey, J. N., Boggero, I. A., Scott, A. B., & Segerstrom, S. C. (2015). Current Directions in Stress and Human Immune Function. *Current Opinion in Psychology*, *5*, 13–17. DOI: 10.1016/j.copsyc.2015.03.007 PMID: 26086030

Motomura, K., Kuwano, A., Tanaka, K., Koga, Y., Masumoto, A., & Yada, M. (2023, August 30). Potential Predictive Biomarkers of Systemic Drug Therapy for Hepatocellular Carcinoma: Anticipated Usefulness in Clinical Practice. *Cancers (Basel)*, *15*(17), 4345. DOI: 10.3390/cancers15174345 PMID: 37686621

Mueller Brown, K., Le Sage, V., French, A. J., Jones, J. E., Padovani, G. H., Avery, A. J., Schultz-Cherry, S., Rosch, J. W., Hiller, N. L., & Lakdawala, S. S. (2022). Secondary infection with *Streptococcus pneumoniae* decreases influenza virus replication and is linked to severe disease. *FEMS Microbes*, *3*, xtac007. Advance online publication. DOI: 10.1093/femsmc/xtac007 PMID: 35392116

Mulangu, S., Dodd, L. E., Davey, R. T.Jr, Tshiani Mbaya, O., Proschan, M., Mukadi, D., Lusakibanza Manzo, M., Nzolo, D., Tshomba Oloma, A., Ibanda, A., Ali, R., Coulibaly, S., Levine, A. C., Grais, R., Diaz, J., Lane, H. C., & Muyembe-Tamfum, J. J.PALM Writing Group. (2019). A Randomized, Controlled Trial of Ebola Virus Disease Therapeutics. *The New England Journal of Medicine*, *381*(24), 2293–2303. DOI: 10.1056/NEJMoa1910993 PMID: 31774950

Müller, R. H., Jacobs, C., & Kayser, O. (2001). Nanosuspensions as particulate drug formulations in therapy: Rationale for development and what we can expect for the future. *Advanced Drug Delivery Reviews*, *47*(1), 3–19. DOI: 10.1016/S0169-409X(00)00118-6 PMID: 11251242

Muteeb, G., Rehman, M. T., Shahwan, M., & Aatif, M. (2023, November 15). Origin of Antibiotics and Antibiotic Resistance, and Their Impacts on Drug Development: A Narrative Review. *Pharmaceuticals (Basel, Switzerland)*, *16*(11), 1615. DOI: 10.3390/ph16111615 PMID: 38004480

Muthukutty, P., MacDonald, J., & Yoo, S. Y. (2024, October 27). Combating Emerging Respiratory Viruses: Lessons and Future Antiviral Strategies. *Vaccines*, *12*(11), 1220. Advance online publication. DOI: 10.3390/vaccines12111220 PMID: 39591123

Nachbagauer, R., Feser, J., Naficy, A., Bernstein, D. I., Guptill, J., Walter, E. B., Berlanda-Scorza, F., Stadlbauer, D., Wilson, P. C., Aydillo, T., Behzadi, M. A., Bhavsar, D., Bliss, C., Capuano, C., Carreño, J. M., Chromikova, V., Claeys, C., Coughlan, L., Freyn, A. W., & Krammer, F. (2021). A chimeric hemagglutinin-based universal influenza virus vaccine approach induces broad and long-lasting immunity in a randomized, placebo-controlled phase I trial. *Nature Medicine*, *27*(1), 106–114. DOI: 10.1038/s41591-020-1118-7 PMID: 33288923

Naidoo, K., & Oliver, S. V. (2024). Gene drives: An alternative approach to malaria control? *Gene Therapy*. Advance online publication. DOI: 10.1038/s41434-024-00468-8 PMID: 39039203

Nainwal, N. (2022). Treatment of respiratory viral infections through inhalation therapeutics: Challenges and opportunities. *Pulmonary Pharmacology & Therapeutics*, *77*, 102170. DOI: 10.1016/j.pupt.2022.102170 PMID: 36240985

Najafi, S., Tan, S. C., Aghamiri, S., Raee, P., Ebrahimi, Z., Jahromi, Z. K., Rahmati, Y., Sadri Nahand, J., Piroozmand, A., Jajarmi, V., & Mirzaei, H. (2022, April). Therapeutic potentials of CRISPR–Cas genome editing technology in human viral infections. *Biomedicine and Pharmacotherapy*, *148*, 112743. DOI: 10.1016/j.biopha.2022.112743 PMID: 35228065

Namasivayam, V., Vanangamudi, M., Kramer, V. G., Kurup, S., Zhan, P., Liu, X., Kongsted, J., & Byrareddy, S. N. (2019). The Journey of HIV-1 Non-Nucleoside Reverse Transcriptase Inhibitors (NNRTIs) from Lab to Clinic. *Journal of Medicinal Chemistry*, *62*(10), 4851–4883. DOI: 10.1021/acs.jmedchem.8b00843 PMID: 30516990

Namba-Nzanguim, C. T., Turon, G., Simoben, C. V., Tietjen, I., Montaner, L. J., Efange, S. M. N., Duran-Frigola, M., & Ntie-Kang, F. (2022). Artificial intelligence for antiviral drug discovery in low resourced settings: A perspective. *Frontiers in Drug Discovery*, *2*, 1013285. Advance online publication. DOI: 10.3389/fddsv.2022.1013285

Naseer, S., Khalid, S., Parveen, S., Abbass, K., Song, H., & Achim, M. V. (2023). COVID-19 outbreak: Impact on global economy. *Frontiers in Public Health*, *10*, 1009393. DOI: 10.3389/fpubh.2022.1009393 PMID: 36793360

Negri, V., Pacheco Torres, J., Calle, D., & López Larrubia, P. (2020). Carbon nanotubes in biomedicine. In *Surface-modified nanobiomaterials for electrochemical and biomedicine applications* (pp. 177-217). DOI: 10.1007/978-3-030-55502-3_6

Neuman, B. W., Kiss, G., Kunding, A. H., Bhella, D., Baksh, M. F., Connelly, S., Droese, B., Klaus, J. P., Makino, S., Sawicki, S. G., Siddell, S. G., Stamou, D. G., Wilson, I. A., Kuhn, P., & Buchmeier, M. J. (2011). A structural analysis of M protein in coronavirus assembly and morphology. *Journal of Structural Biology*, *174*(1), 11–22. DOI: 10.1016/j.jsb.2010.11.021 PMID: 21130884

Neumann, G., Chen, H., Gao, G. F., Shu, Y., & Kawaoka, Y. (2010). H5N1 influenza viruses: Outbreaks and biological properties. *Cell Research*, *20*(1), 51–61. DOI: 10.1038/cr.2009.124 PMID: 19884910

Neumann, G., Noda, T., & Kawaoka, Y. (2009). Emergence and pandemic potential of swine-origin H1N1 influenza virus. *Nature*, *459*(7249), 931–939. DOI: 10.1038/nature08157 PMID: 19525932

Ng, E. W., Shima, D. T., Calias, P., Cunningham, E. T.Jr, Guyer, D. R., & Adamis, A. P. (2006). Pegaptanib, a targeted anti-VEGF aptamer for ocular vascular disease. *Nature Reviews. Drug Discovery*, *5*(2), 123–132. DOI: 10.1038/nrd1955 PMID: 16518379

Nguyen, A., & Trinh, K. (2021). Machine learning and bioinformatics in viral disease research: Future prospects. *Computational Biology and Chemistry*, *88*, 107380.

Nguyen, P. H. A., Fukushi, S., Yamada, S., Harada, S., Yoshikawa, T., Kinoshita, H., Kawahara, M., Ogawa, T., Ebihara, H., Moi, M. L., & Saijo, M. (2024). Analysis of antiviral drug properties of thymidine kinase of herpes B virus using recombinant herpes simplex virus 1. *Microbiology Spectrum, 12*(1), e0309123. DOI: 10.1128/spectrum.03091-23 PMID: 38095468

Nguyen, T., & Li, H. (2021). Mobile health technologies for antiviral medication adherence: A systematic review. *AIDS Patient Care and STDs, 35*(5), 271–283.

Nguyen, T., & Li, H. (2023). mHealth applications in antiviral treatment: A review of benefits and challenges. *Mobile Health Research and Applications, 18*, 45–61.

Nguyen, V. H., Huang, D. Q., Le, M. H., Jin, M., Lee, E. Y., Henry, L., Nerurkar, S. N., Ogawa, E., Thin, K. N., Teng, M. L. P., Goh, K. S., Kai, J. C. Y., Wong, C., Tan, D. J. H., Thuy, L. T. T., Hai, H., Enomoto, M., Cheung, R., & Nguyen, M. H. (2023). Global treatment rate and barriers to direct-acting antiviral therapy: A systematic review and meta-analysis of 146 studies and 1 760 352 hepatitis C virus patients. *Liver International, 43*(6), 1195–1203. DOI: 10.1111/liv.15550 PMID: 36825358

Niakan Kalhori, S. R., Bahaadinbeigy, K., Deldar, K., Gholamzadeh, M., Hajesmaeel-Gohari, S., & Ayyoubzadeh, S. M. (2021). Digital health solutions to control the COVID-19 pandemic in countries with high disease prevalence: Literature review. *Journal of Medical Internet Research, 23*(1), e19473. DOI: 10.2196/19473 PMID: 33600344

Nikzamir, M., Hanifehpour, Y., Akbarzadeh, A., & Panahi, Y. (2021). Applications of dendrimers in nanomedicine and drug delivery: A review. *Journal of Inorganic and Organometallic Polymers and Materials, 31*(6), 2246–2261. DOI: 10.1007/s10904-021-01925-2

Noorbakhsh, F., Abdolmohammadi, K., Fatahi, Y., Dalili, H., Rasoolinejad, M., Rezaei, F., Salehi-Vaziri, M., Shafiei-Jandaghi, N. Z., Gooshki, E. S., Zaim, M., & Nicknam, M. H. (2019). Zika Virus Infection, Basic and Clinical Aspects: A Review Article. *Iranian Journal of Public Health, 48*(1), 20–31. DOI: 10.18502/ijph.v48i1.779 PMID: 30847308

Nourbakhsh, M. (2013). Role of TG Lipases, Arylacetamide Deacetylase and Triacylglycerol Hydrolase, in Hepatitis C Virus Life Cycle.

Nova, N., Athni, T. S., Childs, M. L., Mandle, L., & Mordecai, E. A. (2022). Global change and emerging infectious diseases. *Annual Review of Resource Economics, 14*(1), 333–354. DOI: 10.1146/annurev-resource-111820-024214 PMID: 38371741

Nunes, C., McKee, M., & Howard, N. (2024). The role of global health partnerships in vaccine equity: A scoping review. *PLOS Global Public Health, 4*(2), e0002834. DOI: 10.1371/journal.pgph.0002834 PMID: 38386621

Nyathi, S., Rezende, I. M., Walter, K. S., Thongsripong, P., Mutuku, F., Ndenga, B., Mbakaya, J. O., Aswani, P., Musunzaji, P. S., Chebii, P. K., Maina, P. W., Mutuku, P. S., Ng'ang'a, C. M., Malumbo, S. L., Jembe, Z., Vu, D. M., Mordecai, E. A., Bennett, S., Andrews, J. R., & LaBeaud, A. D. (2024). Molecular epidemiology and evolutionary characteristics of dengue virus 2 in East Africa. *Nature Communications, 15*(1), 7832. DOI: 10.1038/s41467-024-51018-0 PMID: 39244569

O' Kane, R., & Hathorn, E. (2023). Hepatitis C: Recent advances and practical management. *Frontline Gastroenterology, 14*(5), 415–421. DOI: 10.1136/flgastro-2022-102373 PMID: 37581179

Obayomi, O. V., Olaniran, A. F., & Owa, S. O. (2024). Unveiling the role of functional foods with emphasis on prebiotics and probiotics in human health: A review. *Journal of Functional Foods, 119*, 106337. DOI: 10.1016/j.jff.2024.106337

Ochoa, S. A., & Fernández, L. (2021). Advances in wearable technology for monitoring patients undergoing antiviral treatment. *Nature Digital Medicine, 4*(1), 2–9.

Odeyemi, O. E., Williams, S. Y., Adeyemi, S. O., Eyitayo, J. O., Dada, O. E., Adesina, M. A., & Akintayo, A. D. (2020). Mobile Health Technology (Mhealth) in Combating COVID-19 Pandemic: Use, Challenges and Recommendations. *European Journal of Medical and Educational Technologies, 13*(4), em2018. Advance online publication. DOI: 10.30935/ejmets/8572

Offit, P. A. (2005). The Cutter Incident, 50 years later. *The New England Journal of Medicine, 352*(14), 1411–1412. DOI: 10.1056/NEJMp048180 PMID: 15814877

Ogaugwu, C. E., Agbo, S. O., & Adekoya, M. A. (2019). CRISPR in Sub-Saharan Africa: Applications and Education. *Trends in Biotechnology, 37*(3), 234–237. DOI: 10.1016/j.tibtech.2018.07.012 PMID: 30100229

Ogidi, O. I. (2020). A review on the use of Herbal Remedies and Clinical Therapeutics for the Management of Covid-19 Pandemic. *ASIO Journal of Pharmaceuical & Herbal Medicines Research*, 6(2), 68-77. http://doi.org/doilink/10.2020.85722646/

Ogidi, O. I. (2022). Omicron (B.1.1.529): Variant of Concern- A Mini Review. *Biomedical Journal of Scientific & Technical Research,* 41(5), 33106- 33112. https://www.doiDOI: 10.26717/BJSTR.2022.41.006672

Ogidi, O. I. (2024). Therapeutic Role of Biomaterials in The Study of Severe Acute Respiratory Syndrome Corona Virus 2 (Sars-Cov-2). In: Advances in Sustainable Biomaterials: Bioprocessing 4.0, Characterizations and Application. *CRC Press, Taylor and Francis Group, Florida, USA.*

Ogidi, O. I. (2024b). Recent advances in anticancer activity and Bioinformatics Approach from Potential Plants. In: Computational Approaches in Biomaterials, Bioinformatics and Biomedical Engineering Applications. *CRC Press, Taylor and Francis Group, Florida, USA*. DOI: 10.1201/9781003354437-3

Ogidi, O. I., & Ajoko, I. (2024). Neuroprotective Potentials of Phytochemicals. In: M.S. Rajput, T. Sarachana, P.D. Sarkar, M. Raval (eds) NeuroPhytomedicine. *CRC Press, Routledge Taylor and Francis Group.*

Ogidi, O. I., & Emaikwu, N. G. (2024). Plant Phenolic compound isolation and its Bioinformatics Approaches of Molecular Mechanisms in Antimicrobial activities and Resistance. In: Computational Approaches in Biomaterials, Bioinformatics and Biomedical Engineering Applications. *CRC Press, Taylor and Francis Group, Florida, USA*. DOI: 10.1201/9781003354437-8

Ogidi, O.I., Berefagha, W.L., & Okara, E. (2021). Covid-19 vaccination: The pros and cons. *World Journal of Biology Pharmacy and Health Sciences,* 07(01), 015–022. https://doi.org/DOI: 10.30574/wjbphs.2021.7.1.0072

Ogidi, O. I., Barwant, M. M., & Okara, E. (2024). Biomarker Discovery: Protein Phosphorylation State. In *Spatially Variable Genes in Cancer: Development, Progression, and Treatment Response*. IGI Global., DOI: 10.4018/979-8-3693-7728-4.ch005

Ogidi, O. I., & Kumar, P. (2024). Medical and Pharmaceutical Applications of Sustainable Biopolymers. In *Advances in Sustainable Biomaterials: Fundamentals, Modelling and Characterizations* (pp. 431–453). Elsevier., DOI: 10.1016/B978-0-443-13849-2.00023-5

Ogidi, O. I., & Tobia, P. S. (2024). Harnessing Artificial Intelligence (AI) Techniques for the prevention of Microbial Disease in Humans. In *Artificial Intelligence in Microbiology: scopes and Challenges*. Elsevier.

Okafor, N. I., & Nnaji, N. J. (2024). Liposomes, facilitating the encapsulation and improved solubility of zidovudine antiretroviral drug. *BioNanoScience, 3*(5), 1–8. DOI: 10.1007/s12668-024-01504-y

Osterholm, M. T., Moore, K. A., Kelley, N. S., Brosseau, L. M., Wong, G., Murphy, F. A., Peters, C. J., LeDuc, J. W., Russell, P. K., Van Herp, M., Kapetshi, J., Muyembe, J. J., Ilunga, B. K., Strong, J. E., Grolla, A., Wolz, A., Kargbo, B., Kargbo, D. K., Sanders, D. A., & Kobinger, G. P. (2015). Transmission of Ebola viruses: What we know and what we do not know. *mBio*, *6*(2), e00137. DOI: 10.1128/mBio.00137-15 PMID: 25698835

Oster, M. E., Shay, D. K., Su, J. R., Gee, J., Creech, C. B., Broder, K. R., Edwards, K., Soslow, J. H., Dendy, J. M., Schlaudecker, E., Lang, S. M., Barnett, E. D., Ruberg, F. L., Smith, M. J., Campbell, M. J., Lopes, R. D., Sperling, L. S., Baumblatt, J. A., Thompson, D. L., & Shimabukuro, T. T. (2022). Myocarditis cases reported after mRNA-based COVID-19 vaccination in the US from December 2020 to August 2021. *Journal of the American Medical Association*, *327*(4), 331–340. DOI: 10.1001/jama.2021.24110 PMID: 35076665

Painter, G. R., Natchus, M. G., Cohen, O., Holman, W., & Painter, W. P. (2021, October). Developing a direct acting, orally available antiviral agent in a pandemic: The evolution of molnupiravir as a potential treatment for COVID-19. *Current Opinion in Virology*, *50*, 17–22. DOI: 10.1016/j.coviro.2021.06.003 PMID: 34271264

Painter, W. P., Holman, W., Bush, J. A., Almazedi, F., Malik, H., Eraut, N., Morin, M. J., Szewczyk, L. J., & Painter, G. R. (2021, May 1). Human Safety, Tolerability, and Pharmacokinetics of Molnupiravir, a Novel Broad-Spectrum Oral Antiviral Agent with Activity Against SARS-CoV-2. *Antimicrobial Agents and Chemotherapy*, *65*(5), e02428-20. Advance online publication. DOI: 10.1128/AAC.02428-20 PMID: 33649113

Pal, A., & Kundu, R. (2020). Human Papillomavirus E6 and E7: The Cervical Cancer Hallmarks and Targets for Therapy. *Frontiers in Microbiology*, *10*, 3116. DOI: 10.3389/fmicb.2019.03116 PMID: 32038557

Pandey, M., Choudhury, H., Abdul-Aziz, A., Bhattamisra, S. K., Gorain, B., Su, J. S., Tan, C. L., Chin, W. Y., & Yip, K. Y. (2020). Advancement on sustained antiviral ocular drug delivery for herpes simplex virus keratitis: Recent update on potential investigation. *Pharmaceutics*, *13*(1), 1. DOI: 10.3390/pharmaceutics13010001 PMID: 33374925

Pandit, J. A., Radin, J. M., Quer, G., & Topol, E. J. (2022). Smartphone apps in the COVID-19 pandemic. *Nature Biotechnology*, *40*(9), 1013–1022. DOI: 10.1038/s41587-022-01350-x PMID: 35726090

Parambi, D. G. T., Alharbi, K. S., Kumar, R., Harilal, S., Batiha, G. E., Cruz-Martins, N., Magdy, O., Musa, A., Panda, D. S., & Mathew, B. (2022, January). Gene Therapy Approach with an Emphasis on Growth Factors: Theoretical and Clinical Outcomes in Neurodegenerative Diseases. *Molecular Neurobiology*, *59*(1), 191–233. DOI: 10.1007/s12035-021-02555-y PMID: 34655056

Pardi, N., & Krammer, F. (2024). mRNA vaccines for infectious diseases—Advances, challenges and opportunities. *Nature Reviews. Drug Discovery*, *23*(11), 1–24. DOI: 10.1038/s41573-024-01042-y PMID: 39367276

Parker, S., & Patel, A. (2020). Digital health interventions for chronic viral diseases: A systematic review. *BMC Public Health*, *20*, 940. DOI: 10.1186/s12889-020-09468-2

Parkhitko, A. A., Jouandin, P., Mohr, S. E., & Perrimon, N. (2019). Methionine metabolism and methyltransferases in the regulation of aging and lifespan extension across species. *Aging Cell*, *18*(6), e13034. DOI: 10.1111/acel.13034 PMID: 31460700

Park, J., Choi, Y., Chang, H., Um, W., Ryu, J. H., & Kwon, I. C. (2019). Alliance with EPR effect: Combined strategies to improve the EPR effect in the tumor microenvironment. *Theranostics*, *9*(26), 8073–8090. DOI: 10.7150/thno.37198 PMID: 31754382

Parthasarathy, U., Martinelli, R., Vollmann, E. H., Best, K., & Therien, A. G. (2022). The impact of DAMP-mediated inflammation in severe COVID-19 and related disorders. *Biochemical Pharmacology*, *195*, 114847. DOI: 10.1016/j.bcp.2021.114847 PMID: 34801526

Parvathaneni, V., Kulkarni, N. S., Muth, A., & Gupta, V. (2019). Drug repurposing: A promising tool to accelerate the drug discovery process. *Drug Discovery Today*, *24*(10), 2076–2085. DOI: 10.1016/j.drudis.2019.06.014 PMID: 31238113

Pascal, K. E., Coleman, C. M., Mujica, A. O., Kamat, V., Badithe, A., Fairhurst, J., Hunt, C., Strein, J., Berrebi, A., Sisk, J. M., Matthews, K. L., Babb, R., Chen, G., Lai, K. M., Huang, T. T., Olson, W., Yancopoulos, G. D., Stahl, N., Frieman, M. B., & Kyratsous, C. A. (2015). Pre- and postexposure efficacy of fully human antibodies against Spike protein in a novel humanized mouse model of MERS-CoV infection. *Proceedings of the National Academy of Sciences of the United States of America*, *112*(28), 8738–8743. DOI: 10.1073/pnas.1510830112 PMID: 26124093

Patel, A., & Verma, S. (2022). Digital health innovations for chronic disease management: Focus on viral infections. *Health Informatics Journal*, *28*(4), 647–662.

Patel, H., & Desai, M. (2020). Wearable devices for monitoring viral diseases: The next step in personalized antiviral care. *Future Science OA*, *6*(4), FSO542.

Patel, R., Singh, M., & Sharma, P. (2023). Bioinformatics and AI in pandemic preparedness: Applications for pathogen detection and outbreak modeling. *Journal of Biomedical Informatics*, *139*, 104578.

Patel, V. V., Darwal, R., & Rao, A. (2021). Impact of wearable technology on patient care in antiviral therapies. *BMC Medical Informatics and Decision Making*, *21*(1), 141. PMID: 33931058

Paules, C., Subbarao, K., & Fauci, A. S. (2022). Influenza: Progress and remaining challenges. *Clinical Microbiology Reviews*, *35*(2), e00224–e20. DOI: 10.1128/CMR.00224-20

Peci, A., Winter, A. L., Li, Y., Gnaneshan, S., Liu, J., Mubareka, S., & Gubbay, J. B. (2019). Effects of Absolute Humidity, Relative Humidity, Temperature, and Wind Speed on Influenza Activity in Toronto, Ontario, Canada. *Applied and Environmental Microbiology*, *85*(6), e02426–e18. DOI: 10.1128/AEM.02426-18 PMID: 30610079

Peek, N., Sujan, M., & Scott, P. (2020). Digital health and care in pandemic times: Impact of COVID-19. *BMJ Health & Care Informatics*, *27*(1), e100166. DOI: 10.1136/bmjhci-2020-100166 PMID: 32565418

Peeriga, R., Arabath, S. A. M. S., Rao, N. R. A., Yenireddy, A., & Bangaru, G. K. V. (2023). Exploring the coumarins from stem extract of *Ipomea sagittifolia* Burm. f. by high-performance thin layer chromatography. *International Journal of Chemical and Biochemical Science*, *23*(3), 268–273.

Peeriga, R., Balagani, P. K., Raju, G. E., Narayanan, R., Mohammed, J., & Yarlagadda, L. C. (2023). Findings of steroids, coumarins and saponins in *Ipomoea sagittifolia* (Burm. f.) leaf extract by HPTLC fingerprints. *International Journal of Chemical and Biochemical Science*, *24*(4), 132–140.

Peeriga, R., & Balamuruga, S. (2023). Investigation on ethanolic stem extract of *Ipomoea sagittifolia* to explore the presence of saponins by high performance thin layer chromatography. *International Journal of Membrane Science and Technology*, *10*(4), 1012–1020. DOI: 10.15379/ijmst.v10i4.2210

Peeriga, R., Shaik, A., Arabath, S., Chennuru, M. L., & Marimganti, S. N. (2023). Assessment of steroids in ethanolic stem extract of Morning glory by high performance thin layer chromatography. *International Journal of Chemical and Biochemical Science*, *23*(3), 361–366.

Peeri, N. C., Shrestha, N., Rahman, M. S., Zaki, R., Tan, Z., Bibi, S., Baghbanzadeh, M., Aghamohammadi, N., Zhang, W., & Haque, U. (2020). The SARS, MERS and novel coronavirus (COVID-19) epidemics, the newest and biggest global health threats: What lessons have we learned? *International Journal of Epidemiology*, *49*(3), 717–726. DOI: 10.1093/ije/dyaa033 PMID: 32086938

Peiris, J. S., Chu, C. M., Cheng, V. C., Chan, K. S., Hung, I. F., Poon, L. L., Law, K. I., Tang, B. S., Hon, T. Y., Chan, C. S., Chan, K. H., Ng, J. S., Zheng, B. J., Ng, W. L., Lai, R. W., Guan, Y., & Yuen, K. Y.HKU/UCH SARS Study Group. (2003). Clinical progression and viral load in a community outbreak of coronavirus-associated SARS pneumonia: A prospective study. *Lancet*, *361*(9371), 1767–1772. DOI: 10.1016/S0140-6736(03)13412-5 PMID: 12781535

Petersen, L. R., Brault, A. C., & Nasci, R. S. (2013). West Nile virus: Review of the literature. *Journal of the American Medical Association*, *310*(3), 308–315. DOI: 10.1001/jama.2013.8042 PMID: 23860989

Phelps, K. L., Hamel, L., Alhmoud, N., Ali, S., Bilgin, R., Sidamonidze, K., Urushadze, L., Karesh, W., & Olival, K. J. (2019). Bat Research Networks and Viral Surveillance: Gaps and Opportunities in Western Asia. *Viruses*, *11*(3), 240. DOI: 10.3390/v11030240 PMID: 30857374

Pinna, S. M., Lupia, T., Scabini, S., Vita, D., De Benedetto, I., Gaviraghi, A., Colasanto, I., Varese, A., Cattel, F., & De Rosa, F. G. (2021). Monoclonal antibodies for the treatment of COVID-19 patients: An umbrella to overcome the storm? *International Immunopharmacology*, *101*, 108200. DOI: 10.1016/j.intimp.2021.108200 PMID: 34607231

Polack, F. P., Thomas, S. J., Kitchin, N., Absalon, J., Gurtman, A., Lockhart, S., Perez, J. L., Pérez Marc, G., Moreira, E. D., Zerbini, C., Bailey, R., Swanson, K. A., Roychoudhury, S., Koury, K., Li, P., Kalina, W. V., Cooper, D., Frenck, R. W.Jr, Hammitt, L. L., & Gruber, W. C. (2020). Safety and efficacy of the BNT162b2 mRNA Covid-19 vaccine. *The New England Journal of Medicine*, *383*(27), 2603–2615. DOI: 10.1056/NEJMoa2034577 PMID: 33301246

Pradhan, D., Biswasroy, P., Goyal, A., Ghosh, G., & Rath, G. (2021). Recent advancement in nanotechnology-based drug delivery system against viral infections. *AAPS PharmSciTech*, *22*(1), 1–9. DOI: 10.1208/s12249-020-01908-5 PMID: 33447909

Prakash, T. P., Graham, M. J., Yu, J., Carty, R., Low, A., Chappell, A., Schmidt, K., Zhao, C., Aghajan, M., Murray, H. F., Riney, S., Booten, S. L., Murray, S. F., Gaus, H., Crosby, J., Lima, W. F., Guo, S., Monia, B. P., Swayze, E. E., & Seth, P. P. (2014). Targeted delivery of antisense oligonucleotides to hepatocytes using triantennary N-acetyl galactosamine improves potency 10-fold in mice. *Nucleic Acids Research*, *42*(13), 8796–8807. DOI: 10.1093/nar/gku531 PMID: 24992960

Prather, K. A., Wang, C. C., & Schooley, R. T. (2020). Reducing transmission of SARS-CoV-2. *Science*, *368*(6498), 1422–1424. DOI: 10.1126/science.abc6197 PMID: 32461212

PREVAIL II Writing Group. (2016). A randomized, controlled trial of ZMapp for Ebola virus infection. *The New England Journal of Medicine*, *375*(15), 1448–1456. DOI: 10.1056/NEJMoa1604330 PMID: 27732819

Prothero, R. M. Chapter V: Urbanization: A Hot-Bed Of Vector-Borne Diseases. In *Demography and Vector-Borne Diseases* (pp. 59-84). CRC Press.

Przemska-Kosicka, A., Childs, C. E., Enani, S., Maidens, C., Dong, H., Dayel, I. B., Tuohy, K., Todd, S., Gosney, M. A., & Yaqoob, P. (2016). The effect of a synbiotic on the response to seasonal influenza vaccination is strongly influenced by the degree of immunosenescence. *Immunity & Ageing*, *13*(6), 6. Advance online publication. DOI: 10.1186/s12979-016-0061-4 PMID: 26985232

Qahwaji, R., Ashankyty, I., Sannan, N. S., Hazzazi, M. S., Basabrain, A. A., & Mobashir, M. (2024). Pharmacogenomics: A Genetic Approach to Drug Development and Therapy. *Pharmaceuticals (Basel, Switzerland)*, *17*(7), 940. DOI: 10.3390/ph17070940 PMID: 39065790

Quer, J., Colomer-Castell, S., Campos, C., Andrés, C., Piñana, M., Cortese, M. F., González-Sánchez, A., Garcia-Cehic, D., Ibáñez, M., Pumarola, T., Rodríguez-Frías, F., Antón, A., & Tabernero, D. (2022). Next-Generation Sequencing for Confronting Virus Pandemics. *Viruses*, *14*(3), 600. DOI: 10.3390/v14030600 PMID: 35337007

Rabaan, A. A., Al-Ahmed, S. H., Muhammad, J., Khan, A., Sule, A. A., Tirupathi, R., Mutair, A. A., Alhumaid, S., Al-Omari, A., Dhawan, M., Tiwari, R., Sharun, K., Mohapatra, R. K., Mitra, S., Bilal, M., Alyami, S. A., Emran, T. B., Moni, M. A., & Dhama, K. (2021). Role of inflammatory cytokines in COVID-19 patients: A review on molecular mechanisms, immune functions, immunopathology and immunomodulatory drugs to counter cytokine storm. *Vaccines*, *9*(5), 436. DOI: 10.3390/vaccines9050436 PMID: 33946736

Rajan, R., & Suresh, R. (2020). CRISPR-based antiviral strategies for treating viral infections: Current trends. *Frontiers in Microbiology*, *11*, 577523.

Raji, V., Pal, K., Zaheer, T., Kalarikkal, N., Sabu, T., de Souza, F. G., & Asiya, S. I. (2020). Gold nanoparticles against respiratory diseases: Oncogenic and viral pathogens review. *Therapeutic Delivery*, *11*(8), 521–534. DOI: 10.4155/tde-2020-0071 PMID: 32757745

Rajkumar, R. P., Kar, S. K., Chatterjee, S., & Yadav, S. (2021). Telemedicine in the era of COVID-19: Opportunities, challenges, and future directions. *Journal of Medical Systems*, *45*(5), 65.

Raj, V. S., Mou, H., Smits, S. L., Dekkers, D. H., Müller, M. A., Dijkman, R., Muth, D., Demmers, J. A., Zaki, A., Fouchier, R. A., Thiel, V., Drosten, C., Rottier, P. J., Osterhaus, A. D., Bosch, B. J., & Haagmans, B. L. (2013). Dipeptidyl peptidase 4 is a functional receptor for the emerging human coronavirus-EMC. *Nature*, *495*(7440), 251–254. DOI: 10.1038/nature12005 PMID: 23486063

Ralli, G. (2022). Blocking the entry of HIV into host cells through co-receptor inhibition.

Rani, S., & Jain, R. (2021). Future of mHealth in viral disease monitoring: Challenges and opportunities. *Telemedicine Journal and e-Health*, *27*(9), 1010–1018.

Rao, M. R., Ghadge, I. S., Kulkarni, S. P., & Asthana, T. (2023). Computational techniques for drug repurposing: A paradigm shift in drug discovery. *Current Drug Therapy*, *18*(4), 271–284. DOI: 10.2174/1574885518666230207143523

Rao, M. S., Van Vleet, T. R., Ciurlionis, R., Buck, W. R., Mittelstadt, S. W., Blomme, E. A. G., & Liguori, M. J. (2019). Comparison of RNA-Seq and Microarray Gene Expression Platforms for the Toxicogenomic Evaluation of Liver From Short-Term Rat Toxicity Studies. *Frontiers in Genetics*, *9*, 636. DOI: 10.3389/fgene.2018.00636 PMID: 30723492

Rayamajhi, N., Cheng, C. C., & Catchen, J. M. (2022). Evaluating Illumina-, Nanopore-, and PacBio-based genome assembly strategies with the bald notothen, Trematomus borchgrevinki. *G3 (Bethesda, Md.)*, *12*(11), jkac192. Advance online publication. DOI: 10.1093/g3journal/jkac192 PMID: 35904764

Ray, K. K., Landmesser, U., Leiter, L. A., Kallend, D., Dufour, R., Karakas, M., & Wright, R. S. (2020). Inclisiran in patients at high cardiovascular risk with elevated LDL cholesterol. *The New England Journal of Medicine*, *382*(16), 1507–1519. DOI: 10.1056/NEJMoa1912387 PMID: 32187462

Reddy, P., & Singh, M. (2023). The Impact of Research and Development Investments on Drug Development Outcomes and Health Improvements. *international Journal of Responsible Artificial Intelligence, 13*(12), 1-10.

Resop, R. S., Fromentin, R., Newman, D., Rigsby, H., Dubrovsky, L., Bukrinsky, M., Chomont, N., & Bosque, A. (2020). Fingolimod inhibits multiple stages of the HIV-1 life cycle. *PLoS Pathogens, 16*(8), e1008679. DOI: 10.1371/journal.ppat.1008679 PMID: 32790802

Reuters. (2024). China approves CSPC Pharma's mRNA cancer therapy clinical trials. *Reuters*. Retrieved from https://www.reuters.com

Reyed, R. M., & Prabhakar, P. K. (2025). An In-Depth Assessment of Dietary Methods and Nutritional Treatments to Improve Female Athletes. In Chatterjee, A., & Sarkar, T. (Eds.), *Examining Physiology, Nutrition, and Body Composition in Sports Science* (pp. 35–66). IGI Global., DOI: 10.4018/979-8-3693-6317-1.ch002

Reyes, E. D., Kulej, K., Pancholi, N. J., Akhtar, L. N., Avgousti, D. C., Kim, E. T., Bricker, D. K., Spruce, L. A., Koniski, S. A., Seeholzer, S. H., Isaacs, S. N., Garcia, B. A., & Weitzman, M. D. (2017). Identifying Host Factors Associated with DNA Replicated During Virus Infection. *Molecular & Cellular Proteomics, 16*(12), 2079–2097. DOI: 10.1074/mcp.M117.067116 PMID: 28972080

Riccio, G., Ruocco, N., Mutalipassi, M., Costantini, M., Zupo, V., Coppola, D., de Pascale, D., & Lauritano, C. (2020, July 7). Ten-Year Research Update Review: Antiviral Activities from Marine Organisms. *Biomolecules, 10*(7), 1007. Advance online publication. DOI: 10.3390/biom10071007 PMID: 32645994

Ritchie, S. A., van den Hurk, A. F., Smout, M. J., Staunton, K. M., & Hoffmann, A. A. (2022). Mission accomplished? *Wolbachia* mosquito deployments eliminate dengue transmission in North Queensland, Australia. *Gates Open Research, 6*, 54. DOI: 10.12688/gatesopenres.13611.2

Roback, J. D., & Guarner, J. (2020). Convalescent plasma to treat COVID-19: Possibilities and challenges. *Journal of the American Medical Association, 323*(16), 1561–1562. DOI: 10.1001/jama.2020.4940 PMID: 32219429

Roberts, C., & Brown, L. (2021). Telehealth interventions for antiviral treatment in remote areas: A systematic review and meta-analysis. *Telemedicine Journal and e-Health, 27*(6), 123–130.

Robinson, J. C. (2021). Funding of pharmaceutical innovation during and after the COVID-19 pandemic. *Journal of the American Medical Association, 325*(9), 825–826. DOI: 10.1001/jama.2020.25384 PMID: 33443546

Robinson, J. C. (2023). How Industrial Policy Could Accelerate Innovation In The Life Sciences: Article examines how industrial policy could accelerate innovation in the life sciences. *Health Affairs*, *42*(8), 1045–1053. DOI: 10.1377/hlthaff.2023.00047 PMID: 37549321

Rosenblum, J. E., & Chang, A. K. (2019). mHealth for chronic viral infections: A systematic review. *Journal of Mobile Health*, *6*(2), 30–37.

Roy, S., & Singh, B. (2020). The role of artificial intelligence in managing viral outbreaks. *IEEE Access : Practical Innovations, Open Solutions*, *8*, 123589–123596.

Russo, E., Fiorindi, C., Giudici, F., & Amedei, A. (2022). Immunomodulation by probiotics and prebiotics in hepatocellular carcinoma. *World Journal of Hepatology*, *14*(2), 372–385. DOI: 10.4254/wjh.v14.i2.372 PMID: 35317185

Saag, M. S., Gandhi, R. T., Hoy, J. F., Landovitz, R. J., Thompson, M. A., Sax, P. E., Smith, D. M., Benson, C. A., Buchbinder, S. P., Del Rio, C., Eron, J. J.Jr, Fätkenheuer, G., Günthard, H. F., Molina, J.-M., Jacobsen, D. M., & Volberding, P. A. (2020). Antiretroviral drugs for treatment and prevention of HIV infection in adults: 2020 recommendations of the International Antiviral Society–USA Panel. *Journal of the American Medical Association*, *324*(16), 1651–1669. DOI: 10.1001/jama.2020.17025 PMID: 33052386

Sadoff, J., Gray, G., Vandebosch, A., Cárdenas, V., Shukarev, G., Grinsztejn, B., Goepfert, P. A., Truyers, C., Fennema, H., Spiessens, B., Offergeld, K., Scheper, G., Taylor, K. L., Robb, M. L., Treanor, J., Barouch, D. H., Stoddard, J., Ryser, M. F., Marovich, M. A., & Douoguih, M. (2021). Safety and efficacy of single-dose Ad26.COV2.S vaccine against Covid-19. *The New England Journal of Medicine*, *384*(23), 2187–2201. DOI: 10.1056/NEJMoa2101544 PMID: 33882225

Saehlee, S., Seetaha, S., Klankaew, W., Srathong, P., Choowongkomon, K., & Choengpanya, K. (2024). Anti-Human Immunodeficiency Virus-1 Property of Thai Herbal Extract Kerra™. *Pharmaceuticals (Basel, Switzerland)*, *17*(7), 917. DOI: 10.3390/ph17070917 PMID: 39065767

Safarnezhad Tameshkel, F., Karbalaie Niya, M. H., Sohrabi, M., Panahi, M., Zamani, F., Imanzade, F., & Rakhshani, N. (2016). Polymorphism of IL-28B Gene (rs12979860) in HCV Genotype 1 Patients Treated by Pegylated Interferon and Ribavirin. *Iranian Journal of Pathology*, *11*(3), 216–221. PMID: 27799970

Sahu, T., Ratre, Y. K., Chauhan, S., Bhaskar, L. V., Nair, M. P., & Verma, H. K. (2021). Nanotechnology-based drug delivery system: Current strategies and emerging therapeutic potential for medical science. *Journal of Drug Delivery Science and Technology*, *63*, 102487. DOI: 10.1016/j.jddst.2021.102487

Sajid, M. I., Moazzam, M., Cho, Y., Kato, S., Xu, A., Way, J., Lohan, S., & Tiwari, R. K. (2021). siRNA Therapeutics for the Therapy of COVID-19 and Other Coronaviruses. *Molecular Pharmaceutics*, *18*(6), 2105–2121. DOI: 10.1021/acs.molpharmaceut.0c01239 PMID: 33945284

Saleem, M., Ramzan, F., Nazar, M., Khan, M., Mahfooz, A., Zaka, F., Sultan, T., Raza, A., Ahmad, A., & Shakir, A. (2023). Mitigate the Risk of Zoonosis. *Zoonosis, Unique Scientific Publishers, Faisalabad, Pakistan, 1*, 643-658.

Sallam, M., & Khalil, R. (2024). Contemporary Insights into Hepatitis C Virus: A Comprehensive Review. *Microorganisms*, *12*(6), 1035. DOI: 10.3390/microorganisms12061035 PMID: 38930417

Salum, G. M., Dawood, R. M., Abd El-Meguid, M., Ibrahim, N. E., Abdel Aziz, A. O., & El Awady, M. K. (2019). Correlation between IL28B/TLR4 genetic variants and HCC development with/without DAAs treatment in chronic HCV patients. *Genes & Diseases*, *7*(3), 392–400. DOI: 10.1016/j.gendis.2019.05.004 PMID: 32884993

Samira, A., Ehizogie Paul, A., Bisola Oluwafadekemi, A., Adebukola Adejumoke, A., & Emem Henry, U. (2024). Public–private partnerships in health sector innovation: Lessons from around the world. *Magna Scientia Advanced Biology and Pharmacy, 12*(1), 045-059. DOI:DOI: 10.30574/msabp.2024.12.1.0032

Sampat, B. N., & Shadlen, K. C. (2021). The COVID-19 Innovation System: Article describes innovations that emerged during the COVID-19 pandemic. *Health Affairs*, *40*(3), 400–409. DOI: 10.1377/hlthaff.2020.02097 PMID: 33539184

Sanders, R. W.. (2018). HIV-1 neutralizing antibodies: A comprehensive overview. *Annual Review of Immunology*, *36*, 201–232. DOI: 10.1146/annurev-immunol-042617-090014

Sanjuán, R., & Domingo-Calap, P. (2016). Mechanisms of viral mutation. *Cellular and Molecular Life Sciences : CMLS*, *73*(23), 4433–4448. DOI: 10.1007/s00018-016-2299-6 PMID: 27392606

Sanna, V., Satta, S., Hsiai, T., & Sechi, M. (2022). Development of targeted nanoparticles loaded with antiviral drugs for SARS-CoV-2 inhibition. *European Journal of Medicinal Chemistry*, *231*, 114121. DOI: 10.1016/j.ejmech.2022.114121 PMID: 35114539

Sarkar, C., Das, B., Rawat, V. S., Wahlang, J. B., Nongpiur, A., Tiewsoh, I., Lyngdoh, N. M., Das, D., Bidarolli, M., & Sony, H. T. (2023). Artificial intelligence and machine learning technology driven modern drug discovery and development. *International Journal of Molecular Sciences*, *24*(3), 2026. DOI: 10.3390/ijms24032026 PMID: 36768346

Sarkar, C., Mondal, M., Torequl Islam, M., Martorell, M., Docea, A. O., Maroyi, A., Sharifi-Rad, J., & Calina, D. (2020). Potential Therapeutic Options for COVID-19: Current Status, Challenges, and Future Perspectives. *Frontiers in Pharmacology*, *11*, 572870. DOI: 10.3389/fphar.2020.572870 PMID: 33041814

Sarkar, J.. (2022). Needle-free vaccination: A new paradigm in immunization strategies. *Vaccine*, *40*(15), 2201–2212. DOI: 10.1016/j.vaccine.2022.02.011

Sarker, D., Plummer, R., Meyer, T., Sodergren, M. H., Basu, B., Chee, C. E., Huang, K. W., Palmer, D. H., Ma, Y. T., Evans, T. R. J., Spalding, D. R. C., Pai, M., Sharma, R., Pinato, D. J., Spicer, J., Hunter, S., Kwatra, V., Nicholls, J. P., Collin, D., & Habib, N. (2020). MTL-CEBPA, a Small Activating RNA Therapeutic Upregulating C/EBP-α, in Patients with Advanced Liver Cancer: A First-in-Human, Multicenter, Open-Label, Phase I Trial. *Clinical Cancer Research*, *26*(15), 3936–3946. DOI: 10.1158/1078-0432.CCR-20-0414 PMID: 32357963

Satam, H., Joshi, K., Mangrolia, U., Waghoo, S., Zaidi, G., Rawool, S., Thakare, R. P., Banday, S., Mishra, A. K., Das, G., & Malonia, S. K. (2023). Next-Generation Sequencing Technology: Current Trends and Advancements. *Biology (Basel)*, *12*(7), 997. DOI: 10.3390/biology12070997 PMID: 37508427

Schenk, E., García, P., & Capron, A. National Academies of Sciences, E., & Medicine. (2021). Therapeutics. In *Public Health Lessons for Non-Vaccine Influenza Interventions: Looking Past COVID-19*. National Academies Press (US).

Schiano Moriello, N., Pinchera, B., & Gentile, I. (2024). Personalized care approaches to hepatitis C therapy: Recent advances and future directions. *Expert Review of Anti-Infective Therapy*, *22*(4), 139–151. DOI: 10.1080/14787210.2024.2328336 PMID: 38459735

Schneider, A. R., & Busse, R. (2020). The integration of digital health tools into the healthcare system. *Healthcare Policy = Politiques de Santé*, *16*(4), 9–20.

Schwartz, D. A., & Graham, A. L. (2020). Potential maternal and infant outcomes from coronavirus 2019-nCoV (SARS-CoV-2) infecting pregnant women: Lessons from SARS, MERS, and other human coronavirus infections. *Viruses*, *12*(2), 194. DOI: 10.3390/v12020194 PMID: 32050635

Sha, A., Liu, Y., & Hao, H. (2023). Current state-of-the-art and potential future therapeutic drugs against COVID-19. *Frontiers in Cell and Developmental Biology*, *11*, 1238027. DOI: 10.3389/fcell.2023.1238027 PMID: 37691829

Shah, A., Aftab, S., Nisar, J., Ashiq, M. N., & Iftikhar, F. J. (2021). Nanocarriers for targeted drug delivery. *Journal of Drug Delivery Science and Technology*, *62*, 102426. DOI: 10.1016/j.jddst.2021.102426

Shah, N. J., Aloysius, M. M., Sharma, N. R., & Pallav, K. (2021). Advances in treatment and prevention of hepatitis B. *World Journal of Gastrointestinal Pharmacology and Therapeutics*, *12*(4), 56–78. DOI: 10.4292/wjgpt.v12.i4.56 PMID: 34316384

Sharif-Yakan, A., & Kanj, S. S. (2014). Emergence of MERS-CoV in the Middle East: Origins, transmission, treatment, and perspectives. *PLoS Pathogens*, *10*(12), e1004457. DOI: 10.1371/journal.ppat.1004457 PMID: 25474536

Sharma, A., Mishra, R. K., Goud, K. Y., Mohamed, M. A., Kummari, S., Tiwari, S., Li, Z., Narayan, R., Stanciu, L. A., & Marty, J. L. (2021). Optical biosensors for diagnostics of infectious viral disease: A recent update. *Diagnostics (Basel)*, *11*(11), 2083. DOI: 10.3390/diagnostics11112083 PMID: 34829430

Sharma, A., Patel, V., & Gupta, K. (2021). Digital health interventions to improve adherence to antiretroviral therapy: A meta-analysis. *JMIR mHealth and uHealth*, *9*(2), e24102.

Sharma, P., & Garg, S. (2010). Pure drug and polymer-based nanotechnologies for the improved solubility, stability, bioavailability and targeting of anti-HIV drugs. *Advanced Drug Delivery Reviews*, *62*(4–5), 491–502. DOI: 10.1016/j.addr.2009.11.019 PMID: 19931328

Sharma, P., & Yadav, K. (2020). Integration of digital tools in chronic viral disease management: A future perspective. *Future Medicine*, *22*(6), 251–260.

Shenge, J. A., & Opayele, A. V. (2020). The impact and control of emerging and re-emerging viral diseases in the environment: an African perspective. *Current Microbiological Research in Africa: Selected Applications for Sustainable Environmental Management*, 185-202.

Shen, X., & Li, J. (2020). Virtual consultations in the management of viral infections: A review of current applications. *Telemedicine Journal and e-Health*, *26*(10), 1235–1243.

Shi, J., Xiao, Y., Zhang, Y., Geng, D., Cong, D., Shi, K. X., & Knapp, R. J. (2021, May). Challenges of drug development during the COVID-19 pandemic: Key considerations for clinical trial designs. *British Journal of Clinical Pharmacology*, *87*(5), 2170–2185. DOI: 10.1111/bcp.14629 PMID: 33119136

Shimabukuro, T. T.. (2021). Reports of Anaphylaxis and Myocarditis After mRNA COVID-19 Vaccination. *Journal of the American Medical Association*, *325*(11), 1101–1102. DOI: 10.1001/jama.2021.1967 PMID: 33576785

Shi, Y., van der Meel, R., Chen, X., & Lammers, T. (2020). The EPR effect and beyond: Strategies to improve tumor targeting and cancer nanomedicine treatment efficacy. *Theranostics*, *10*(17), 7921–7924. DOI: 10.7150/thno.49577 PMID: 32685029

Shrestha, B., Tang, L., & Romero, G. (2019). Nanoparticles-mediated combination therapies for cancer treatment. *Advanced Therapeutics*, *2*(11), 1900076. DOI: 10.1002/adtp.201900076

Shyr, Z. A., Cheng, Y. S., Lo, D. C., & Zheng, W. (2021, October). Drug combination therapy for emerging viral diseases. *Drug Discovery Today*, *26*(10), 2367–2376. DOI: 10.1016/j.drudis.2021.05.008 PMID: 34023496

Siddiqui, D., Badar, U., Javaid, M., Farooqui, N., Shah, S. A., Iftikhar, A., Sultan, F., Mir, F., Furqan, S., Mahmood, S. F., & Abidi, S. H. (2023). Genetic and antiretroviral drug resistance mutations analysis of reverse transcriptase and protease gene from Pakistani people living with HIV-1. *PLoS One*, *18*(8), e0290425. DOI: 10.1371/journal.pone.0290425 PMID: 37616294

Sinclair, F., Begum, A. A., Dai, C. C., Toth, I., & Moyle, P. M. (2023). Recent advances in the delivery and applications of nonviral CRISPR/Cas9 gene editing. *Drug Delivery and Translational Research*, *13*(5), 1500–1519. DOI: 10.1007/s13346-023-01320-z PMID: 36988873

Singh, K. B. (2023). Impact of Patent (Amendment) Act, 2005 on Indian Pharmaceutical Industry. *Journal of Intellectual Property Rights*, *27*(6), 397–414. DOI: 10.56042/jipr.v27i6.63793

Singh, R., Gupta, A., & Sharma, P. (2021). Privacy concerns in digital health platforms for antiviral therapy: A systematic review. *Journal of Medical Internet Research*, *24*(9), e34123.

Singh, R., & Kaushik, A. (2021). CRISPR technology and its role in advancing antiviral therapies: Future directions. *Bioengineering & Translational Medicine*, *6*(3), e10239. PMID: 34589605

Singh, S., Kumar, R., Payra, S., & Singh, S. K. (2023). Artificial Intelligence and Machine Learning in Pharmacological Research: Bridging the Gap Between Data and Drug Discovery. *Cureus, 15*(8), e44359. DOI: 10.7759/cureus.44359 PMID: 37779744

Sinumvayo, J. P., Munezero, P. C., Tope, A. T., Adeyemo, R. O., Bale, M. I., Nyandwi, J. B., Haakuria, V. M., Mutesa, L., & Adedeji, A. A. (2024, July 3). Advancing Vaccinology Capacity: Education and Efforts in Vaccine Development and Manufacturing across Africa. *Vaccines, 12*(7), 741. Advance online publication. DOI: 10.3390/vaccines12070741 PMID: 39066380

Sirnaomics. (2023). STP707 Phase I clinical study results. *Sirnaomics*. Retrieved from https://sirnaomics.com

Sizar, O., Khare, S., & Goyal, A.. (2024 Jan). Vitamin D Deficiency. [Updated 2023 Jul 17] In *StatPearls* [Internet]. StatPearls Publishing., Available from https://www.ncbi.nlm.nih.gov/books/NBK532266/

Smith, A. E., & Brown, R. D. (2023). Next-generation sequencing: Transforming pathogen identification and resistance monitoring in low-resource settings. *International Journal of Infectious Diseases, 128*, 95–101.

Smith, R., & Jones, T. (2023). Advancements in antiviral therapies and the role of digital health. *Journal of Clinical Virology, 120*, 15–23.

Smith, S. J., Zhao, X. Z., Passos, D. O., Lyumkis, D., Burke, T. R.Jr, & Hughes, S. H. (2021). Integrase Strand Transfer Inhibitors Are Effective Anti-HIV Drugs. *Viruses, 13*(2), 205. DOI: 10.3390/v13020205 PMID: 33572956

Song, Y., Zhang, M., Yin, L., Wang, K., Zhou, Y., Zhou, M., & Lu, Y. (2020). COVID-19 treatment: Close to a cure? A rapid review of pharmacotherapies for the novel coronavirus (SARS-CoV-2). *International Journal of Antimicrobial Agents, 56*(2), 106080. DOI: 10.1016/j.ijantimicag.2020.106080 PMID: 32634603

Soriano, V., Moreno-Torres, V., de Mendoza, C., Fernández-Montero, J. V., Treviño, A., Corral, O., de Jesús, F., & Barreiro, P. (2023). Pre-exposure prophylaxis of non-HIV viral infections and the role of long-acting antivirals. *AIDS Reviews, 25*(4). Advance online publication. DOI: 10.24875/AIDSRev.M23000066 PMID: 38092029

Soriano, V., Treviño, A., de Mendoza, C., Moreno-Torres, V., Pintos, I., Barreiro, P., & Corral, O. (2022). Pre-Exposure Prophylaxis for viral infections other than HIV. *Le Infezioni in Medicina, 30*(3), 362. PMID: 36148176

Souri, M., Soltani, M., Kashkooli, F. M., & Shahvandi, M. K. (2022). Engineered strategies to enhance tumor penetration of drug-loaded nanoparticles. *Journal of Controlled Release, 341*, 227–246. DOI: 10.1016/j.jconrel.2021.11.024 PMID: 34822909

Stebbing, J., Krishnan, V., de Bono, S., Ottaviani, S., Casalini, G., Richardson, P. J., Monteil, V., Lauschke, V. M., Mirazimi, A., Youhanna, S., Tan, Y. J., Baldanti, F., Sarasini, A., Terres, J. A. R., Nickoloff, B. J., Higgs, R. E., Rocha, G., Byers, N. L., Schlichting, D. E., & Corbellino, M.Sacco Baricitinib Study Group. (2020). Mechanism of baricitinib supports artificial intelligence-predicted testing in COVID-19 patients. *EMBO Molecular Medicine, 12*(8), e12697. DOI: 10.15252/emmm.202012697 PMID: 32473600

Stefanicka-Wojtas, D., & Kurpas, D. (2023, February 21). Personalized Medicine-Implementation to the Healthcare System in Europe (Focus Group Discussions). *Journal of Personalized Medicine, 13*(3), 380. DOI: 10.3390/jpm13030380 PMID: 36983562

Stern, A., & Andino, R. (2016). Viral evolution: it is all about mutations. In *Viral pathogenesis* (pp. 233–240). Elsevier. DOI: 10.1016/B978-0-12-800964-2.00017-3

Stojanovic, J., Melano, R., & Petrovic, B. (2020). Application of telemedicine and remote health monitoring in treating chronic viral diseases. *Computer Methods and Programs in Biomedicine, 196*, 105713.

Strasfeld, L., & Chou, S. (2010). Antiviral drug resistance: Mechanisms and clinical implications. *Infectious Disease Clinics of North America, 24*(2), 413–437. DOI: 10.1016/j.idc.2010.01.001 PMID: 20466277

Strauss, J. H., & Strauss, E. G. (2012). Emerging and reemerging viral diseases. Viruses and Human Disease, 325.

Strauss, M., Mičetić Turk, D., Lorber, M., Pogačar, M. Š., Koželj, A., Tušek Bunc, K., & Fijan, S. (2023). The Multi-Strain Probiotic OMNi-BiOTiC® Active Reduces the Duration of Acute Upper Respiratory Disease in Older People: A Double-Blind, Randomized, Controlled Clinical Trial. *Microorganisms, 11*(7), 1760. DOI: 10.3390/microorganisms11071760 PMID: 37512932

Suardi, C., Cazzaniga, E., Graci, S., Dongo, D., & Palestini, P. (2021). Link between Viral Infections, Immune System, Inflammation and Diet. *International Journal of Environmental Research and Public Health, 18*(5), 2455. DOI: 10.3390/ijerph18052455 PMID: 33801527

Sudevan, N., Manrai, M., Tilak, T. V. S. V. G. K., Khurana, H., & Premdeep, H. (2024). Chronic hepatitis B and occult infection in chemotherapy patients - evaluation in oncology and hemato-oncology settings: The CHOICE study. *World Journal of Virology, 13*(1), 89104. DOI: 10.5501/wjv.v13.i1.89104 PMID: 38616860

Sugo, T., Terada, M., Oikawa, T., Miyata, K., Nishimura, S., Kenjo, E., Ogasawara-Shimizu, M., Makita, Y., Imaichi, S., Murata, S., Otake, K., Kikuchi, K., Teratani, M., Masuda, Y., Kamei, T., Takagahara, S., Ikeda, S., Ohtaki, T., & Matsumoto, H. (2016). Development of antibody-siRNA conjugate targeted to cardiac and skeletal muscles. *Journal of controlled release: official journal of the Controlled Release Society, 237*, 1–13. https://doi.org/DOI: 10.1016/j.jconrel.2016.06.036

Suhaimi, N. A., Ahmad, S., Husna, S. M., Sarmiento, M. E., Acosta, A., Norazmi, M. N., Ibrahim, J., Mohamud, R., & Kadir, R. (2022). Application of liposomes in the treatment of infectious diseases. *Life Sciences, 305*, 120734. DOI: 10.1016/j.lfs.2022.120734 PMID: 35760094

Su, J., Yang, L., Sun, Z., & Zhan, X. (2024). Personalized Drug Therapy: Innovative Concept Guided With Proteoformics. *Molecular & Cellular Proteomics, 23*(3), 100737. DOI: 10.1016/j.mcpro.2024.100737 PMID: 38354979

Suman, A., & Mishra, S. (2020). Role of CRISPR technology in managing viral infections: A comprehensive review. *Frontiers in Microbiology, 11*, 1093.

Sun, D., Gao, W., Hu, H., & Zhou, S. (2022, July). Why 90% of clinical drug development fails and how to improve it? *Acta Pharmaceutica Sinica. B, 12*(7), 3049–3062. DOI: 10.1016/j.apsb.2022.02.002 PMID: 35865092

Sun, J., Zhuang, Z., Zheng, J., Li, K., Wong, R. L., Liu, D., Huang, J., He, J., Zhu, A., Zhao, J., Li, X., Xi, Y., Chen, R., Alshukairi, A. N., Chen, Z., Zhang, Z., Chen, C., Huang, X., Li, F., & Zhao, J. (2020). Generation of a Broadly Useful Model for COVID-19 Pathogenesis, Vaccination, and Treatment. *Cell, 182*(3), 734–743.e5. DOI: 10.1016/j.cell.2020.06.010 PMID: 32643603

Sun, S. H., Chen, Q., Gu, H. J., Yang, G., Wang, Y. X., Huang, X. Y., Liu, S. S., Zhang, N. N., Li, X. F., Xiong, R., Guo, Y., Deng, Y. Q., Huang, W. J., Liu, Q., Liu, Q. M., Shen, Y. L., Zhou, Y., Yang, X., Zhao, T. Y., & Wang, Y. C. (2020). A Mouse Model of SARS-CoV-2 Infection and Pathogenesis. *Cell Host & Microbe, 28*(1), 124–133.e4. DOI: 10.1016/j.chom.2020.05.020 PMID: 32485164

Surur, A. S., Fekadu, A., Makonnen, E., & Hailu, A. (2020). Challenges and opportunities for drug discovery in developing countries: The example of cutaneous leishmaniasis. *ACS Medicinal Chemistry Letters, 11*(11), 2058–2062. DOI: 10.1021/acsmedchemlett.0c00446 PMID: 33214808

Su, S., & Kang, P. M. (2020). Recent advances in nanocarrier-assisted therapeutics delivery systems. *Pharmaceutics, 12*(9), 837. DOI: 10.3390/pharmaceutics12090837 PMID: 32882875

Sutherland, K. (2014). *Contribution of Gag and protease to variation in susceptibility to protease inhibitors between different strains of HIV-1* UCL (University College London)].

Su, Y., & Wang, X. (2019). Personalized medicine in antiviral treatment: The role of artificial intelligence. *Frontiers in Immunology, 10*, 1833.

Szabat, M., Lorent, D., Czapik, T., Tomaszewska, M., Kierzek, E., & Kierzek, R. (2020). RNA Secondary Structure as a First Step for Rational Design of the Oligonucleotides towards Inhibition of Influenza A Virus Replication. *Pathogens (Basel, Switzerland), 9*(11), 925. DOI: 10.3390/pathogens9110925 PMID: 33171815

Szliszka, E., Czuba, Z. P., Domino, M., Mazur, B., Zydowicz, G., & Krol, W. (2009, February 13). Ethanolic extract of propolis (EEP) enhances the apoptosis- inducing potential of TRAIL in cancer cells. *Molecules (Basel, Switzerland), 14*(2), 738–754. DOI: 10.3390/molecules14020738 PMID: 19223822

Szyk, P., Czarczynska-Goslinska, B., Mlynarczyk, D. T., Ślusarska, B., Kocki, T., Ziegler-Borowska, M., & Goslinski, T. (2023). Polymer-based nanoparticles as drug delivery systems for purines of established importance in medicine. *Nanomaterials (Basel, Switzerland), 13*(19), 2647. DOI: 10.3390/nano13192647 PMID: 37836288

Tabrizi, S. J., Leavitt, B. R., Landwehrmeyer, G. B., Wild, E. J., Saft, C., Barker, R. A., & Smith, A. V. (2019). Targeting huntingtin expression in patients with Huntington's disease. *The New England Journal of Medicine, 380*(24), 2307–2316. DOI: 10.1056/NEJMoa1900907 PMID: 31059641

Tan, E. L., Ooi, E. E., Lin, C. Y., Tan, H. C., Ling, A. E., Lim, B., & Stanton, L. W. (2004). Inhibition of SARS coronavirus infection in vitro with clinically approved antiviral drugs. *Emerging Infectious Diseases, 10*(4), 581–586. DOI: 10.3201/eid1004.030458 PMID: 15200845

Tang, H., Abouleila, Y., & Mashaghi, A. (2020). Lassa hemorrhagic shock syndromeon-a-chip. *Biotechnology and Bioengineering*. Advance online publication. DOI: 10.1002/bit.27636 PMID: 33241859

Tang, Y., Zhang, H., Lu, X., Jiang, L., Xi, X., Liu, J., & Zhu, J. (2015). Development and evaluation of a dry powder formulation of liposome-encapsulated oseltamivir phosphate for inhalation. *Drug Delivery, 22*(5), 608–618. DOI: 10.3109/10717544.2013.863526 PMID: 24299495

Tazerji, S. S., Nardini, R., Safdar, M., Shehata, A. A., & Duarte, P. M. (2022). An overview of anthropogenic actions as drivers for emerging and re-emerging zoonotic diseases. *Pathogens (Basel, Switzerland), 11*(11), 1376. DOI: 10.3390/pathogens11111376 PMID: 36422627

Telenti, A., Hodcroft, E. B., & Robertson, D. L. (2022). The evolution and biology of SARS-CoV-2 variants. *Cold Spring Harbor Perspectives in Medicine, 12*(5), a041390. DOI: 10.1101/cshperspect.a041390 PMID: 35444005

Temereanca, A., & Ruta, S. (2023). Strategies to overcome HIV drug resistance-current and future perspectives. *Frontiers in Microbiology, 14*, 1133407. DOI: 10.3389/fmicb.2023.1133407 PMID: 36876064

The Complex Trait Consortium. (2018). The Collaborative Cross, a community resource for the genetic analysis of complex traits. *Nature Genetics, 36*, 1133–1137. PMID: 15514660

Thomas, S. J., & Yoon, I. K. (2019). A review of Dengvaxia®: Development to deployment. *Human Vaccines & Immunotherapeutics, 15*(10), 2295–2314. DOI: 10.1080/21645515.2019.1658503 PMID: 31589551

Tian, L., Qiang, T., Liang, C., Ren, X., Jia, M., Zhang, J., Li, J., Wan, M., YuWen, X., Li, H., Cao, W., & Liu, H. (2021). RNA-dependent RNA polymerase (RdRp) inhibitors: The current landscape and repurposing for the COVID-19 pandemic. *European Journal of Medicinal Chemistry, 213*, 113201. DOI: 10.1016/j.ejmech.2021.113201 PMID: 33524687

Tian, W. J., & Wang, X. J. (2023). Broad-Spectrum Antivirals Derived from Natural Products. *Viruses, 15*(5), 1100. DOI: 10.3390/v15051100 PMID: 37243186

Tian, Y., Hu, D., Li, Y., & Yang, L. (2022). Development of therapeutic vaccines for the treatment of diseases. *Molecular Biomedicine, 3*(1), 40. DOI: 10.1186/s43556-022-00098-9 PMID: 36477638

Tian, Y., Ran, H., Wen, X., Fu, G., Zhou, X., Liu, R., & Pan, T. (2023). Probiotics improve symptoms of patients with COVID-19 through gut-lung axis: A systematic review and meta-analysis. *Frontiers in Nutrition, 10*, 1179432. DOI: 10.3389/fnut.2023.1179432 PMID: 37284648

Tiwana, M. H., & Smith, J. (2024). Faith and vaccination: A scoping review of the relationships between religious beliefs and vaccine hesitancy. *BMC Public Health, 24*(1), 1806. DOI: 10.1186/s12889-024-18873-4 PMID: 38971784

Tiwari, P., & Rana, V. (2020). The future of mHealth in managing chronic diseases: Insights from COVID-19. *Telemedicine Journal and e-Health, 26*(11), 1426–1432.

Tiwari, R., & Singh, M. (2023). Advancements in Antiviral Drug Discovery: A Comprehensive Review. *Recent Trends in Pharmaceutical Technology and Industries, 5*(2), 26–33.

Tompa, D. R., Immanuel, A., Srikanth, S., & Kadhirvel, S. (2021, March 1). Trends and strategies to combat viral infections: A review on FDA approved antiviral drugs. *International Journal of Biological Macromolecules, 172*, 524–541. DOI: 10.1016/j.ijbiomac.2021.01.076 PMID: 33454328

Treanor, J. J., Hayden, F. G., Vrooman, P. S., Barbarash, R., Bettis, R., Riff, D., & Singh, S. (2019). Efficacy and safety of the oral neuraminidase inhibitor oseltamivir in treating acute influenza. *The Journal of Infectious Diseases, 220*(4), 651–661. DOI: 10.1093/infdis/jiy684

Tsai, M. C., Hung, C. H., Lu, S. N., Wang, J. H., Chen, C. H., Kee, K. M., Chang, K. C., Chao, T. L., & Hu, T. H. (2021). The incidence of resistance-associated variants to NS5A in HCV subtypes 1a and 1b in Taiwan. *Biomedical Journal, 44*(6, Suppl 1), S126–S131. DOI: 10.1016/j.bj.2020.08.004 PMID: 35123932

Tse, L. V., Meganck, R. M., Graham, R. L., & Baric, R. S. (2020). The Current and Future State of Vaccines, Antivirals and Gene Therapies Against Emerging Coronaviruses. *Frontiers in Microbiology, 11*, 658. DOI: 10.3389/fmicb.2020.00658 PMID: 32390971

Tshiani Mbaya, O., Mukumbayi, P., & Mulangu, S. (2021). insights on current FDA-approved monoclonal antibodies against Ebola virus infection. *Frontiers in Immunology, 12*, 721328. DOI: 10.3389/fimmu.2021.721328 PMID: 34526994

Uddin, T. M., Chakraborty, A. J., Khusro, A., Zidan, B. R., Mitra, S., Emran, T. B., Dhama, K., Ripon, M. K., Gajdács, M., Sahibzada, M. U., Hossain, M. J., & Koirala, N. (2021). Antibiotic resistance in microbes: History, mechanisms, therapeutic strategies and future prospects. *Journal of Infection and Public Health, 14*(12), 1750–1766. DOI: 10.1016/j.jiph.2021.10.020 PMID: 34756812

Udeze, A. O., Olaleye, D. O., & Odaibo, G. N. (2020). Polymorphisms and drug resistance analysis of HIV-1 isolates from patients on first line antiretroviral therapy (ART) in Southeastern Nigeria. *PLoS One, 15*(4), e0231031. DOI: 10.1371/journal.pone.0231031 PMID: 32267869

Ullah, A. A., & Ferdous, J. (2022). *The post-pandemic world and global politics*. Springer. DOI: 10.1007/978-981-19-1910-7

Ul-Rahman, A., Sbabbir, M., Rasheed, M., Bano, N., Shahid, M., Habib, M., Mehmood, R., Shafi, N., Khera, H., & Kabeer, S. (2023). Zoonotic and reverse zoonotic transmission of SARS-CoV-2 virus: A perspective on human-animal interface. *Zoonosis, Unique Scientific Publishers, Faisalabad, Pakistan, 3*, 269-282.

Vakili, B., Shoaei, P., Shahzamani, K., Siadat, S. D., Shojaei, H., Esfandiari, Z., Nasri, E., Shabani, S., Zamani Moghadam, A., & Ataei, B. (2024). Gut-Lung Microbiota Characterization in Patients with Non-Small Cell Lung Carcinoma and COVID-19 Coinfection. *Archives of Iranian Medicine, 27*(2), 62–71. DOI: 10.34172/aim.2024.11 PMID: 38619029

van Griensven, J., De Weiggheleire, A., Delamou, A., Smith, P. G., Edwards, T., Vandekerckhove, P., Bah, E. I., Colebunders, R., Herve, I., Lazaygues, C., Haba, N., & Lynen, L. (2016). The use of Ebola convalescent plasma to treat Ebola virus disease in resource-constrained settings: A perspective from the field. *Clinical Infectious Diseases, 62*(1), 69–74. DOI: 10.1093/cid/civ680 PMID: 26261205

Van Poelvoorde, L. A. E., Saelens, X., Thomas, I., & Roosens, N. H. (2020). Next-Generation Sequencing: An Eye-Opener for the Surveillance of Antiviral Resistance in Influenza. *Trends in Biotechnology, 38*(4), 360–367. DOI: 10.1016/j.tibtech.2019.09.009 PMID: 31810633

Van Tran, V., Moon, J. Y., & Lee, Y. C. (2019). Liposomes for delivery of antioxidants in cosmeceuticals: Challenges and development strategies. *Journal of Controlled Release, 300*, 114–140. DOI: 10.1016/j.jconrel.2019.03.003 PMID: 30853528

Varahachalam, S. P., Lahooti, B., Chamaneh, M., Bagchi, S., Chhibber, T., Morris, K., Bolanos, J. F., Kim, N. Y., & Kaushik, A. (2021). Nanomedicine for the SARS-CoV-2: State-of-the-Art and Future Prospects. *International Journal of Nanomedicine, 16*, 539–560. DOI: 10.2147/IJN.S283686 PMID: 33519200

Varghese, N., Majeed, A., Nyalakonda, S., Boortalary, T., Halegoua-DeMarzio, D., & Hann, H. W. (2024). Review of Related Factors for Persistent Risk of Hepatitis B Virus-Associated Hepatocellular Carcinoma. *Cancers (Basel), 16*(4), 777. DOI: 10.3390/cancers16040777 PMID: 38398168

Vecchio, R., Gentile, L., Tafuri, S., Costantino, C., & Odone, A. (2024). Exploring future perspectives and pipeline progression in vaccine research and development. *Annali di igiene : medicina preventiva e di comunità, 36*(4), 446–461. PMID: 38436081

Veiga, N., Goldsmith, M., Granot, Y., Rosenblum, D., Dammes, N., Kedmi, R., Ramishetti, S., & Peer, D. (2018). Cell specific delivery of modified mRNA expressing therapeutic proteins to leukocytes. *Nature Communications, 9*(1), 4493. DOI: 10.1038/s41467-018-06936-1 PMID: 30374059

Verger, P., Botelho-Nevers, E., Garrison, A., Gagnon, D., Gagneur, A., Gagneux-Brunon, A., & Dubé, E. (2022). Vaccine hesitancy in health-care providers in Western countries: A narrative review. *Expert Review of Vaccines*, *21*(7), 909–927. DOI: 10.1080/14760584.2022.2056026 PMID: 35315308

Villar, L. M., Del Campo, J. A., Ranchal, I., Lampe, E., & Romero-Gomez, M. (2013). Association between vitamin D and hepatitis C virus infection: A meta-analysis. *World Journal of Gastroenterology*, *19*(35), 5917–5924. DOI: 10.3748/wjg.v19.i35.5917 PMID: 24124339

von Delft, A., Hall, M. D., Kwong, A. D., Purcell, L. A., Saikatendu, K. S., Schmitz, U., Tallarico, J. A., & Lee, A. A. (2023). Accelerating antiviral drug discovery: Lessons from COVID-19. *Nature Reviews. Drug Discovery*, *22*(7), 585–603. DOI: 10.1038/s41573-023-00692-8 PMID: 37173515

Voshavar, C. (2019). Protease inhibitors for the treatment of HIV/AIDS: Recent advances and future challenges. *Current Topics in Medicinal Chemistry*, *19*(18), 1571–1598. DOI: 10.2174/1568026619666190619115243 PMID: 31237209

Voysey, M.. (2021). Safety and efficacy of the ChAdOx1 nCoV-19 vaccine (AZD1222) against SARS-CoV-2: A preliminary report of a phase 3 trial. *Lancet*, *397*(10269), 99–111. DOI: 10.1016/S0140-6736(20)32661-1 PMID: 33306989

Wang, H. (2024). Practical updates in clinical antiviral resistance testing. *Journal of Clinical Microbiology*, *62*(8), e0072823. DOI: 10.1128/jcm.00728-23 PMID: 39051778

Wang, H., Zhao, Y., Pei, Z., Zhao, J., Zhang, P., Zhang, X., Zhang, Z., & Lu, W. (2024). Effect of Mixed Probiotics on Alleviating H1N1 Influenza Infection and Regulating Gut Microbiota. *Foods*, *13*(19), 3079. DOI: 10.3390/foods13193079 PMID: 39410114

Wang, J., & Liu, M. (2020). Digital health solutions for COVID-19 management: A comprehensive review. *Journal of the American Medical Informatics Association : JAMIA*, *27*(9), 1415–1428.

Wang, J., Peng, Y., Zhang, L., & Xu, Y. (2022). Virus-like particle vaccinology, from bench to bedside. *Cellular & Molecular Immunology*, *19*(4), 456–475. DOI: 10.1038/cmi.2014.72 PMID: 34992277

Wang, J., Zhu, Q., Xing, X., & Sun, D. (2024). A Mini-Review on the Common Antiviral Drug Targets of Coronavirus. *Microorganisms*, *12*(3), 600. DOI: 10.3390/microorganisms12030600 PMID: 38543651

Wang, L., Yang, Z., & Zhou, Y. (2021). A comprehensive review of AI in digital health and antiviral therapy. *Journal of Medical Systems*, *45*(4), 61.

Wang, P.. (2021). Neutralizing antibody responses to SARS-CoV-2 variants. *Nature*, *593*(7857), 133–138. DOI: 10.1038/s41586-021-03398-2 PMID: 33684923

Wang, R., Gu, Y., Chen, H., Tian, B., & Li, H. (2025). Uracil base PCR implemented for reliable DNA walking. *Analytical Biochemistry*, *696*, 115697. DOI: 10.1016/j.ab.2024.115697 PMID: 39442604

Wang, R., Xiong, K., Wang, Z., Wu, D., Hu, B., Ruan, J., Sun, C., Ma, D., Li, L., & Liao, S. (2023). Immunodiagnosis - the promise of personalized immunotherapy. *Frontiers in Immunology*, *14*, 1216901. DOI: 10.3389/fimmu.2023.1216901 PMID: 37520576

Wang, Y. H., Limaye, A., Liu, J. R., & Wu, T. N. (2022). Potential probiotics for regulation of the gut-lung axis to prevent or alleviate influenza in vulnerable populations. *Journal of Traditional and Complementary Medicine*, *13*(2), 161–169. DOI: 10.1016/j.jtcme.2022.08.004 PMID: 36970463

Wang, Y., Ling, L., Zhang, Z., & Marin-Lopez, A. (2022). Current Advances in Zika Vaccine Development. *Vaccines*, *10*(11), 1816. DOI: 10.3390/vaccines10111816 PMID: 36366325

Wang, Y., Wei, L., Jiang, D., Cong, X., Fei, R., Xiao, J., & Wang, Y. (2005). In vitro resistance to interferon of hepatitis B virus with precore mutation. *World Journal of Gastroenterology*, *11*(5), 649–655. DOI: 10.3748/wjg.v11.i5.649 PMID: 15655815

Wang, Y., & Zhang, Y. (2021). Artificial intelligence in antiviral drug discovery: Recent advances and future perspectives. *Frontiers in Pharmacology*, *12*, 654.

Wang, Z., & Liu, Y. (2021). Artificial intelligence applications in antiviral drug discovery: A comprehensive review. *Computational Biology and Chemistry*, *89*, 107358.

Wei, J., Alfajaro, M. M., DeWeirdt, P. C., Hanna, R. E., Lu-Culligan, W. J., Cai, W. L., Strine, M. S., Zhang, S. M., Graziano, V. R., Schmitz, C. O., Chen, J. S., Mankowski, M. C., Filler, R. B., Ravindra, N. G., Gasque, V., de Miguel, F. J., Patil, A., Chen, H., Oguntuyo, K. Y., & Wilen, C. B. (2021). Genome-wide CRISPR Screens Reveal Host Factors Critical for SARS-CoV-2 Infection. *Cell*, *184*(1), 76–91.e13. DOI: 10.1016/j.cell.2020.10.028 PMID: 33147444

Weinreich, D. M.. (2021). mRNA COVID-19 Vaccines Efficacy and Safety. *The New England Journal of Medicine*, *384*(20), 1–15.

Weinreich, D. M., Sivapalasingam, S., Norton, T., Ali, S., Gao, H., Bhore, R., Xiao, J., Hooper, A. T., Hamilton, J. D., Musser, B. J., Rofail, D., Hussein, M., Im, J., Atmodjo, D. Y., Perry, C., Pan, C., Mahmood, A., Hosain, R., Davis, J. D., & Yancopoulos, G. D. (2021). REGEN-COV antibody combination and outcomes in outpatients with Covid-19. *The New England Journal of Medicine*, *385*(23), e81. DOI: 10.1056/NEJMoa2108163 PMID: 34587383

Weiss, R. A., & Sankaran, N. (2022). Emergence of epidemic diseases: Zoonoses and other origins. *Faculty Reviews*, *11*, 11. DOI: 10.12703/r/11-2 PMID: 35156099

Wei, Y., Zhao, Z., & Ma, X. (2022). Description of CRISPR-Cas9 development and its prospects in human papillomavirus-driven cancer treatment. *Frontiers in Immunology*, *13*, 1037124. DOI: 10.3389/fimmu.2022.1037124 PMID: 36479105

Wen, F.. (2021). Adverse events following immunization: Current perspectives. *Drug Safety*, *44*(7), 781–794.

Whitten-Bauer, C., Chung, J., Gómez-Moreno, A., Gomollón-Zueco, P., Huber, M. D., Gerace, L., & Garaigorta, U. (2019). The Host Factor Erlin-1 is Required for Efficient Hepatitis C Virus Infection. *Cells*, *8*(12), 1555. DOI: 10.3390/cells8121555 PMID: 31810281

WHO Digital Health Technical Advisory Group. (2019). Global strategy on digital health 2020–2025. World Health Organization. Retrieved from https://www.who.int/docs/default-source/documents/gsdh2020-2025-en.pdf

WHO. (2015). *Summary of Probable SARS Cases with Onset of Illness from 1 November 2002 to 31 July 2003*. World Health Organization.

WHO. (2017). *Global Hepatitis Report, 2017*. WHO.

WHO. (2020). *Ebola in Democratic Republic of Congo—North Kivu, Ituri 2018–2020*. WHO.

WHO. (2020). *Latest HIV Estimates and Updates on HIV Policies Uptake, July 2020*. WHO.

WHO. (2022). *Avian Influenza A (H5N1)—United States of America*. WHO.

WHO. (2022). *Coronavirus Disease (COVID-19) Dashboard*. WHO.

WHO. (2022). *Ebola Virus Disease-Democratic Republic of Congo*. WHO.

WHO. (2022). *Middle East Respiratory Syndrome Coronavirus (MERS-CoV)—Saudi Arabia*. WHO.

Widagdo, W., Okba, N. M. A., Stalin Raj, V., & Haagmans, B. L. (2016). MERS-coronavirus: From discovery to intervention. *One Health*, *3*, 11–16. DOI: 10.1016/j.onehlt.2016.12.001 PMID: 28616497

Wilson, S. L., & Wiysonge, C. (2020). Social media and vaccine hesitancy. *BMJ Global Health*, *5*(10), e004206. DOI: 10.1136/bmjgh-2020-004206 PMID: 33097547

Winkler, D. A. (2024). Computational repurposing of drugs for viral diseases and current and future pandemics. *Journal of Mathematical Chemistry*, *62*(11), 2844–2879. DOI: 10.1007/s10910-023-01568-3

Woodson, S. E., & Morabito, K. M. (2024). Continuing development of vaccines and monoclonal antibodies against Zika virus. npj. *Vaccines*, *9*(1), 91. DOI: 10.1038/s41541-024-00889-x PMID: 38789469

World Health Organization (WHO). (2022). *Challenges and solutions in global vaccine access*. Retrieved from https://www.who.int

World Health Organization (WHO). (2023). Hepatitis B Vaccines. Retrieved from https://www.who.int

World Health Organization. (2019). *Measles*. https://www.who.int/news-room/fact-sheets/detail/measles

World Health Organization. (2020). "Digital Health." WHO. Retrieved from https://www.who.int/health-topics/digital-health

World Health Organization. (2020). History of smallpox vaccination. Retrieved March 28, 2025, from https://www.who.int/news-room/spotlight/history-of-vaccination/history-of-smallpox-vaccination

World Health Organization. (2021). *Ebola virus disease*. https://www.who.int/news-room/fact-sheets/detail/ebola-virus-disease

World Health Organization. (2021, April 7). *A push to vaccinate every child, everywhere, ended polio in India*. Retrieved March 28, 2025, from https://www.who.int/india/news/feature-stories/detail/a-push-to-vaccinate-every-child-everywhere-ended-polio-in-india

World Health Organization. (2022). Global progress report on HIV, viral hepatitis, and sexually transmitted infections, 2022. Retrieved from https://www.who.int/publications/i/item/9789240053779

World Health Organization. (2022). *Monkeypox*. https://www.who.int/news-room/fact-sheets/detail/monkeypox

World Health Organization. (2022). *Rabies*. Retrieved from WHO Website.

World Health Organization. (2022). *Viral diseases epidemiology*. https://www.who.int/news-room/fact-sheets

World Health Organization. (2023). *2023 WHO Global Vaccine Market Report*. Retrieved March 28, 2025, from https://www.who.int/publications/i/item/B09022

World Health Organization. (2023). *Dengue and severe dengue*. https://www.who.int/news-room/fact-sheets/detail/dengue-and-severe-dengue

World Health Organization. (2023). *Influenza*. Retrieved from WHO Website.

World Health Organization. (2024). *COVID-19 dashboard*. Retrieved from https://data.who.int/dashboards/covid19/data

World Health Organization. (2024). *COVID-19 vaccines: A guide for healthcare providers*. Retrieved from WHO Website.

World Health Organization. (2024). Global immunization efforts have saved at least 154 million lives over the past 50 years. Retrieved March 28, 2025, from https://www.who.int/news/item/24-04-2024-global-immunization-efforts-have-saved-at-least-154-million-lives-over-the-past-50-years

World Health Organization. (2024). *Influenza (seasonal)*. Retrieved from https://www.who.int/news-room/fact-sheets/detail/influenza-(seasonal)

World Health Organization. (2024). Urban health. Retrieved March 28, 2025, from https://www.who.int/news-room/fact-sheets/detail/urban-health

World Health Organization. (2025). European Region reports highest number of measles cases in more than 25 years. Retrieved March 28, 2025, from https://www.who.int/europe/news/item/13-03-2025-european-region-reports-highest-number-of-measles-cases-in-more-than-25-years---unicef--who-europe

World Health Organization. (2025, January 7). Disease outbreak news: Trends of acute respiratory infection, including human metapneumovirus, in the Northern Hemisphere. Retrieved from https://www.who.int/emergencies/disease-outbreak-news/item/2025-DON550

World Health Organization. (n.d.). *COVID-19 vaccine misinformation and disinformation: An urgent public health challenge*. Retrieved March 28, 2025, from https://www.who.int/news-room/spotlight/ten-threats-to-global-health-in-2019

World Health Organization. Food and Agriculture Organization of the United Nations, & World Organisation for Animal Health. (2018). *Zero by 30: The global strategic plan to end human deaths from dog-mediated rabies by 2030*. https://www.who.int/publications/i/item/9789241513838

Wright, P. F.. (2021). Universal influenza vaccines: Progress and challenges. *Nature Reviews. Immunology, 21*(6), 374–390. DOI: 10.1038/s41577-020-00473-6

Wynendaele, E., Furman, C., Wielgomas, B., Larsson, P., Hak, E., Block, T., Van Calenbergh, S., Willand, N., Markuszewski, M., Odell, L. R., Poelarends, G. J., & De Spiegeleer, B. (2021). Sustainability in drug discovery. *Medicine in Drug Discovery, 12*, 100107. Advance online publication. DOI: 10.1016/j.medidd.2021.100107

Xiao, B., Wu, L., Sun, Q., Shu, C., & Hu, S. (2024). Dynamic analysis of SARS-CoV-2 evolution based on different countries. *Gene, 916*, 148426. DOI: 10.1016/j.gene.2024.148426 PMID: 38575101

Xuan, Q., Wang, J., Nie, Y., Fang, C., & Liang, W. (2024). Research Progress and Application of Miniature CRISPR-Cas12 System in Gene Editing. *International Journal of Molecular Sciences, 25*(23), 12686. DOI: 10.3390/ijms252312686 PMID: 39684395

Xu, Z., Su, C., Xiao, Y., & Wang, F. (2022). Artificial intelligence for COVID-19: Battling the pandemic with computational intelligence. *Intelligent Medicine, 2*(1), 13–29. DOI: 10.1016/j.imed.2021.09.001 PMID: 34697578

Yamey, G., Garcia, P. J., Hassan, F., Mao, W., McDade, K. K., Pai, M., & Zhao, F. (2023). It is not too late to achieve global COVID-19 vaccine equity. *BMJ (Clinical Research Ed.), 376*, e070650. DOI: 10.1136/bmj-2022-070650 PMID: 35331982

Yang, J. W., Yang, L., Luo, R. G., & Xu, J. F. (2020). Corticosteroid administration for viral pneumonia: COVID-19 and beyond. *Clinical microbiology and infection: the official publication of the European Society of Clinical Microbiology and Infectious Diseases, 26*(9), 1171–1177. https://doi.org/DOI: 10.1016/j.cmi.2020.06.020

Yang, L., Han, Y., Nilsson-Payant, B. E., Gupta, V., Wang, P., Duan, X., Tang, X., Zhu, J., Zhao, Z., Jaffré, F., Zhang, T., Kim, T. W., Harschnitz, O., Redmond, D., Houghton, S., Liu, C., Naji, A., Ciceri, G., Guttikonda, S., & Chen, S. (2020). A Human Pluripotent Stem Cell-based Platform to Study SARS-CoV-2 Tropism and Model Virus Infection in Human Cells and Organoids. *Cell Stem Cell, 27*(1), 125–136.e7. DOI: 10.1016/j.stem.2020.06.015 PMID: 32579880

Yang, L., Pei, R. J., Li, H., Ma, X. N., Zhou, Y., Zhu, F. H., He, P. L., Tang, W., Zhang, Y. C., Xiong, J., Xiao, S. Q., Tong, X. K., Zhang, B., & Zuo, J. P. (2021). Identification of SARS-CoV-2 entry inhibitors among already approved drugs. *Acta Pharmacologica Sinica*, *42*(8), 1347–1353. DOI: 10.1038/s41401-020-00556-6 PMID: 33116249

Yang, Y., Zheng, Y., Zou, Q., Li, J., & Feng, H. (2024). Overcoming CRISPR-Cas9 off-target prediction hurdles: A novel approach with ESB rebalancing strategy and CRISPR-MCA model. *PLoS Computational Biology*, *20*(9), e1012340. DOI: 10.1371/journal.pcbi.1012340 PMID: 39226304

Yao, Z. Q., Schank, M. B., Zhao, J., El Gazzar, M., Wang, L., Zhang, Y., Hill, A. C., Banik, P., Pyburn, J. S., & Moorman, J. P. (2024). The potential of HBV cure: An overview of CRISPR-mediated HBV gene disruption. *Frontiers in Genome Editing*, *6*, 1467449. DOI: 10.3389/fgeed.2024.1467449 PMID: 39444780

Yeh, Y. C., Huang, T. H., Yang, S. C., Chen, C. C., & Fang, J. Y. (2020). Nano-based drug delivery or targeting to eradicate bacteria for infection mitigation: A review of recent advances. *Frontiers in Chemistry*, *8*, 286. DOI: 10.3389/fchem.2020.00286 PMID: 32391321

Yetisgin, A. A., Cetinel, S., Zuvin, M., Kosar, A., & Kutlu, O. (2020, May 8). Therapeutic nanoparticles and their targeted delivery applications. *Molecules (Basel, Switzerland)*, *25*(9), 2193. DOI: 10.3390/molecules25092193 PMID: 32397080

Yount, B., Curtis, K. M., Fritz, E. A., Hensley, L. E., Jahrling, P. B., Prentice, E., Denison, M. R., Geisbert, T. W., & Baric, R. S. (2003). Reverse genetics with a full-length infectious cDNA of severe acute respiratory syndrome coronavirus. *Proceedings of the National Academy of Sciences of the United States of America*, *100*(22), 12995–13000. DOI: 10.1073/pnas.1735582100 PMID: 14569023

Yuan, S., Chu, H., Chan, J. F., Ye, Z. W., Wen, L., Yan, B., Lai, P. M., Tee, K. M., Huang, J., Chen, D., Li, C., Zhao, X., Yang, D., Chiu, M. C., Yip, C., Poon, V. K., Chan, C. C., Sze, K. H., Zhou, J., & Yuen, K. Y. (2019). SREBP-dependent lipidomic reprogramming as a broad-spectrum antiviral target. *Nature Communications*, *10*(1), 120. DOI: 10.1038/s41467-018-08015-x PMID: 30631056

Yudkin, J. S., Hodes, R. M., Sandler-Loeff, A., & Messiah, S. E. (2022). Needs assessment and best practices for digital trainings for health professionals in Ethiopia using the RE-AIM framework: COVID-19 case study. *Disaster Medicine and Public Health Preparedness*, *2022*, 1–9. PMID: 36226522

Yu, Z. J., Mosher, E. P., & Bumpus, N. N. (2021). Pharmacogenomics of Antiretroviral Drug Metabolism and Transport. *Annual Review of Pharmacology and Toxicology*, *61*(1), 565–585. DOI: 10.1146/annurev-pharmtox-021320-111248 PMID: 32960701

Zahedipour, F., Zahedipour, F., Zamani, P., Jaafari, M. R., & Sahebkar, A. (2024). Harnessing CRISPR technology for viral therapeutics and vaccines: From preclinical studies to clinical applications. *Virus Research*, *341*, 199314. DOI: 10.1016/j.virusres.2024.199314 PMID: 38211734

Zakharova, M. A., & Chudinov, M. V. (2024). PROTAC® technology and potential for its application in infection control. *Fine Chemical Technologies*, *19*(3), 214–231. DOI: 10.32362/2410-6593-2024-19-3-214-231

Zeng, M. C., Jia, Q. J., & Tang, L. M. (2021). *rpoB* gene mutations in rifampin-resistant *Mycobacterium tuberculosis* isolates from rural areas of Zhejiang, China. *The Journal of International Medical Research*, *49*(3), 300060521997596. DOI: 10.1177/0300060521997596 PMID: 33715498

Zhang, H., Wang, J., Zhang, K., Shi, J., Gao, Y., Zheng, J., He, J., Zhang, J., Song, Y., Zhang, R., Shi, X., Jin, L., & Li, H. (2024). Association between heavy metals exposure and persistent infections: The mediating role of immune function. *Frontiers in Public Health*, *12*, 1367644. DOI: 10.3389/fpubh.2024.1367644 PMID: 39104887

Zhang, J., & Cheng, Z. (2020). Data-driven insights into the impact of digital health in managing viral infections. *Journal of Medical Systems*, *44*(4), 68. PMID: 32072331

Zhang, K., Wan, P., Wang, L., Wang, Z., Tan, F., Li, J., Ma, X., Cen, J., Yuan, X., Liu, Y., Sun, Z., Cheng, X., Liu, Y., Liu, X., Hu, J., Zhong, G., Li, D., Xia, Q., & Hui, L. (2024). Efficient expansion and CRISPR-Cas9-mediated gene correction of patient-derived hepatocytes for treatment of inherited liver diseases. *Cell Stem Cell*, *31*(8), 1187–1202.e8. DOI: 10.1016/j.stem.2024.04.022 PMID: 38772378

Zhang, M.-L., Li, H.-B., & Jin, Y. (2024). Application and perspective of CRISPR/Cas9 genome editing technology in human diseases modeling and gene therapy. *Frontiers in Genetics*, *15*, 1364742. DOI: 10.3389/fgene.2024.1364742 PMID: 38666293

Zhang, M., Wang, P., Luo, R., Wang, Y., Li, Z., Guo, Y., Yao, Y., Li, M., Tao, T., Chen, W., Han, J., Liu, H., Cui, K., Zhang, X., Zheng, Y., & Qin, J. (2020). Biomimetic Human Disease Model of SARS-CoV-2-Induced Lung Injury and Immune Responses on Organ Chip System. *Advanced Science (Weinheim, Baden-Wurttemberg, Germany)*, *8*(3), 2002928. DOI: 10.1002/advs.202002928 PMID: 33173719

Zhang, M., Wang, S., Wilffert, B., Tong, R., van Soolingen, D., van den Hof, S., & Alffenaar, J. W. (2018). The association between the NAT2 genetic polymorphisms and risk of DILI during anti-TB treatment: A systematic review and meta-analysis. *British Journal of Clinical Pharmacology*, *84*(12), 2747–2760. DOI: 10.1111/bcp.13722 PMID: 30047605

Zhang, Q., Dong, G., Meng, W., Chen, Z., Cao, Y., & Zhang, M. (2022). Perceived Stress and Psychological Impact Among Healthcare Workers at a Tertiaty Hospital in China During the COVID-19 Outbreak: The Moderating Role of Resilience and Social Support. *Frontiers in Psychiatry*, *12*, 570971. DOI: 10.3389/fpsyt.2021.570971 PMID: 35281206

Zhang, W. F., Stephen, P., Thériault, J. F., Wang, R., & Lin, S. X. (2020). Novel Coronavirus Polymerase and Nucleotidyl-Transferase Structures: Potential to Target New Outbreaks. *The Journal of Physical Chemistry Letters*, *11*(11), 4430–4435. DOI: 10.1021/acs.jpclett.0c00571 PMID: 32392072

Zhang, Y., Wang, M., Huang, M., & Zhao, J. (2024). Innovative strategies and challenges mosquito-borne disease control amidst climate change. *Frontiers in Microbiology*, *15*, 1488106. DOI: 10.3389/fmicb.2024.1488106 PMID: 39564491

Zhang, Y., & Wang, X. (2021). Machine learning in the development of mRNA vaccines. *Frontiers in Immunology*, *12*, 654.

Zhang, Z., & Liu, J. (2021). Insights into digital health technologies for the treatment of viral diseases. *International Journal of Medical Informatics*, *145*, 104300.

Zhen, S., Qiang, R., Lu, J., Tuo, X., Yang, X., & Li, X. (2023). CRISPR/Cas9-HPV-liposome enhances antitumor immunity and treatment of HPV infection-associated cervical cancer. *Journal of Medical Virology*, *95*(1), e28144. DOI: 10.1002/jmv.28144 PMID: 36121194

Zhou, J., Lazar, D., Li, H., Xia, X., Satheesan, S., Charlins, P., O'Mealy, D., Akkina, R., Saayman, S., Weinberg, M. S., Rossi, J. J., & Morris, K. V. (2018). Receptor-targeted aptamer-siRNA conjugate-directed transcriptional regulation of HIV-1. *Theranostics*, *8*(6), 1575–1590. DOI: 10.7150/thno.23085 PMID: 29556342

Zhou, Y., Fletcher, N. F., Zhang, N., Hassan, J., & Gilchrist, M. D. (2021). Enhancement of antiviral effect of plastic film against SARS-CoV-2: Combining nanomaterials and nanopatterns with scalability for mass manufacturing. *Nano Letters*, *21*(24), 10149–10156. DOI: 10.1021/acs.nanolett.1c02266 PMID: 34881894

Zhou, Y., Peng, S., Wang, H., Cai, X., & Wang, Q. (2024). Review of Personalized Medicine and Pharmacogenomics of Anti-Cancer Compounds and Natural Products. *Genes, 15*(4), 468. DOI: 10.3390/genes15040468 PMID: 38674402

Zhou, Y., Vedantham, P., Lu, K., Agudelo, J., Carrion, R.Jr, Nunneley, J. W., Barnard, D., Pöhlmann, S., McKerrow, J. H., Renslo, A. R., & Simmons, G. (2015). Protease inhibitors targeting coronavirus and filovirus entry. *Antiviral Research, 116*, 76–84. DOI: 10.1016/j.antiviral.2015.01.011 PMID: 25666761

Zhu, N., Zhang, D., Wang, W., Li, X., Yang, B., Song, J., Zhao, X., Huang, B., Shi, W., Lu, R., Niu, P., Zhan, F., Ma, X., Wang, D., Xu, W., Wu, G., Gao, G. F., & Tan, W. (2020). A Novel Coronavirus from Patients with Pneumonia in China, 2019. *The New England Journal of Medicine, 382*(8), 727–733. DOI: 10.1056/NEJMoa2001017 PMID: 31978945

Žigrayová, D., Mikušová, V., & Mikuš, P. (2023). Advances in Antiviral Delivery Systems and Chitosan-Based Polymeric and Nanoparticulate Antivirals and Antiviral Carriers. *Viruses, 15*(3), 647. DOI: 10.3390/v15030647 PMID: 36992356

Zi, Y., Yang, K., He, J., Wu, Z., Liu, J., & Zhang, W. (2022). Strategies to enhance drug delivery to solid tumors by harnessing the EPR effects and alternative targeting mechanisms. *Advanced Drug Delivery Reviews, 188*, 114449. DOI: 10.1016/j.addr.2022.114449 PMID: 35835353

Zou, G., Xu, H. Y., Qing, M., Wang, Q. Y., & Shi, P. Y. (2011). Development and characterization of a stable luciferase dengue virus for high-throughput screening. *Antiviral Research, 91*(1), 11–19. DOI: 10.1016/j.antiviral.2011.05.001 PMID: 21575658

About The Contributors

Muhammad Shahzad Aslam is recognized as 1% top reviewer according to web of science in 2017, 2018 and 2019 in multi-discipline, cross discipline and clinical medicine respectively. He has served as honorary academic editor in several journals and had experience of more than 8 years in peer review and academic editor. He is authored/editor of three books title as Cases on Teaching Pharmacology to Complementary and Alternative Medicine Students, Multidisciplinary Applications of Natural Science for Drug Discovery and Integrative Medicine and, Artificial Intelligence Applications Using ChatGPT in Education: Case Studies and Practices. He is currently working in School of Traditional Chinese Medicine, Xiamen University Malaysia. He had completed his Ph.D. from School of Bioprocess Engineering, Universiti Malaysia Perlis (UniMAP) and did worked on natural product chemistry. He was awarded scholarship by the Ministry of higher education (MOHE), Malaysia under the Malaysian International Scholarship (MIS) to pursue his Ph.D. He did Master of Philosophy from Bahauddin Zakariya University, Pakistan in Pharmaceutical Chemistry. He has completed his graduation from Baqai Medical University, Pakistan in Doctor of Pharmacy degree (often abbreviated as Pharm.D. or PharmD). He has started his career as a production supervisor and quality control analyst in different pharmaceutical companies and then in 2013 he has moved his academic career, as a lecturer in Lahore pharmacy college, Pakistan where he had taught pharmaceutical chemistry, instrumentation, and organic chemistry. He has served as assistant professor in Rashid Latif college of Pharmacy and Ziauddin University after his Ph.D. During his research work, Dr. Muhammad Shahzad Aslam had worked on the metabolomics analysis, fragmentation analysis of bioactive compounds, polyherbal formulation and their interactions in wound healing. His area of expertise is phytochemistry, Pharmacognosy, green extraction technique, wound healing studies, animal modeling, novel plant extraction methods, and green solvents. He had published more than nighty five (95) articles and have an h-index=16. Currently, he is working on public health issues, new extraction techniques, green extraction methods and identifying new material in the healing wound.

Bancha Yingngam, an Associate Professor at Ubon Ratchathani University in Thailand, brings a wealth of knowledge and experience in Pharmaceutical Sciences and Technology. His academic path started in 2005, when he graduated with a bachelor's degree in pharmacy, receiving second-class honors. Driven by his passion for phytomedicines and community pharmacy, he furthered his studies, obtaining a Ph.D. in 2011 from the same university. Between 2013 and 2021, during his postdoctoral fellowship at Karl-Franzens-Universität Graz, Austria, he extended his research focus to include optimization and machine learning algorithms. This focus has significantly influenced his field, as exemplified by his Scopus H-index of 9. Throughout his career, Dr. Yingngam has written over twenty scholarly articles, book chapters, and conference papers. His research largely explores optimization algorithms and the integration of machine learning in the pharmaceutical sciences. He is currently authoring book chapters that delve into artificial intelligence applications for predicting biopharmaceutical properties and making decisions in pharmaceutical sciences. His research has been acknowledged and published in various esteemed journals, and his contributions have enriched several key publications in his field. His exceptional work has twice earned him the "Outstanding Research Award in Health Science" from Ubon Ratchathani University in 2020 and 2022. Beyond his extensive research, Dr. Yingngam excels as an educator and mentor, having guided 10 master's students in Austria toward achieving their academic and professional aspirations. Furthermore, his dedication to his field is showcased in his review of more than 70 articles listed in the Scopus database. Through his work in research and education, Dr. Yingngam continues to advance Pharmaceutical Sciences and Technology.

Tariq Javed is a distinguished academic and researcher with extensive experience in the field of pharmaceutical sciences. He currently serves as the Incharge and Chairperson of the Department of Pharmaceutical Sciences at Government College University (GCU), Lahore, a position he has held since 2022. Prior to this, Dr. Javed was an Associate Professor at Lahore Pharmacy College, Lahore Medical and Dental College (LMDC), affiliated with the University of Health Sciences (UHS) Lahore, where he also served as the Principal of Ruth Pfau College Life Sciences. Dr. Javed completed his Ph.D. in Molecular Medicine from the University of the Punjab, Lahore, where his research focused on the therapeutic potential of medicinal plants against Hepatitis C virus. He holds an M.Phil. in Pharmacy and a B.Pharm., both from Bahauddin Zakariya University, Multan. With a career spanning over two decades, Dr. Javed has made significant contributions to pharmaceutical research, particularly in phytochemistry, antiviral drug development, and pharmaceutical microbiology. His work is widely published in renowned international journals, including studies on Alzheimer's disease, anticancer potential of phytochemicals, and antiviral properties of medicinal plants. He is also a recognized Ph.D. supervisor, approved by the Higher Education Commission (HEC) of Pakistan. Dr. Javed is an active member of several professional societies, including the American Society of Microbiology, the American Society of Pharmacognosy, and the Pakistan Biological Safety Association. He is a life member of the Pakistan Biological Safety Association and a global assessor for the Royal Society of Tropical Medicine and Hygiene (RSTMH), UK. His administrative roles are equally notable, having served on various ethical committees, technical committees, and boards of studies in different academic institutions. Dr. Javed is also involved in collaborative research projects with pharmaceutical industries and diagnostic laboratories across Pakistan. In addition to his academic and research achievements, Dr. Javed has actively participated in numerous workshops and conferences worldwide, focusing on cutting-edge topics such as CRISPR technology, nanotechnology, and biosafety. His commitment to advancing pharmaceutical sciences is evident through his continuous efforts in research, education, and professional development. Dr. Tariq Javed's contributions to the field of pharmaceutical sciences are not only recognized nationally but also on the international stage, making him a leading figure in his area of expertise.

Sana Riaz is a Senior Research Officer at the Food and Biotechnology Research Center (FBRC), PCSIR Lahore, where she has been serving since 2022. She brings a wealth of experience in microbiology, molecular genetics, and biotechnology, with a distinguished career that spans over a decade in academia and research. Prior to her current role, Dr. Riaz held positions as an Assistant Professor at Lahore Institute of Biomedical and Allied Health Sciences and as a Research Scientist at the School of Biomedicine, KU Leuven, Belgium. Dr. Riaz completed her Ph.D. in Microbiology and Molecular Genetics at The Women University Multan and KU Leuven, Belgium, where her research focused on the biochemical and molecular characterization of potential anticancer phytochemicals. She also holds an M.Phil. in Molecular Biology from the University of the Punjab, Lahore, and an MSc in Biochemistry from Government College University, Lahore. Her research expertise encompasses a broad range of topics, including gene editing, bioinformatics, phytochemical drug development, and antimicrobial resistance. Dr. Riaz has successfully led multiple R&D projects, including the development of rapid detection kits for microbes, RNA extraction kits from viruses, and microbial production of active pharmaceutical ingredients (APIs). She has also contributed significantly to the development of innovative products, such as a herbal topical gel and an edible, sustainable food coating biofilm, for which she holds several patents. Dr. Riaz is an active member of various scholarly societies, including the American Society of Microbiology, the European Society of Clinical Microbiology and Infectious Diseases, and the Pakistan Biological Safety Association. She has published extensively in reputable journals and has authored several books on topics ranging from anticancer medicinal plants to the production of interferon alpha 2b. In addition to her research and publications, Dr. Riaz has organized and participated in numerous workshops and conferences globally, focusing on cutting-edge technologies such as CRISPR/Cas9, biosafety, and halal science. Her commitment to advancing the field of biotechnology and her contributions to research and innovation have earned her several fellowships and awards, including a Responsible Research Fellowship from the American Society of Microbiology and Health Security Partner USA. Dr. Sana Riaz's dedication to research, coupled with her leadership in innovative projects, makes her a prominent figure in the fields of microbiology, biotechnology, and pharmaceutical sciences.

***

Minal Hussain pursuing an M.Phil in Microbiology from Cholistan University of Veterinary and Animal Sciences, Bahawalpur, Pakistan.

Abhiruj Navabhatra is an assistant professor at Rangsit University in Thailand, specializing in pharmaceutical sciences and technology. His journey in academia began in 2005 when he earned a bachelor's degree in pharmacy. His deep interest in pharmacology and toxicology led him to further his education, culminating in a Ph.D. from Chulalongkorn University in 2016. During his research fellowship and internship from 2014 to 2016, he worked at the Toxicogenomics Laboratory within the Department of Clinical Pharmacology at Nagoya University's Graduate School of Medicine, located in Nagoya, Aichi, Japan. Here, he broadened his research horizons to encompass the study of enzymes' mechanisms, specifically cytochrome P450 (CYP), with a focus on drug-drug and drug-herb interactions (DDIs). Dr. Navabhatra's concentrated research has significantly impacted his field, earning him a Scopus H-index of 5. Over his career, he has published more than ten scholarly articles and conference papers. His research predominantly examines pharmacology and toxicology, optimization algorithms, and the integration of machine learning in the realm of pharmaceutical sciences. His research, recognized and published in several esteemed journals, has enriched numerous pivotal publications within his field. Through his dedicated efforts in research and education, Dr. Navabhatra continues to be a progressive force within the field of pharmaceutical sciences and technology.

Odangowei Inetiminebi Ogidi is a Lecturer and Acting Head, Department of Biochemistry, Bayelsa Medical University, Yenagoa, Bayelsa State, Nigeria. He holds a Bachelor of Science degree in Biochemistry, Master of Science and Doctor of Philosophy degrees in Biotechnology. He has also served as a Lecturer, Head of Departments of Science Laboratory Technology and Biochemistry, and Deputy Director, Centre for Continuing Education in Federal Polytechnic Ekowe, Bayelsa State. He has authored over 80 peer-reviewed academic articles in reputable local and international journals, published over 40 book chapters and presented over 35 conference papers. He is a member of various Journal Editorial Boards. He is indexed in World Scientist and University Ranking (AD Scientific Index ID 4306173) amongst other international bodies. He have supervised over 30 ND and HND students and currently supervising over 10 undergraduate students. His research interest has been Environmental Biotechnology, Phytomedicine, Phytochemistry, Human and Environmental health risk assessments and ecotoxicology amongst others. He is a member of West Africa Society of Toxicology, Nigerian Society of Biochemistry and Molecular Biology, Foundation for African Development through International Biotechnology and Nigerian Environmental Society.

Raveesha Peeriga is a Professor and Head of the Department of Pharmacognosy at V.V. Institute of Pharmaceutical Sciences, Gudlavalleru. With over thirteen years of her experience, She is a dedicated professional with a profound understanding of phytochemistry and herbal medicine. She has received several awards for her teaching excellence and Associate Fellow from the Andhra Pradesh Akademi of Sciences. She presented her research work at numerous National and International Conferences as well as served as a convener and coordinator for several such events. She has published more than 40 Research Articles in reputed Journals and delivered Guest Lectures. She also authored 2 Books titled "Herbal Drug Technology" and "Concise Pharmacognosy. She also acted as a Chairperson for BOS for Paramedicals, Krishna University. She is a Life Member of ISTE, APTI, IPA and OPF.

Pranav Kumar Prabhakar is currently affiliated with the Department of Biotechnology, School of Engineering and Technology, Nagaland University, Meriema Campus, Kohima, Nagaland, India. He is one amongst the World's Top 2% Scientists (list published by Stanford University, USA, 2021, 2022, 2023, and 2024). He has completed his PhD in Biotechnology from IIT Madras. His main area of research interest focuses Elucidate Molecular Mechanism and Strategies for Oral Insulin Delivery, Mimic the signaling pathways in Metabolic Disorders (diabetes) with Natural Products. He is member of Royal Society of Chemistry, and Asia-Pacific Chemical, Biological& Environmental Engineering Society. He is also serving as editorial board member and reviewer for many reputed national and International Journals. Dr. Pranav received honors includes recipient of travel grant towards attending ATTD 2009 in Greece from Indian Institute of Technology Madras & Council for scientific and Industrial research (CSIR) and approved by Department of Science and Technology (DST). He has published 75+ research articles in journals, authored 3 books, and 14 book chapters. He also delivered 9 oral and poster presentations in scientific meeting.

Reyed M. Reyed is a distinguished Associate Professor of Microbial Biotechnology at the Genetic Engineering and Biotechnology Research Institute within the City of Scientific Research and Technological Applications (SRTA-City), located in New Borg El-Arab, Alexandria, Egypt. His academic journey spans multiple disciplines, from microbial genomics and computational enzyme research to synthetic biology and environmental sustainability. Holding a range of prestigious qualifications, including Dia.M.C, RN, PAND, PhD, MSc, and BSc in Microbial Biotechnology, Dr. Reyed's expertise is comprehensive, making him a leading figure in his field. Dr. Reyed's career has been defined by a commitment to solving critical global challenges. His early research (2008–2019) focused on biofilm formation and marine biofouling in reverse osmosis systems, driving innovations in water treatment technologies. This was followed by his leadership in the Egyptian-Indian joint project on water treatment (2020–2023) in collaboration with the Environmental Biotechnology and Genomics Division (EBGD) at CSIR-NEERI, India. His work provided practical solutions to pressing water sustainability issues and showcased his expertise in environmental biotechnology. In the realm of food microbiology, Dr. Reyed has advanced food safety through groundbreaking research. He also established a winter school in food microbiology, offering a platform to train future scientists. His research extends to the human gut microbiome, where he combines bioinformatics and machine learning to explore personalized probiotics aimed at enhancing gut health. This demonstrates Dr. Reyed's visionary approach to human nutrition and health. In computational enzyme research, Dr. Reyed has integrated synthetic biology and systems biology to develop innovative solutions for environmental sustainability. His interdisciplinary expertise has led to significant advancements in microbial biotechnology, where computational tools intersect with biological systems, driving cutting-edge research in this domain. Dr. Reyed currently serves as the Program Coordinator at SRTA-City, overseeing critical research initiatives. His leadership is also evident in his role as Co-Chair of the Organizing Committee and Chair of the Scientific and Research Board for the Climate Change 2025 Conference, scheduled to take place in Paris, France, under the theme "Green Technologies for a Green Future." Additionally, he is a member of the Advisory Board of Advances in Agriculture and Biology (ISSN: 3079-174X) and serves as the Editor-in-Chief of the Journal of Climate Change and Renewable Energy, as well as Academic Editor for the International Journal of Environment and Climate Change. Through these roles, Dr. Reyed contributes significantly to the global scientific discourse on environmental challenges and sustainability. Dr. Reyed's qualifications extend beyond academia. He is a Registered Clinical Nutritionist and a Registered Physical and Health Nutritionist in Egypt, demonstrating his commitment to bridging the gap between biotechnology and health sciences. His certifications include membership in several professional syndicates such as the Egyptian Syndicate of Scientific Professions and the General Syndicate of Egyptian Ambulance Workers. His volunteer work in the UAE further underscores his dedication to public health and environmental sustainability. Dr. Reyed has been invited to represent Egypt at leading international conferences, including the International Conference on Bacteria, Viruses, and Infectious Diseases in Dubai (2018) and the Food Safety Exhibition in Abu Dhabi (2023). In May 2024, he will serve on the Research Evaluation Committee at the 8th International Conference on Genetic Disorders in Dubai, underscoring his esteemed reputation in the scientific community. In parallel to his professional work, Dr. Reyed has engaged in various specialized workshops and training programs to further his expertise. He completed the CRISPR/Cas9 Genome Editing course at Helwan University and attended the 4th Computer-Assisted Drug Design Workshop (CADD). His participation in the European Summer School on Nutrigenomics in Italy and numerous bioinformatics courses at the European Bioinformatics Institute (EMBL-EBI) showcases his commitment to emerging technologies in his field. His extensive training in bioinformatics, computational biology, and machine learning further highlights Dr. Reyed's commitment to professional development. In 2024, he will attend the Introduction to Bioinformatics (IBT) course at H3ABioNet, Alexandria University, reinforcing his foundation in bioinformatics. He has also completed the Introduction to Deep Learning course via Mahara Tech & iTi Platform, where he explored deep learning techniques in bioinformatics. His engagement in Google AI Courses, including Intro to Machine Learning: Image Processing, Intro to Machine Learning: Natural Language Processing, and Exploring API (Qwiklabs), further complements his research interests. Dr. Reyed's active participation in key workshops continued in 2021, where he attended the NCBI Virtual Workshop on cloud computing for biologists, the Data Science Summer School 2022 at the Hertie School Data Science Lab in Berlin, and the Virtual VEuPathDB Bioinformatics Workshop, focusing on infectious disease research. He was also involved in the Illumina Workshop on Rare Genetic Disorders, which emphasized genome sequencing in healthcare applications. His dedication to data integrity was further reflected in his participation in the H3ABioNet Bioinformatics Training (IBT), Research Data Management Course, and the Cancer Clinical Genetics in Africa series. These experiences illustrate his interdisciplinary approach, combining bioinformatics, healthcare, and environmental sustainability to address global challenges. In 2020, Dr. Reyed contributed to the GED-PerMed Project on big data and precision medicine, an initiative aimed at harnessing the power of data to address health disparities. His involvement in the EG-CompBio Project also played a key role in empowering computational biology and bioinformatics research in Egypt. Through these and other initiatives, Dr. Reyed has played a pivotal role in shaping the future of bioinformatics and computational biology in his region and beyond. Dr. Reyed's ongoing contributions to research, sustainability, and education demonstrate his unwavering commitment to advancing science for the betterment of humanity and the planet. His work continues to inspire the next generation of scientists and professionals in the fields of biotechnology, health, and environmental sustainability.

Index

A

Antiviral drug discovery 135, 139, 140, 141, 144, 145, 147, 148, 150, 160, 164, 166, 170, 171, 173, 210, 211, 215, 216
Antiviral drugs 2, 4, 8, 17, 32, 33, 34, 36, 37, 38, 40, 42, 43, 44, 46, 47, 49, 50, 53, 60, 63, 72, 75, 76, 79, 80, 81, 86, 87, 89, 93, 94, 95, 96, 100, 108, 110, 113, 123, 139, 140, 141, 142, 144, 145, 153, 157, 160, 161, 162, 163, 167, 169, 173, 209, 211, 228, 245, 256, 257, 263, 264, 267, 271, 272, 274, 275, 276, 282, 286, 287, 289, 294
Antiviral Resistance 17, 63, 123, 135, 249, 257, 290
Antiviral therapeutics 38, 56, 59, 217, 219, 225, 261, 262, 263, 271, 282, 284, 285, 287
Antiviral Therapy 26, 31, 33, 34, 37, 38, 39, 40, 42, 43, 45, 46, 48, 49, 51, 53, 54, 63, 64, 65, 66, 67, 68, 69, 75, 76, 77, 78, 80, 81, 82, 84, 86, 87, 90, 91, 92, 93, 96, 99, 100, 101, 103, 104, 108, 109, 110, 111, 114, 120, 123, 124, 142, 178, 180, 182, 185, 186, 187, 195, 196, 199, 203, 204, 205, 206, 207, 209, 210, 214, 215, 235, 281, 282

B

Biomarkers 42, 63, 67, 82, 83, 128, 187, 188, 189, 190, 196, 199

C

Centers for Disease Control and Prevention 15, 23, 24
Clinical Trials 19, 48, 58, 63, 77, 83, 100, 141, 144, 147, 152, 154, 217, 232, 233, 234, 236, 237, 244, 255, 278

D

Digital Health 125, 155, 170, 176, 177, 179, 186, 190, 191, 192, 194, 195, 196, 197, 198, 199, 203, 204, 205, 206, 207, 208, 209, 210, 211, 212, 213, 214, 215, 216
Drug Efficacy 59, 78, 82, 86, 91, 117, 197, 208, 274, 283
Drug Repurposing 57, 126, 145, 159, 166, 225, 261, 277, 278, 285, 292
Drugs 2, 4, 8, 17, 32, 33, 34, 35, 36, 37, 38, 40, 42, 43, 44, 46, 47, 48, 49, 50, 51, 52, 53, 57, 58, 59, 60, 61, 63, 65, 67, 72, 75, 76, 78, 79, 80, 81, 82, 83, 86, 87, 88, 89, 92, 93, 94, 95, 96, 100, 108, 110, 113, 115, 116, 123, 133, 136, 139, 140, 141, 142, 143, 144, 145, 146, 147, 148, 149, 152, 153, 154, 157, 159, 160, 161, 162, 163, 164, 167, 169, 172, 173, 186, 209, 211, 217, 228, 230, 235, 237, 242, 245, 256, 257, 258, 263, 264, 266, 267, 271, 272, 273, 274, 275, 276, 277, 278, 280, 282, 283, 284, 286, 287, 289, 291, 292, 293, 294

E

Ebola 3, 4, 5, 7, 8, 26, 29, 79, 85, 97, 98, 121, 124, 140, 155, 160, 161, 164, 169, 217, 218, 222, 224, 225, 238, 240, 247, 248, 251, 253, 254, 257, 258, 261, 263, 265, 266, 267, 268, 269, 270, 277, 278, 279, 280, 282, 288, 289, 290, 294, 295
Emerging economies 149, 162, 163, 164

G

Genetic Screening 63, 64, 85, 91, 92, 93, 94, 95, 96, 110, 111
Global health 4, 5, 8, 13, 14, 21, 22, 25, 29, 121, 139, 140, 142, 147, 150, 151, 163, 164, 170, 171, 194, 199, 208, 210, 218, 239, 261, 262, 266, 292

H

HCV 53, 64, 87, 114, 125, 128, 132, 134, 159, 201, 217, 218, 224, 225, 232, 240
HIV 3, 4, 7, 17, 19, 20, 23, 24, 28, 29, 32, 38, 44, 45, 46, 47, 48, 49, 50, 51, 52, 57, 59, 61, 65, 66, 69, 70, 72, 75, 78, 79, 80, 81, 84, 85, 88, 89, 90, 91, 92, 94, 96, 110, 111, 113, 114, 116, 117, 118, 119, 122, 124, 128, 129, 131, 133, 134, 139, 140, 148, 149, 150, 160, 164, 168, 176, 178, 179, 180, 181, 182, 183, 184, 185, 186, 187, 189, 190, 191, 192, 193, 195, 196, 197, 200, 201, 203, 204, 205, 207, 208, 209, 210, 217, 218, 224, 225, 231, 238, 240, 245, 258, 259, 261, 263, 264, 265, 266, 268, 272, 273, 274, 275, 276, 278, 281, 283, 286, 287, 288, 290, 291, 292, 293, 294, 295

I

Influenza 3, 4, 5, 7, 8, 9, 10, 17, 19, 20, 24, 27, 30, 44, 45, 47, 49, 58, 64, 72, 75, 79, 80, 81, 85, 88, 89, 93, 96, 97, 98, 100, 102, 104, 105, 107, 108, 121, 122, 124, 125, 126, 127, 129, 130, 131, 135, 136, 140, 144, 149, 150, 160, 176, 178, 182, 188, 191, 208, 217, 218, 219, 221, 222, 223, 224, 225, 238, 240, 242, 245, 249, 252, 256, 257, 258, 264, 265, 269, 274, 276, 278, 281, 290, 293

L

Lifestyle Factors 63, 97, 99, 102

M

Measles 2, 3, 6, 8, 9, 10, 11, 12, 16, 17, 21, 23, 24, 29, 30
MERS-CoV 143, 146, 165, 217, 221, 240, 250, 253, 255, 258, 270
Monkeypox 4, 29
Mumps 3, 6, 9, 10, 11, 12

N

Nanocarriers 33, 34, 35, 36, 37, 39, 40, 41, 43, 46, 47, 53, 54, 55, 57, 59, 61
Nanotechnology 31, 32, 33, 34, 35, 36, 41, 43, 46, 47, 49, 53, 54, 56, 57, 58, 60, 65, 111, 175, 208, 210, 211, 215, 282, 283, 287

P

Personalized Medicine 63, 68, 76, 81, 85, 90, 93, 95, 118, 133, 138, 215
Pharmacogenomics 63, 64, 65, 66, 67, 78, 79, 86, 87, 90, 92, 110, 111, 124, 131, 137, 138
Precision Medicine 31, 33, 63, 74, 80, 92, 93, 115, 118, 121

R

Resource-limited settings 122, 139, 157, 267, 280
Respiratory Infection 3, 136
Rubella 3, 6, 9, 10, 11, 12

S

SARS-CoV 3, 4, 5, 18, 23, 27, 28, 60, 61, 62, 64, 72, 78, 79, 81, 84, 86, 89, 93, 96, 111, 114, 115, 116, 119, 120, 127, 136, 143, 144, 146, 147, 148, 149, 151, 159, 160, 161, 165, 169, 171, 173, 217, 218, 220, 221, 223, 224, 228, 236, 240, 242, 243, 247, 248, 249, 250, 251, 252, 256, 257, 258, 259, 266, 267, 269, 270, 277, 279, 281, 283, 289, 290, 293, 294, 295

T

Targeted Drug Delivery 31, 35, 37, 40, 57, 59, 61
Technology 9, 16, 19, 20, 21, 26, 45, 60, 61, 64, 65, 68, 70, 71, 72, 74, 76, 77, 82, 83, 85, 91, 108, 111, 112, 121, 129, 132, 137, 153, 166, 171, 173,

174, 177, 187, 190, 191, 192, 194, 195, 198, 199, 200, 203, 204, 206, 207, 209, 211, 212, 213, 214, 215, 234, 249, 281, 293
Telemedicine 176, 177, 178, 179, 180, 181, 183, 186, 189, 190, 191, 192, 193, 195, 196, 197, 198, 199, 201, 202, 203, 204, 205, 207, 208, 209, 210, 211, 212, 214, 215
Treatment 1, 3, 4, 5, 7, 8, 16, 17, 24, 25, 26, 32, 33, 34, 36, 37, 40, 42, 43, 45, 46, 47, 49, 50, 51, 52, 53, 54, 57, 59, 61, 64, 65, 66, 67, 69, 70, 72, 74, 75, 76, 77, 78, 79, 80, 81, 82, 83, 84, 85, 86, 87, 88, 89, 90, 91, 92, 93, 94, 95, 96, 97, 99, 100, 101, 102, 103, 104, 105, 106, 109, 110, 111, 113, 114, 115, 116, 120, 122, 124, 126, 127, 132, 134, 136, 137, 138, 140, 141, 143, 148, 149, 155, 158, 161, 171, 175, 176, 177, 178, 179, 180, 181, 182, 183, 184, 185, 186, 187, 188, 189, 190, 191, 192, 193, 194, 195, 196, 197, 198, 199, 200, 201, 202, 203, 204, 205, 206, 207, 208, 209, 210, 211, 212, 214, 215, 216, 227, 229, 230, 232, 235, 236, 237, 239, 243, 244, 246, 248, 253, 255, 256, 263, 266, 267, 268, 273, 274, 275, 276, 277, 279, 280, 281, 282, 283, 284, 288, 291, 292, 293, 295

V

Vaccines 1, 2, 3, 8, 9, 10, 11, 12, 13, 14, 15, 16, 17, 18, 19, 20, 21, 22, 23, 24, 25, 26, 28, 30, 57, 84, 89, 91, 93, 114, 115, 116, 119, 134, 135, 136, 137, 140, 146, 147, 149, 151, 153, 167, 169, 171, 172, 173, 208, 209, 210, 211, 215, 216, 222, 223, 224, 230, 231, 232, 233, 234, 235, 239, 257, 263, 265, 267, 271, 279, 281, 282, 283, 284, 286, 287, 289, 291, 292
Viral Diseases 1, 2, 3, 4, 5, 7, 20, 21, 22, 29, 32, 34, 37, 42, 46, 47, 64, 65, 66, 68, 72, 74, 75, 77, 83, 84, 88, 99, 103, 104, 105, 106, 107, 108, 110, 111, 112, 132, 133, 136, 142, 149, 170, 175, 176, 195, 206, 207, 209, 210, 213, 215, 216, 217, 218, 227, 239, 270, 271, 276, 288, 290, 291, 293, 294
Viruses 2, 3, 5, 8, 16, 17, 20, 25, 32, 33, 45, 47, 57, 64, 66, 67, 68, 69, 71, 72, 75, 76, 77, 80, 86, 90, 94, 98, 101, 104, 105, 107, 114, 115, 116, 118, 119, 120, 122, 124, 127, 131, 133, 134, 138, 140, 141, 143, 144, 149, 152, 159, 160, 161, 168, 169, 170, 171, 175, 182, 208, 217, 218, 219, 220, 221, 222, 223, 224, 226, 227, 228, 235, 236, 237, 240, 242, 243, 245, 249, 250, 252, 253, 254, 261, 262, 263, 264, 265, 267, 268, 269, 270, 271, 272, 273, 274, 275, 276, 277, 281, 282, 283, 284, 286, 287, 289, 293, 294, 296

W

Wearable devices 176, 179, 185, 186, 188, 189, 190, 191, 198, 199, 202, 203, 204, 205, 210, 211, 213
World Health Organization 13, 14, 15, 25, 29, 30, 71, 136, 176, 205, 215, 240, 257, 284

Z

Zika 3, 5, 6, 19, 71, 72, 80, 85, 89, 90, 93, 97, 98, 102, 117, 119, 125, 129, 135, 136, 144, 217, 221, 224, 225, 249, 265, 267, 268, 270, 281, 286, 290
Zoonotic pathogens 261, 262, 268

www.ingramcontent.com/pod-product-compliance
Ingram Content Group UK Ltd.
Pitfield, Milton Keynes, MK11 3LW, UK
UKHW052001051026
15624UKWH00003B/503

9 798369 392768